MW00762869

Subsea
Pipeline
Engineering

Subsea
Pipeline
Engineering

Andrew C. Palmer and Roger A. King

Copyright © 2004 by
PennWell Corporation
1421 S. Sheridan Road/P.O. Box 1260
Tulsa, Oklahoma 74101

800.752.9764
+1.918.831.9421
sales@pennwell.com
www.pennwellbooks.com
www.pennwell.com

Managing Editor: Marla Patterson
Production Editor: Sue Rhodes Dodd
Cover Designer: Beth Caissie
Book Designer: Clark Bell

Library of Congress Cataloging-in-Publication Data available on request

Subsea Pipeline Engineering / by Andrew C. Palmer and Roger A. King
    p. cm
Includes index
  ISBN 1-59370-013-X

Printed in the United States of America.
1 2 3 4 5   08 07 06 05 04

For Emily and Hannah

Contents

List of Figures

List of Tables

Chapter 7

Chapter 8

Chapter 15

Appendix A

Appendix B

Acknowledgments

I am grateful to many people who have helped me over the years, but I owe a special debt of gratitude to two people, Bob Brown and the late Jack Ells. Jack Ells was British Petroleum's pipeline guru, and it was he who saw that an academic specialist in applied mechanics who had never heard of a lay barge might nevertheless have something to contribute. He had immense insight into what might be practicable and what the difficulties might be, and I miss his wise advice. Bob Brown hired me five years later. He was ready to take on someone who had never worked outside a university environment and was in many ways innocent about consulting. He is an extraordinarily creative and charismatic person, full of energy and new ideas, many of which have become influential parts of marine pipeline practice. It was a privilege to work in his team: being with him is never boring.

I have benefited enormously from the environments I have worked in and from the companionship and advice of many colleagues in Cambridge, Brown, Liverpool, University of Manchester Institute of Science and Technology (UMIST) and Harvard Universities, at R. J. Brown and Associates, Andrew Palmer and Associates, Science Applications International Corporation (SAIC), in many customer companies, and in specialist courses. My work at Cambridge is supported by a generous benefaction from the Jafar Foundation to the University and to Churchill College, and I am extremely grateful to Hamid Jafar for his support and encouragement.

There are too many individuals to mention all of them by name, but in addition to Jack Ells, Bob Brown, and Hamid Jafar, I would particularly like to thank Jack Apgar, Steve Booth, David Bruton, Chris Calladine, Malcolm Carr, Phil Corbishley, Nigel Curson, the late Dan Drucker, Shon Ffowcs-Williams, Pat Fontaine, Gary Harrison, Willot Heerde, Jacques Heyman, the late George Hinkle, David Kaye, John Kenny, Ibrahim Konuk, Carl Langner, BJ Lowe, Alan Niedoroda, Marinke Overwater, Amanda Pyatt, Allan Reece, Jim Rice, Andrew Schofield, the late Ted Schultz, Simon Shaw, Lawrence Tebboth, the late Stu Tholan, John Tiratsoo, David Walker, and Ron Watkins.

I am grateful to my co-author Roger King for his unbreakable good humor and patience. Roger and I first thought of this book on a flight from Jakarta to Singapore, and it has been with us through many journeys. It was completed during a sabbatical year at Harvard University, and I am grateful to Cambridge for allowing me to take leave and to Harvard for its calm but stimulating ambience.

The photographs in chapter 12 were generously provided by Heerema, Allseas, Technip-Coflexip, Saipem, and Global Industries; and Roger King and myself are grateful for them. The cover is based on a photograph by Bob Brown of the Technip-Coflexip reelship "Deep Blue."

I would like to thank the staff of PennWell, and above all Marla Patterson and Sue Rhodes Dodd, for their hard work in discussing the plan for the book with us and in removing obscurities and infelicities.

Most of all I would like to thank my wife Jane and my daughter Emily for their constant support and for putting up with my idiosyncrasies.

Preface

Submarine pipelines are the arteries of the oil and gas industry, important in engineering practice and full of excitement and interest, though at first sight less spectacular than platforms and floating production systems. The technology has advanced dramatically. Projects that were dreams 20 years ago are now becoming reality.

I came into this field by lucky chance. I was a university lecturer in Cambridge; and in the summer of 1970 was invited to spend the summer at Brown University, where I had completed my Ph.D. five years earlier. The Alaska pipeline was in the news. Former colleagues in the Division of Engineering remembered that I had worked on permafrost as a sideline to my Ph.D., and they had some money for research on geotechnical aspects of environmental problems. They told me that if I could link my research to the pipeline, they could support me for the summer. I was keen to go and networked through a colleague in Cambridge to locate the people in BP in London who were working on the problem. I told that them that I was not looking for money but was looking for an interesting problem. Impressed by this refreshingly naive approach, they gave me a problem to do with differential settlement on thawing permafrost; I researched that topic and was taken to talk to the pipeline team in Houston.

A year later, BP called me again and said that there were questions about a planned North Sea pipeline, Forties, the first large-diameter line in what was then thought of as deep water, 125 m deep.

"That sounds interesting," I said, "but I don't know anything about underwater pipelines."

"We think that's an advantage," they replied. "You'll bring a fresh mind to it. Come and see us tomorrow."

That phone call changed my life. Over the next few years, I studied a number of problems, among them pipelaying mechanics, buckle propagation, interaction with sand waves, modeling and surface tie in. Since then, most of my work has been in pipelines. In 1975 I went to work for the leading consultant in underwater pipelines, R. J. Brown and Associates, and worked on projects in Arctic Canada, the North Sea, and the Middle East. It was in that period that I first met my friend and co-author Roger King, and later we were both at UMIST. Ten years later, I

started my own consulting company, Andrew Palmer and Associates, and we worked on exciting projects in every continent.

There are many books, and each one needs a reason for its existence. Our justification for the present book is that we believe—and, more importantly, others have confirmed to us—that there is no other book that covers the same ground, though there are excellent books on land pipelines and on outfalls. There are many technical papers, but most of them are in the conference literature and not readily accessible. Roger and I thought it useful to bring together in one volume the state of the art as it stands in the summer of 2003, in a form that would be accessible to the non-specialist and, at the same time, would provide references to topics that we did not have space to follow up in detail. We are primarily concerned with oil and gas pipelines that operate at relatively high pressures and sometimes in very deep water, but most of the ideas are equally applicable to lines that carry water and sewage.

No doubt there are mistakes. They are our responsibility, and we hope that readers will tell us about them.

Andrew Palmer
Deer Isle, Maine
10 July 2003

1 Introduction

1.1 Introduction

Humankind needs to move fluids from place to place. Some fluids have to be moved in huge quantities and over long distances: water, oil, natural gas, and carbon dioxide are examples. Other fluids have to be moved in smaller quantities or over shorter distances: steam, ethylene, blood, milk, wine, helium, mercury, nitroglycerin, and petrochemicals are examples.

There are essentially three ways of moving fluids. The first is to pour the fluid into a tank, move the filled tank to where the fluid is needed, and empty the tank. The essential components of that method are a tank that can be moved and a way of filling and emptying it. The second way is to construct a pipe from where the fluid is to where the fluid needs to be and to pump the fluid along the pipe. The third method, sometimes used in combination with the other two, is to transform the fluid into a solid or into another fluid that can be transported easily.

The tank option is flexible and often has lower capital costs but higher operating costs. This option invariably is used for small volumes of high-value fluids such as mercury,

wine, blood, and helium. The tank protects the fluid against contamination from outside. The risk that the transported fluid might escape and damage the environment depends on the integrity of the tank. The method is used widely to transport oil and liquefied natural gas (an example of the transformation option) by sea because of the flexibility it allows. The same very large crude carrier (VLCC) can transport oil from the Middle East to Japan on one voyage and oil from Alaska to California on the next voyage; and during the voyages, the cargo can be sold and resold, diverted to a different destination, and partly offloaded at yet another location. Oil is transported over long distances by rail where no pipelines exist. For example, oil from central Russia goes to the Russian Far East by rail, and oil from the Wytch Farm field in southern England was exported by rail until production increased to the level that justified building a pipeline.

The pipeline option is relatively inflexible by comparison. A pipeline is a fixed asset with large capital costs. Once the pipeline is in place, though, the operation and maintenance costs are relatively small; and the pipeline has an operating life of 40 years or more. A land pipeline may be vulnerable to damage by war or terrorist attack or to interruption of service by political interference from one of the countries it crosses. It is this factor that has limited the application of pipelines to carry oil from the Middle East to Europe and so far has prevented the construction of a gas pipeline from Nigeria to Europe. Pipeline transportation consumes little energy. The most dramatic example is the comparison between, on the one hand, pipeline transportation of natural gas and, on the other hand, liquefaction followed by sea transportation by tanker followed by regasification. Even over a long distance, operation of a pipeline uses no more than 10% of the energy content of the gas, whereas the liquefied gas option uses more than 30%.

This book is about pipelines under water, which are used in various contexts. More and more oil and gas are being produced from fields that lie under the sea. The product has to be carried to shore; and that is usually, though not always, done by pipeline. Intrafield pipelines carry oil and gas from wellheads and manifolds to platforms and from one manifold to another. Sometimes gas from one field is transported to another field to be injected to maintain reservoir pressure. Then treated seawater is injected to displace the oil, and occasionally carbon dioxide is separated from the gas and reinjected, as occurs in the Sleipner field in the North Sea and, in the future, in the Natuna field off Indonesia.

Many pipelines that are primarily on land also have to cross seas, straits between islands, and river estuaries. Chapter two describes two examples, the first, a pipeline to Vancouver Island in western Canada and the second, a pipeline from Algeria to Spain. Several ambitious long distance projects currently under study include crossings of this type: examples are the pipeline from Papua New Guinea to Australia and the pipeline from Qatar to India and Pakistan.

These examples refer to gas and oil, but transportation of water is also important. It has been argued that, as world population grows, water will be a greater problem than energy and that it will become a major source of conflict. Several projects are examining long distance water pipelines, which are necessarily very large if they are to be useful.

There is growing concern about carbon dioxide released into the atmosphere by burning fossil fuels and about its effect on the global climate. One mitigation option is to capture carbon dioxide at large point sources such as power plants and cement plants, and then to store it in the deep ocean or in geological formations under the ocean.[1] If these schemes come to pass, very large quantities of carbon dioxide will have to be handled for the beneficial impact on climate to be significant, because the present rate of release is 6 billion tonnes a year. It follows that any storage scheme that involves the oceans necessarily will generate a large additional demand for pipelines.

It is also necessary to dispose of water. Outfalls carry treated sewage and storm water into the seas and rivers. Coastal power stations draw in cooling from intakes and dispose of warmer water from outfalls. Outfalls have also been used to dispose of materials such as mine tailings and raw sewage, but there are often substantial environmental and legal objections to these disposal options, making them less and less often appropriate.

Section 1.2 presents the book's organization. Section 1.3 is a brief historical introduction to pipelines.

1.2 How This Book is Organized

The chapter themes relate to areas of knowledge, which broadly match groups of decisions that a pipeline engineer has to make.

The first task of the pipeline system designer is to choose the pipeline route. Sometimes this task is straightforward: If the seabed is smooth and featureless, a straight line between the end points is the

shortest and most economical route. More often, there are various obstructions and interferences that compel the designer to select a more complex route. The factors involved may be physical, environmental, political, or related to other human uses of the seabed.

The designer then has to consider materials. Many pipelines are made from low-alloy carbon steel, which is a robust and inexpensive material. However, frequently the fluid in the pipe corrodes carbon steel too rapidly for its use to be acceptable. The designer then has to consider a corrosion-resistant alloy, a flexible, or a composite. Usually, each of these alternatives is more expensive than a low-alloy carbon steel; and an alternate material may present other problems such as difficulty in making connections, but it may lend itself to an economical overall solution. The pipes have to be joined together, and careful decisions have to be made if the pipes are to be welded.

The next task is to decide on the diameter of the pipeline. That decision primarily rests on the hydraulics. If the diameter is too small, the pressure drop between the ends will be excessively large, but if the diameter is too large the cost will be unnecessarily great, and undesirable flow modes may occur in two-phase flow. Detailed decisions about the composition and specification of the material involve many interacting factors of corrosion resistance, weldability, strength, fracture toughness, and cost.

The next choice is the wall thickness of the pipeline. This is primarily an issue of structural engineering in which the designer has to ensure that the pipe is strong enough to resist many kinds of loading, among them internal pressure, external pressure, bending and fatigue during construction, concentrated loads, and impact.

Almost all underwater pipelines have an external coating to protect them against corrosion, complemented by a cathodic protection system that prevents external corrosion if the coating is damaged. Many pipelines have additional concrete weight coating to provide stability against waves and currents and to give the anti-corrosion coating protection against mechanical damage. Some pipelines have one or more additional layers of thermal insulation, required to maintain the fluid contents at a high temperature. Still other pipelines have internal coatings as corrosion protection or to provide a smooth inner surface to reduce the resistance to flow.

A pipeline must be constructible. The designer needs to know the limitations of the available construction systems, and he has to design the pipeline so that it can be built safely and economically. Many pipelines are trenched or buried in order to provide shelter against hydrodynamic forces, to protect them against mechanical damage, or to provide thermal insulation and resistance to upheaval buckling.

Pipelines do not always rest continuously in contact with the seabed, and there may be spans in which the pipeline bridges across low points in the profile. Spans can give rise to various structural problems and may need to be corrected. Uneven seabed profiles can also initiate upheaval buckling in which the pipeline arches above the bottom. That situation also has to be guarded against and, if necessary, corrected. Pipelines can also buckle sideways.

Pipelines in service may be subject to damage by chemical and microbiological corrosion. The designer needs to know how to suppress corrosion as far as possible and how to allow for it in the choice of wall thickness. The operator needs to know what operating practices are likely to minimize corrosion and how to monitor it.

A pipeline is at risk of damage, and repairs may become necessary. A chapter on risk, accidents, and repairs examines the general principles that apply to repair and presents case studies of incidents that have required intervention and the repair techniques used in each instance. Ultimately, a pipeline has to be decommissioned when its operating life has ended or when it is no longer needed. Decommissioning is an evolving subject of which there is little experience as yet, but it is an increasing concern to operators, to regulatory authorities, and to the wider community. It engages various factors; among them, national and international law, environmental protection, the safety of other seabed users, and engineering.

Finally, neither design nor construction is a mature or dead technology. New ideas are coming forward all the time, There are new understandings that overturn established notions and new techniques that make it possible to build pipelines more cheaply, more quickly, and more safely than ever before. It is famously difficult to foresee the future, but the final chapter discusses some promising ideas.

References are indicated in the text by superscripts and are listed at the end of each chapter.

1.3 Historical Background

The oldest marine pipelines were outfalls, which have been used since the nineteenth century when it was realized that simply dumping sewage into rivers and onto beaches created harmful and unpleasant pollution.[2] The earliest marine pipelines for oil were short loading and

unloading lines, generally constructed by building them on shore and winching them into the water. This remains the usual way of constructing pipelines between the shore and offshore loading points and of constructing river and estuary crossings. Alternative versions of the pull technique for installation are various kinds of tow, where the pipe is built on shore then towed or dragged to its final location.

The offshore petroleum industry is relatively recent historically. The first well out of sight of land was drilled in the Gulf of Mexico in 1947 at Kerr–McGee's Ship Shoal block 32, 17 km from shore and in 6 m of water. The earliest petroleum pipelines date from before 1947 and were set in very shallow water in the Gulf of Maracaibo and the Caspian Sea off Azerbaijan. In time, there came to be a need for pipelines over longer distances. One urgent need arose during the Second World War when the Allies realized that they would have to invade the continent of Europe and that that the ports would have been destroyed during the invasion. The Allies also knew that an invading army would require vast quantities of gasoline. The military authorities called in Anglo-Iranian, the forerunner of British Petroleum, and asked the company if it would be possible to lay pipelines from England to France across the English Channel. Two kinds of pipeline were devised for the project, which was called Pipe–Line Under The Ocean (PLUTO). One was rather like a submarine cable with no central core, formed from a lead tube wound with armor layers of steel tape and plastic. The other was a welded steel pipe with no anti-corrosion coating. Among the impressive aspects of the project was that the first trials were carried out just a week after the first meeting, something that few oil companies could manage nowadays.

Conventional small cable ships laid the cable-like pipe. The steel pipe was made up in great lengths and then wound onto floating reels (conundrums). Tugs towed the drums across the Channel, unwinding the pipe as they went. By this means, a pipe could be laid from the Isle of Wight to the Cotentin Peninsula in ten hours (and that, too, could not be accomplished now). It is fair to add that the value of PLUTO is disputed. Some authors argue that it had little impact in comparison to tanker transportation into newly constructed harbors and imply that it was a waste of effort, but others are more positive.[3,4]

Reel ship technology was picked up and developed in the Gulf of Mexico and led to the construction of a series of reel barges and reel ships, which continue to earn a significant share of the pipelaying market.

A different technology was developed in parallel. It is related to land pipeline construction, where lengths of pipe are laid out along the right of way, welded together length by length, and progressively lowered into a trench. The lay barge system consists essentially of a barge with a fore-and-aft line of rollers (firing line) along which are spaced a number of welding stations. Lengths of pipe are brought to the bow of the barge and welded to the end of the length that has already been completed. The barge moves forward one length at a time. Successive weld passes at the stations complete the welds. The pipe leaves the stern of the barge over a curved support called a stinger, and then lifts off the stinger and curves downward until it reaches the bottom.

Another development pulled pipelines in various ways, sometimes along the seabed, sometimes just above the seabed, and sometimes at the sea surface supported by pontoons. This group of methods is widely used for shore crossings but can be applied in deeper water and to longer pipelines.

Lay barge technology was developed in the Gulf of Mexico and then brought to the North Sea and the Arabian Gulf. The earliest barges had simple rectangular box-like hulls, but later barges had ship-form hulls. Some barges applied the semi-submersible principle, which reduces motion in a seaway by setting most of the buoyancy well below the waterline, with the barge topsides supported on comparatively slender surface-piercing columns. Different combinations of larger hulls and semi-submersible hull forms made it possible to increase lay productivity in exposed locations such as the North Sea.

Most marine pipelines are laid by the lay barge method, which has proved itself flexible and versatile. The first barges were positioned by long mooring lines stretching from winches to anchors on the seabed; but many recent barges are dynamically positioned by thrusters, which liberate them from the potential problems inherent in complex mooring systems.

Early pipelines were all within water depths accessible to divers, and many construction operations were carried out with diver help, particularly when connections were needed. The demands of offshore petroleum production have taken the requirement to construct pipelines far beyond the maximum depth that divers can reach, which is about 300 m; and the past 25 years have seen an enormous development of diverless operations. Depths of 1000 m are now routine. A pipeline across the Black Sea in depths down to 2200 m has been completed and is in operation, and several other projects in depths between 1500 and 2500 m are now in progress.

8

References

1 Palmer, A.C., and B. Ormerod. (1997) "Global Warming: the Ocean Solution." *Science and Public Affairs*. Autumn, 49–51.

2 Grace, R.A. (1978) *Marine Pipeline Systems*. Englewood Cliffs, NJ: Prentice-Hall.

3 Yergin, D. (1991) *The Prize: The Epic Quest for Oil, Money and Power*. New York: Simon & Schuster.

4 Searle, A. (1995) *PLUTO (Pipe-Line Under the Ocean): the Definitive Story*. Shanklin, England

2 Route Selection

2.1 Introduction

An early task for the pipeline designer is to select the route of the pipeline.

Route selection is a critical activity. A poorly chosen route can be much more expensive than a well chosen route. It can lead to costly surprises and delays at later stages, particularly if unexpected geotechnical or marine conditions are encountered or if the route leads to conflicts with public authorities, with environmental interests, or with other operators. It is not an exaggeration to say that a few days (and a few thousand dollars) spent on sensitive and thoughtful evaluation of the pipeline route can save months and millions later. Several notorious examples illustrate the problems that can arise.

In some instances, the choice of route is straightforward. This is so in many areas of the North Sea and the Gulf of Mexico where more than 30 years' experience has led to a good understanding of the seabed geotechnics and the oceanographic conditions: knowledge of the locations of geotechnically uniform and smooth

seabed, free of obstructions or existing pipelines and not in conflict with other fields, existing or planned subsea installations. In many other parts of the world, the choice of route is much more difficult. In areas such as the Arabian Gulf, northwest Australia, Indonesia, and the approaches to the Norwegian coast, the seabed has a heavily uneven topography and is hard to trench.

Even in developed areas, a surprising number of factors have to be taken into account, among them:

- Politics
- Environment
- Approaches to existing platforms and risers
- Avoidance of zones exposed to anchor damage
- Avoidance of zones exposed to dropped-object damage
- Crossings of existing pipelines
- Cables
- Areas of very hard seabed
- Areas of very soft seabed
- Boulder fields
- Pockmarks
- Iceberg plow marks
- Submarine exercise areas
- Fishing
- Mine fields
- Dumping grounds
- Dredging
- Wrecks

A rational choice of route cannot be made without information about the seabed topography and geotechnics. It is never sensible to embark on a marine survey without carrying out a desk study first. Much information can be gathered from charts, geological maps, fishing charts, aerial photography, satellite photography and synthetic-aperture radar, other operators, navigation authorities and navies, local inquiry, and sometimes from more obscure sources. In one instance, a book by a nineteenth century bird-watcher gave invaluable information about a proposed pipeline route through the Pechora Delta in Arctic Russia.

Landfalls are particularly complicated. Many of the famous disasters of pipeline construction have been in very shallow water in shore approaches. A bad choice of landfall location can be very expensive indeed and can lead to devastating cost overruns and interminable legal disputes. The designer needs to have a perceptive understanding of the geomorphologic factors that decide the form of the coast and of the way in which the construction of a pipeline will interact with environmental factors such as wave refraction, longshore sediment transport, wave breaking, and seabed geology.

Section 2.2 discusses physical factors such as seabed geotechnics. Section 2.3 discusses route selection implications prompted by interaction with other users of the seabed. Section 2.4 is concerned with political and environmental factors. Section 2.5 describes two case studies.

2.2 Physical Factors

A pipeline rests on or in the seabed. From the pipeline point of view, the ideal seabed is level and smooth so that no spans are formed and is composed of stable medium clay. The pipe settles into the clay and gains enhanced lateral stability. If the seabed is not smooth but uneven and rocky, there will be many free spans where the pipeline bridges above hollows, some of them long enough to need correction. At high points, the concentrated forces between the bed and the pipe may damage the external coating. Hard seabed is difficult and expensive to trench. If the seabed is very soft, on the other hand, a pipeline will sink into it and may be difficult to reach for inspection and for operations such as tie ins to other pipelines or possible repairs.

Some seabeds are highly mobile and include sandwaves (which may be 15 m high and 100 m long) and smaller ripple features (which range in size on many scales from millimeters to meters high).[1,2] These features move significantly during the life of a pipeline so that a pipeline supported on the crest of a sand wave when it was constructed may later be left unsupported when the wave has moved on. The movements are irregular and difficult to predict confidently. For these reasons, it is better to avoid sand wave fields whenever possible. The route of the first Forties pipeline in the North Sea, for instance, was shifted southward to avoid a sand wave field. Sometimes, it is impossible to avoid sand waves and mega-ripples, and then a trench is dredged to a level at or below the troughs of the sand waves before the pipe is laid along the trench. This technique is called *presweeping*. It has been used for several pipelines in the southern North

Sea where a combination of shallow water; high tidal currents; high waves induced by northeasterly gales; and mobile, loose sand creates a complicated and changing seabed topography. The wider issue of seabed mobility and its implications for pipeline stability is considered in chapter 11. Chapter 12 covers trenching, and chapter 14 spans.

The complexity of marine geotechnics is at last being recognized. The variety of geomorphology and topography on the seabed is as great as it is on land. There are many features that influence the choice of pipeline route.

Submarine landslides occur when high sedimentation rates overload and oversteepen slopes. This happens particularly in deltas, such as the Mississippi and Fraser deltas in North America; but there have also been large submarine slides in the North Sea. An earthquake may trigger a slide in a marginally stable slope. A slide across a pipeline can lead to very large movements of the line and can easily induce tensile forces large enough to break it. A slide along a pipeline is less serious because the forces it induces are smaller.

Parts of the seabed of the Norwegian sector of the North Sea are covered with boulders and cobbles, some of them lying on the surface and some partially or wholly buried in clay. The boulders are erratics that fell out of melting icebergs. They may be more than 1 m in diameter and are a significant obstacle to most kinds of trenching machines. Pockmark depressions in the seabed are formed when shallow gas escapes to the surface, reducing the pore pressure and allowing loose sediments to collapse.

In tropical seas, coral forms large humps on the seabed or coral pinnacles that can be 15 m high. Coral is fracture-resistant and extremely difficult to trench. Moreover, it is, of course, undesirable to damage coral because of its ecological significance. The beds of tropical seas are often carbonate sands (rather that the silica sands met in temperate climates); and if the sand is not disturbed by storms, chemical processes progressively harden it into a tough and trenching-resistant rock.

In Arctic seas, pipelines are faced with many other features. In the Arctic spring, the rivers thaw while the sea is still frozen; and river water floods out across the sea ice. If the sea ice includes a hole or a crack, fresh water flows downward through the hole and forms a whirlpool vortex and a rotating jet below the surface. The jet can excavate deep holes in the bottom, called *strudel scours* (after the German word for *whirl*).[3] Large ice masses drift into shallow water, ground, and are driven forward by wind and by the pressure of ice sheets and ice packs. The ice gouges into the seabed and makes grooves, which can be 10 m deep and 100 m wide. Design and route selection to minimize the risk of gouge damage is the major challenge of Arctic marine pipeline design.[3,4] Relic gouges left by drifting icebergs in the last glaciation are found in the northern North Sea.

Turning now to the water above the seabed, hydrodynamic factors too can influence the choice of pipeline route. It is desirable to avoid high currents, which can sweep a pipeline sideways and complicate pipelaying. High tidal currents occur in shallow seas in macrotidal areas, in estuaries, and in straits between islands. It will often be better not to cross the narrowest part of a strait and instead to take a longer route that has weaker currents.

It is also desirable to avoid areas where waves are particularly high because of the adverse effect on stability of wave-induced water movements and because high waves slow or stop pipelaying. Wave-induced velocities are negligibly small at depths greater than half a wave length, but velocities must increase as the water depth decreases. The effect is still more severe if the waves break. For this reason, it is good to avoid long sections in shallow water and better to choose a route in deeper water (even if it is longer).

A wave moving in from deep water offshore slows down and steepens as it comes into shallower water.[1,2] If the propagation direction of the wave is not at right angles to the depth contours, the propagation direction changes so that it becomes closer to the direction at right angles to the contours. This directional change is a refraction effect analogous to the change in direction of a ray of light when it reaches at an angle an interface between air and glass. The wave crests turn so that they become more nearly parallel to the depth contours, an effect that can be seen on most beaches. It is often wise to include a refraction analysis in the pipeline design process.[2,5] If the coast includes headlands and if the depth contours are convex away from the shore, this refraction tends to focus wave energy on the headlands and away from the bays in between. The waves are higher off the headlands, and this is a good reason for a pipeline route to avoid them. Another reason is that the headlands often indicate the presence of harder rock because the softer rock has been preferentially eroded to form the bays and the harder rock will be more difficult to dredge or excavate.

The water column may include sharp discontinuities of density, where lighter, less saline water is on top of denser more saline water. These conditions occur in the Strait of Gibraltar, Southeast Asia, and elsewhere. Internal waves can occur on the interface and can induce high velocities at the bottom.

2.3 Interactions with Other Users of the Seabed

A remarkably large number of other human activities engage with the seabed, and the choice of route must take into account potential interference with them.

Oil and gas exploration and production are the most obvious conflicting activities. It is prudent to keep pipelines away from platforms unless they have to be connected to the platforms because of the possibility of damage by dropped objects; an increased risk of anchor damage from supply vessels and construction vessels; and the remote possibility that fire, explosion, or structural failure on the platform might involve the pipeline. For the same reasons, it is good to keep away from existing wellheads and manifolds and wise to find out where future subsea activities might occur.

Existing pipelines are the most common problem for designers. One pipeline can cross another one, but it is not practicable simply to lay the second pipeline across the first. A crossing has to be carefully designed so that neither pipeline damages the other, so that there is no undesirable interference between the two cathodic protection systems, and so that neither is overstressed or destabilized by hydrodynamic forces. One simple option is to trench the first pipeline more deeply at the crossing point, to lay mattresses over the first line to provide physical separation, to prevent coating damage, and to separate the cathodic protection systems, and then carefully to lay the second line over the crossing point. Sometimes, much more elaborate designs are required, particularly if the first line cannot be lowered because the seabed is too hard to trench, requiring the second line to cross by a bridge-like structure made of mattresses, concrete units, or rock dump. It follows that the number of crossings should be kept to a minimum. At the very least, a decision to cross another operator's pipeline necessarily implies time-consuming discussions with the operator and the regulators about the design of the construction and the measures that need to be taken during construction.

Vulnerable submarine cables crisscross many areas of the sea floor. The usual way of crossing a cable with a pipeline is to sever the cable first, to lay the pipeline through the gap, and then to splice the cable and lower it back over the pipeline. This is an expensive operation to be avoided if possible.

Fishing for demersal fish that live at the bottom is an important activity in many shallow seas such as the North Sea. The heightened level of exploitation of fishery resources leads fishermen into new geographical areas and deeper water and often leads to the use of larger and heavier bottom gear. Though fish are attracted to pipelines, a politically influential fishing industry perceives the invasion of fishing grounds by pipelines as another threat to its livelihood and may wish to see a pipeline route changed to skirt the best fishing grounds.

Military activities use the seabed in various ways. Mines are laid in wars and are not recovered or deactivated afterwards. The explosives and detonators remain sensitive, and the mines may drift about and lodge against pipelines. This is a major problem in the North Sea and the Arabian Gulf (Persian Gulf). Submarines exercise and may navigate close to the bottom in order to hide from sonar. Some areas are used for artillery and bombing practice. Various kinds of magnetic, acoustic, and electrical sensors are laid on the seabed to detect submarines and are linked by cables. Munitions left over after wars are dumped at sea and are sometimes picked up by fishing trawls and then dropped overboard so that the contaminated area widens with time.

Ships rarely try to anchor in the open sea but do anchor in the approaches to ports, sometimes in carefully designated anchorage areas but sometimes indiscriminately. Port approaches are dredged to increase depths in shipping channels, and the construction of new ports and existence of larger ships may lead to much extended dredging during the design life of a pipeline. Ships have sunk onto marine pipelines and damaged them in at least two instances, one in Singapore and the other in the Netherlands sector of the North Sea. The beds of many seas are littered with shipwrecks, often not properly recorded and only to be found by survey.

The seabed is also a potential source of minerals. The most important seabed mining activity is dredging for sand and gravel. This activity is becoming more significant as land sites become harder to find, because of intensive land use and environmental restrictions.

Finally, in the past the seabed was used as a dumping ground for all kinds of materials, from sewage sludge through chemical and nuclear waste to obsolete equipment such as ships and nuclear reactors. Only recently has it been recognized how shortsighted and undesirable indiscriminate dumping is, though it may be an environmentally sound solution in a few cases. Material dumped in the past remains in place, and there may be damaging consequences if it is disturbed.

2.4 Environmental and Political Factors

The importance and sensitivity of the environment cannot be overstated, particularly after the controversies surrounding the Alaska pipeline, the abandonment of the Brent Spar, the pipeline to Point of Air in North Wales, pipelines from the Norwegian sector to North Germany, the Arctic National Wildlife Reserve, and marine pipelines currently proposed for the Alaskan Beaufort Sea. An ill-judged scheme that appears to ignore environmental factors can be fatal to a project. Even if it is not fatal, it can incite opposition and provide ammunition for people opposed to the petroleum industry on broad political grounds. The effect is to tie the project down in years of public hearings and mountains of experts' reports, inevitably conflicting with each other and leading to further controversy.

Each case is different, and there can be no simple answer. The best solution is a strategy that includes being aware of the possibility of environmental impact from the very beginning and consulting widely with interested organizations and individuals.

In deep water, environmental concerns have generally been absent. This may be in the process of change, particularly in parts of the world where the environmental movement is strong and active. Concern has been expressed about construction noise, disturbance to marine mammals, damage to coral and tubeworm colonies, damage to benthic life in the seabed (though that damage is minimal by comparison with trawling damage), and disturbance to heavy metals such as cadmium and mercury previously deposited in sediments by human activities.

In shallow water and at landfalls, on the other hand, there are almost invariably concerns about environmental factors. Shallow water is biologically productive, and its immensely complex food web engages bacteria, plankton, plants, invertebrates, fish, birds, and marine mammals so that damage to one component may have far-reaching consequences. The response is to study and quantify the effects and to look for route alternatives and mitigation measures that eliminate or minimize them. Much can be done by scheduling the construction period at an appropriate season.

Environmental and political issues often become mixed, but there are political issues that have nothing to do with the environment. It is usually a good idea to try to minimize the number of different regulatory and political organizations that have to be dealt with and to avoid unnecessary incursions across national or state boundaries or into offshore leases that will necessitate negotiations with other operators. A glance at a pipeline map of the North Sea reveals several instances where routes have been selected so as to skirt national boundaries.

2.5 Case Studies

Two case studies show the interaction between different factors.

British Columbia Hydro and Power Authority wished to build a pipeline across the Strait of Georgia, from the Fraser Delta area south of Vancouver to a landing on Vancouver Island. The pipeline project is described in greater detail by Park.[5,6] The pipeline diameter was a nominal 10 in (273.05 mm outside diameter), chosen because the expected demand would justify that size. The plan was to construct two lines to obtain additional security. Figure 2–1 shows the area and some of the factors that determined the route.

The political boundary between Canada and the USA at 49 °N lies just south of the delta, cuts across the Point Roberts peninsula, and extends westward to the middle of the Strait of Georgia. An early decision was not to cross that boundary because to do so would bring part of the pipeline under the jurisdiction of United States federal authorities, as well as numerous state and local authorities, and would make it liable to challenge in the United States legal system.

The land portion of the delta consists of low-lying islands. The delta is fronted by a tidal flat, Roberts Bank, which is mostly dry at low tide and covered by about 1 m of water at high tide. Landward sections of the bank are covered by grasses and seaweeds and are fish spawning areas. Therefore, it was agreed to schedule construction to avoid the spawning season. The top of the bank is almost level, but the foreslope on the seaward side is relatively steep. British Columbia has fewer earthquakes than California to the south or Alaska to the north, but there are occasional large earthquakes. Oscillatory shear stresses induced by an earthquake might liquefy the loose and geologically recent sand and silt sediments of the bank, and parts of the foreslope could then become unstable, liquefy, and slide downhill into deeper water. The risk of liquefaction is least where the gradient of the foreslope is smallest, about mid-way between the South Arm of the Fraser and Canoe Pass. So it was decided that the pipeline should traverse the foreslope at that point. The route runs straight down the slope (down the *fall line*), so that if a flowside should occur, the sand

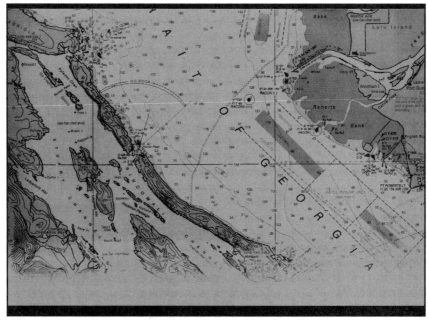

Fig 2–1 *Strait of Georgia between Fraser Delta and Vancouver Island*

would flow along the pipeline rather than across it. This was the decision because a pipeline can withstand very large forces applied along its length but only much smaller forces applied across its length.

In the deeper water in the center of the Strait, the maximum depth is about 380 m. At the time the route was being investigated, larger pipelines had already been constructed across the Strait of Messina (between Sicily and mainland Italy) in 380 m and across the Strait of Sicily (between Tunisia and Sicily) in 615 m. With that information, it was reasonable to conclude that the deep water would not be an obstacle.

On the western side of the deep water, a submerged ridge of sandstone, Galiano Ridge, lies parallel to Galiano and Valdes Islands. The ridge is bordered by near-vertical cliffs up to 20 m high. On either side of the ridge, the seabed is silt, too soft to support construction equipment or rock embankments. At first it was thought that there were no gaps in the ridge. Various construction alternatives were studied, and the studies concluded that the best option was to lay the pipeline to the top of the cliffs, to restart laying at the foot of the cliffs, and to join the ends by a length of pipe (spool piece), preshaped to fit the cliff and connected by hyperbaric welding, which is feasible because the depth to the top of the ridge is about 160 m. Later, however, another marine survey found that a gently sloping curved valley led up onto the ridge and that there was enough space to construct two pipelines through the valley. The project was called the valley Valdes Gap because it was a gap in the ridge opposite Valdes Island. In retrospect, it would have been wiser to call it Valdes Valley because the word gap allowed people opposed to the project to imply that it is a steep-sided canyon feature through which it would be difficult to lay pipelines. In reality, its sides have quite modest slopes.

The route then continued onto Valdes Island and across Stuart Channel to Vancouver Island.

That route was selected by the Public Utilities Commission of British Columbia after lengthy public hearings, but ultimately the pipeline was constructed by Pacific Energy Corporation along a route much further to the north. It crosses from the mainland to the south end of Texada Island, continues to the north end of the island, and then crosses westward to Vancouver Island, with a smaller spur to Powell River on the mainland. At the time of this writing, 15 years later, gas demand of the island has markedly increased; and an application to construct a second pipeline along a more southerly route north of San Juan Island is being considered by the National Energy Board.

The second example is the gas pipeline from Algeria to Spain. The shortest marine crossing is the Strait of Gibraltar, but laying the pipeline there means traversing Morocco; and for a long time, political differences between Algeria and Morocco made that route inaccessible. Direct crossings from Algeria to Spain were considered; but if Moroccan waters are to be avoided, the greatest water depth is at least 2500 m, and at the time that was thought technically impracticable for a large-diameter pipeline. That position has now changed.

In the late 1980s, reconciliation between the two countries made a route through Morocco practicable. The next question was where to cross the Strait. A crossing at the easterly end of the Strait encounters very deep water, more than 900 m to the south of Gibraltar, and also has to avoid Gibraltar which is disputed between Britain and Spain. Further west, the water is much shallower because of a submarine mountain range, called by oceanographers the Camarinal Sill, which runs north-south in an irregular curve. Depths on the top of the range are much less, between 300 and 400 m. However, a route along the range encounters two difficulties. The first is that the range has a rough and broken topography (as one might expect from a similar range on land) so that a pipeline that followed the crest of the range would have many long spans. The second is related to the complex flow of water through the Strait. Dense saline Mediterranean water flows from east to west through the bottom of the Strait. Less saline and less dense Atlantic water flows from west to east through the top of the Strait. The level of the interface changes with the tides in the Atlantic and is influenced by large-scale oceanographic changes. Internal waves form on the interface. The consequence is that currents are strong and highly variable and that they change markedly within a few minutes. These effects are at their most severe on the crest of the Camarinal Sill and are less strong in the deeper water on either side. That is another reason to avoid the crest of the range. The route finally chosen skirted the ridge to the west. The project is described in detail in the proceedings of a conference.[7]

References

[1] Sleath, J. F. A. (1984) *Sea Bed Mechanics*. New York: John Wiley & Sons.

[2] Komar, P. D. (1976) *Beach Processes and Sedimentation*. New York: Prentice-Hall.

[3] Palmer, A. C. (2000). "Are We Ready to Construct Submarine Pipelines in The Arctic?" Paper 12183 presented at 32nd Annual Offshore Technology Conference. Houston, TX.

[4] Woodworth-Lynas, C. M. L., J. D. Nixon, R. Phillips, A. C. Palmer. (1996) "Subgouge Deformations and the Security of Arctic Marine Pipelines." Paper 8222 presented at 28th Annual Offshore Technology Conference, Houston, TX.

[5] United States Army Corp of Engineers. (1973) *Shore Protection Manual*. Coastal Engineering Research Center.

[6] Park, C. A., A. C. Palmer, R. McGovern, and J. P. Kenny. (1986) "The Proposed Pipeline Crossing to Vancouver Island." Paper presented at European Seminar on Offshore Oil and Gas Pipeline Technology, Paris.

[7] Conference Proceedings. (1995) *Gibraltar Submarine Gas Pipeline: Meeting the Challenge*

3 Carbon-Manganese Steels

3.1 Introduction

For economic reasons, carbon-manganese steels are used whenever possible for the fabrication of pipelines for production and transmission of oil and gas and also for water injection systems.[1] Pipeline engineers need to be familiar with the modern methods of fabrication of pipe and of the limitations of particular steels to the type of product that can be safely transported in pipelines construct of those steels. This chapter describes the manufacture of carbon-manganese steel pipelines and the compositions and fabrication methods of the steel plate used for forming pipe. Advice is given on preparing a specification for pipeline material and on altering the specification correctly if the pipe is to be used for the transport of sour product. Corrosion, calculation of corrosion allowances, and corrosion limitations of the carbon-manganese steels are discussed in the section on internal corrosion.

Joining together short lengths of pipe, termed *pipe lengths* or *joints*, forms a pipeline. Early pipelines were constructed using screwed, bell and spigot, or flanged connections, as suitable steel and welding techniques were not available. Mechanical connections are still used sometimes nowadays, but almost all flowlines and transmission pipelines for the oil and gas industry are constructed from pipe joints fused together by arc welding. The steels used to form the pipe joints are low carbon–manganese steels. The higher strength grades are microalloyed and may be referred to as high-strength low-alloy (HSLA) steels. Similar types of steels are used for ships, pressure vessels, pump bodies, and oil country tubular goods (OCTG).

From a materials and corrosion viewpoint, it is generally the case that pipeline service is becoming more severe—both for new pipelines and pipelines in service. For example, there are new submarine multiphase pipelines in deep water operating at temperatures above 150° C at very high shut-in pressures and high concentrations of carbon dioxide. There are changes of use of existing pipelines in the Middle East where much of the crude and gas is sour. In this region, many companies producing from extremely extensive fields had a policy—now abandoned—to shut–in wells when significant water was produced. The pipeline production systems operated with dry hydrocarbons, so they avoided corrosion problems. As a consequence of the more severe service, a higher quality of pipe is required both for new fields and for replacements in the older Middle East fields, which are now in the refurbishment phase. To meet these demands, the steel and pipe production processes have become much more complicated.

3.2 Materials Specifications

In most parts of the world, the pipe joints for oil and gas will conform to the American Petroleum Institute API Specification 5L; for submarine pipelines the relevant specification is PSL 2.[2] In 1999 this specification was converted into an international standard, International Organization of Standardization (ISO) 3183, which covers the selection and use of seamless, longitudinal welded and helical (spiral) welded line pipe.[3] ISO 3183 covers a wider range of pipe material compositions than the API 5L that it replaced. ISO 3183 is in three parts with the various steel grades divided between Parts 1 and 2. Part 3 is based, in part, on the Engineering Equipment and Materials Users Association (EEMUA) Publication 166 and deals with both compositional and sour service requirements and is relevant to submarine pipelines.[4] The previous API 5L documents identified the steel grade by the yield strength as X42 to X80 where the number refers to the yield strength in thousands of pounds per square inch, psi. Hence, grade X52 has a yield strength of 52 psi or 52,000 pounds per square inch (psi), or in metric units, 358.5 MPa, since 145.04 psi is1 MPa.

Units are discussed further in Appendix 4. The range of standard pipeline strengths is given in Table 3–1. The more recent revisions and the ISO standard now refer to international system of units, abbreviated as *SI* units throughout.

***Table 3–1** Standard Strength Grades and Yield to Tensile Ratios*

Grade	Minimum Yield Strength		Minimum Tensile Strength		YS/TS
	lb/in²	*MPa*	*lb/in²*	*MPa*	ratio
A25	25,000	172	45,000	310	0.556
A	30,000	207	48,000	331	0.625
B	35,000	241	60,000	413	0.583
X42	42,000	289	60,000	413	0.700
X46	46,000	317	63,000	434	0.730
X52	52,000	358	66,000	455	0.788
X56	56,000	386	71,000	489	0.789
X60	60,000	413	75,000	517	0.800
X65	65,000	448	77,000	530	0.844
X70	70,000	482	82,000	565	0.854
X80	80,000	551	90,000	620	0.889

Though the API 5L Specification dated back to the 1920s, it became the basic international specification in 1948. At that time, the highest strength grades was X42. The ISO Standard now includes pipe grades up to X80. Despite the recent conversion to SI units, it remains common parlance in the oil and gas industry to use the foot-pound-second (FPS) units for general discussion. To accommodate previous and present design terminology, reference is made here to API 5L though it is to be understood that the same comments relate to ISO 3183. European Standard EN 10208, derived from British Standard EN 1028, Parts 1 & 2, is related to the new ISO 3183. EN 10208 gives compositional specifications for pipe usually as the maximum compositional values only; if considered relevant, minimum values need to be additionally specified. EN 10208 is only used within Europe, and elsewhere ISO 3183 is the primary document.

API Specification 5L is the minimum requirement to provide serviceable pipes, and most contractors and operators impose additional requirements. The specification gives a very limited chemical composition for the steel, and the purchaser is expected to negotiate the details with the pipe producer or supplier to obtain pipe suitable for its purpose. Typical formulations for the higher-grade steels are given in Table 3–2 and are discussed in more detail in the section that follows. API Specification 5L details the mechanical properties of the steel, methods for producing the pipe and for testing finished pipe. One of the more important functions of the API Specification 5L is the classification of dimensions and tolerances of the pipe joints, including standard diameters, wall thickness (termed schedule), lengths of joints, ovality, and out of straightness.

Table 3–2 Typical Compositions of Pipeline Steels

Pipeline Grade /wall	C	Mn	Si	Al $\times10^2$	Ca $\times10^3$	Ni	N $\times10^2$	Cu	V $\times10^2$	Nb $\times10^2$	Ti $\times10^2$	B $\times10^3$	P $\times10^2$	S $\times10^3$
Typical Formulations														
Basic API 5L	0.31	1.80											3	03
API 5L	0.16	1.56	0.35	4			1.2		7	5		1	3	15
Sweet Onshore	0.11	1.56	0.35	4		0.2	1	0.25	8	5			2.5	10
Sweet Offshore	0.08	1.56	0.30	4	3	0.2	0.8	0.25	8	4			1.5	5
Sour Offshore	0.05	1.00	0.30	4	5	0.2	0.7	0.25	6	5		4	1.5	1
Examples of Actual Pipeline Steels														
X65 16 mm	0.02	1.59	0.14							4	1.7	1	1.8	3
X65 25 mm	0.03	1.61	0.16			0.17				5	1.6	1	1.6	3
X65 25 mm	0.06	1.35				0.25		0.33	7	4	1.8		2.5	5
X70 20 mm	0.03	1.91	0.14							5		1	1.8	3
X70 20 mm	0.08	1.60						0.04	7					

Most national standards impose similar requirements. The Norwegian Offshore Standard OS-F101: Submarine Pipeline Systems published by Det norske Veritas in 2000 is beginning to be widely used internationally, not only in the North Sea. It describes specification requirements in great detail. For the first time, it includes an additional optional, tighter specification requirement, supplementary requirement U, whose objective is to penalize materials with a high variability of yield stress. The characteristic yield and tensile strengths to be applied in design are the specified minimum yield and tensile strengths multiplied by a material strength factor α_U, which is 0.96 for materials with a normal degree of variability. The factor is increased to 1.00 for materials that meet supplementary requirement U, which ensures increased confidence in yield strength.

3.3 Material Properties

3.3.1 Strength requirement

A pipeline steel must have high strength while retaining ductility, fracture toughness, and weldability.[5] There is some conflict between these properties. Strength is the ability of the pipe steel (and associated welds) to resist the longitudinal and transverse tensile forces imposed on the pipe in service and during installation. Ductility is the ability of the pipe to absorb overstressing by deformation. Toughness is the ability of the pipe material to withstand impacts or shock loads. Metallic engineering materials are generally tough and fail in a ductile manner, i.e., they yield before they break. In comparison brittle materials are glass-like and fail suddenly by fracture. Weldability is the ability and ease of production of a quality weld and heat-effected zone of adequate strength and toughness. Most metals can be welded but not all have good weldability. For example, the parts of an aluminum alloy airplane are held together with bolts, rivets, and adhesive, rather than by welding. For submarine pipelines, the prime factor driving the need for good weldability is economic. The largest percentage cost of a submarine pipeline is the installation because of the high cost of operating the lay barge. The faster the pipe can be welded, the faster it can be installed and the shorter the period of use of the lay barge.

The balance of properties (strength, toughness, and weldability) required depends on the intended use of the pipeline. An example of a severe service pipeline would be a high-pressure sour gas/condensate pipeline in Arctic conditions. Such a pipe would require heavy wall thickness with high toughness at low temperatures while having resistance to sulphide stress cracking. The heavy wall thickness would complicate the welding process. Obtaining both high strength and toughness without sacrificing weldability requires limited alloying combined with complex thermo-mechanical treatment of the steel combined with micro alloying.

Yield strength is a primary design parameter. As the yield strength increases, the wall thickness requirement decreases. A thinner pipe wall reduces material costs, transportation costs, load on the lay barge stinger, and welding costs. Up to the 1950s, the diameters and operating pressures of oil and gas pipelines were modest; and most lines were fabricated from seamless pipe. Product was relatively clean because problem gas was flared or the field not produced. In the mid 1960s, increased utilization of oil and gas led to a requirement for significant increases in the diameters of pipe for long gas transmission pipelines, which would operate at high pressures to reduce transportation costs. The increased diameter and higher operating pressures necessitated thicker wall pipe. Conventional stick welding of thick wall pipe led to hydrogen cracking problems at the weldment, termed *cold cracking*. Cold cracking could be overcome by altering the welding process to include the use of low hydrogen welding rods and by preheat and post weld heat treatments of the weld area. Some attempt was made to modify the composition of the steel but without notable success. It was soon realized that the most cost-effective approach was to increase the strength of the steel to permit use of thinner wall pipe by using heat treatment techniques while allowing a reduction in the alloying content, and thereby preserving weldability.

Submarine pipelines are generally designed in strength grades up to X65. The reason for this limitation on strength, compared to the highest strengths available, relates to the other pipeline design requirements that can mandate the provision of a thicker wall than that required for pressure containment. These requirements may include on-bottom stability, resistance to buckling during installation and stresses imposed during reeling. Resistance to bending and buckling depends primarily on the pipeline diameter/wall thickness ratio (D/t) and to a much lesser extent on material properties such as yield strength and elastic modulus. The reasons for this are discussed in chapter 10. It follows that if buckling is the factor that governs the wall-thickness design, there is no economic advantage to providing the additional steel as high-grade steel, rather than lower cost lower-grade steel. Welding of higher strength steels requires close control, and this may also be a factor in limiting strength in some cases. However, it is generally the case that higher-grade steel is cost effective because the reduction in tonnage of material more than makes up for the increased cost per tonne.

Over the last decade, many land pipelines have been constructed with grade X70 and X80 steels; and confidence has grown in the use and welding of these steels. Consideration is now being given to use of pipe to grade X100.[6] It is expected that the higher strength steels, X70 and X80, will be more widely used for submarine pipelines in the future. For example for S-lay installation of larger diameter pipelines in very deep water, the suspended weight becomes a limiting factor. The risk of buckling can be reduced by installing the pipeline partially flooded with seawater.[7] Use of X80 steel would reduce suspended weight while meeting the reduced D/t ratio. Pipelines enclosed in bundles are fabricated onshore, and the welding reservations do not apply. As the bundle is required to have minimum submerged weight, the use of the higher-grade steels and consequent minimum wall thickness and lower submerged weight becomes beneficial.

3.3.2 Improving strength

Steel strength can be increased by the use of one or a combination of the following mechanisms:

- Solid solution strengthening by addition of alloying elements C, Si, Mn, etc.
- Grain refining
- Precipitation strengthening by microalloying with Nb, V, Ti
- Transformation strengthening by formation of martensite and bainite
- Dislocation strengthening by work hardening

Steps in the historical use of these mechanisms in pipeline and OCTG production are illustrated in Figure 3–1. Some of these strengthening mechanisms also have disadvantages for material for submarine pipelines, and these are noted where relevant. For API 5L type pipeline steel with 1.5% manganese, the typical percentage effects of the strengthening processes are given in Table 3–3.

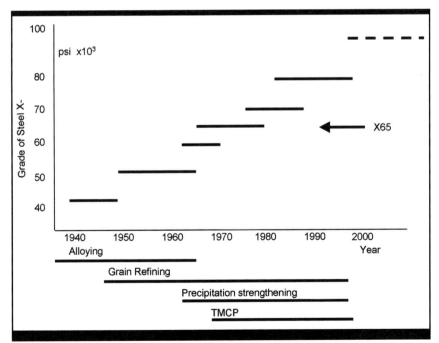

Fig. 3–1 *Application of Strengthening Mechanisms to Pipeline Steels*

Table 3–3 *Percentage Effect of Strengthening Mechanisms*

Strengthening Option	Mechanism	% Effect on Strength
Base Line strength		18
Addition of silicon and nitrogen	solid solution	8
Addition of manganese	solid solution	12
Ferrite grain size	Grain refining	45
Micro-alloying	Precipitation	17

3.3.3 Solid solution strengthening

The oldest and simplest method of increasing the strength of a metal is to increase the alloying elements: this is termed *solid solution strengthening*. When molten steel solidifies, it does so by forming very small crystals of iron, which are intimately interlocked. These small crystals are termed *grains*. When the metal is stressed, the stresses are relieved within the crystals by slippage (dislocation) of the atoms within the grains and by transference of stresses across the grain boundaries by slippage between grains. The dislocations cluster around the areas of high stress and reduce further slippage and atom movement within the metal. The dislocations may be seen as causing a localized work hardening of the steel. These processes may be reversible on the macro scale. If the stress applied deforms the steel within this reversible region, the steel recovers elastically. If the stress on the steel is increased further, there comes a point when the dislocation movements become irreversible; and permanent alteration is made to the steel. This deformation occurs at the yield point of the steel. Further increasing the stress will lead to the failure of cohesion of the material, and a tensile failure will occur at the tensile strength of the material. When a material is fatigued, the clusters of dislocations steadily grow to produce micro-voids within the steel that eventually lead to fracture of the material.

The introduction of alloying elements distorts the crystal lattice and reduces the freedom of movement of the dislocations. It takes a greater applied stress to move the dislocations through the distorted crystal lattice. Though material is strengthened, the ductility is reduced. If the alloying is increased further, there comes a point where slippage of dislocations does not occur; and the metal loses all ductility and becomes brittle. Cast iron and ductile iron are examples. The carbon content is high (~ 3%) so that the material has good strength (210 to 480 MPa) but has lost almost all its ductility. Another example of the effect of high alloying on steel is silicon iron (~ 14% Si), which again has moderately high strength (380 MPa) but is brittle. Table 3–4 gives the strengthening relationships for the more common alloying elements used in structural steels.

Table 3–4 *Solid Solution Strengthening*

Element	Strengthening MPa per wt%	Element	Strengthening MPa per weight %
Carbon	5,500	Copper	40
Nitrogen	5,500	Manganese	30
Vanadium	1,500	Molybdenum	11
Phosphorous	700	Nickel	0
Silicon	80	Chromium	-31

Of the available solid solution alloying agents, carbon, silicon and manganese are cost effective. There are upper limits to the strength that can be obtained by solid solution strengthening, and this process on its own is inadequate to produce high-grade pipeline steels. Moreover, though strength is increased by the addition of the solid solution strengthening elements, many of them cause a loss of toughness and weldability. The elements that contribute most to the strengthening of the steel are detrimental to toughness, e.g., carbon and nitrogen. Manganese appears to be a most beneficial element because it also helps to reduce the grain size and is involved in the distribution of precipitation strengthening particles. Manganese sulphide, present as a tramp compound, is, however, a prime reason for reduction of resistance of steel to hydrogen-induced cracking (HIC), which is important in sour service.

Carbon is a primary alloying element and also causes a major effect during welding. When the molten steel solidifies, the crystal lattice formed is face-centered cubic (FCC), also termed *γ*, *austenite*, or *austenitic*, in which an iron atom is situated at each corner of a cube with other atoms in each face of the cube. On further cooling, the crystal lattice spontaneously alters to body centered cubic (BCC), also termed *α*, *ferrite*, or *ferritic*, in which an iron atom is situated at each corner of a cube and one atom is in the center of the cube. In the FCC form, the steel dissolves considerable carbon. In the BCC form the solubility is reduced, and excess carbon precipitates as a separate iron-carbon phase.

Steels with an equilibrium state between iron and carbon are produced only in laboratories because it takes an impracticably long time to achieve. In practice, the steel is not at equilibrium because some excess carbon and other alloying elements will be held in the steel or will have been shed in a non-equilibrium form. The quantity and form of the carbon-steel interactions alter the properties of the steel. These are discussed in the section 3.4.7 on transformation strengthening.

Very rapid cooling results in the carbon being trapped in the ferrite grains, producing an acicular structure called martensite, which is very hard and prone to cracking. Slower cooling allows some equilibrium material to be formed interspersed with a non-equilibrium structure and is termed *pearlite*. Pearlite is alternate layers of ferrite and iron carbide and has a glossy appearance at high visual magnification. Use of carbon as a strengthening agent, therefore, requires subsequent heat treatment of the steel to obtain a suitable distribution of phases.

Tempering involves the re-heating of the steel to a temperature below the austenitizing temperature (specific to the formulation and about 740–760 °C). *Normalizing* is a term indicating the heated steel is cooled in still air. Normalizing of steels is used for grades up to X52. *Annealing* indicates the steel was very slowly cooled in a furnace and produces very ductile, low strength steel and is not a technique used for production of steel plate for pipe fabrication. *Quenching* is the term reserved for the rapid cooling of steel and generally involves using oil or water (sometimes forced air or brine is used). During quenching, the high carbon phases formed with the ferrite may be martensitic and very fine grained. On tempering, the martensite transforms to ferrite and pearlite. Quench and temper is typically used for steels up to X70. Time–temperature–transformation (TTT) diagrams or continuous cooling transformation (CCT) diagrams are the practical diagrams used to decide the properties resulting from a heat treatment.[8]

3.3.4 Weldability

Alloying elements affect the weldability of the steel; this subject is discussed in more detail in chapter 5. The weld itself is a cast structure, and the steel adjacent to the weld is heated into the austenite region and then annealed by the subsequent weld passes. This action results in changes in the composition and morphology of the steel grains in this region. During the welding, some hydrogen gas will be dissolved into the austenite and will attempt to escape when the material converts to ferrite. If martensitic or bainitic material has been formed, the escape attempt can lead to hydrogen cracking. Empirical formulae have been developed to provide guidance on the level of alloying that can be accepted while maintaining weldability and avoiding hydrogen cracking. The two most important formulae are the International formula (used in API Specification 5L) and the Ito Bessyo formula, also known as the parameter of crack measurement (PCM) formula. These equations use empirical factors to adjust each alloying element to a carbon equivalent so that a single relative number is obtained. The equations are as follows:

International equation:

Carbon Equivalent = *C* + *Mn*/6 + (*Cr* + *Mo* + *V*)/5 + (*Cu* + *Ni*)/15   (3.1)

Ito Bessyo (parameter of crack measurement) equation:

PCM = *C* + *Si*/30 + (*Mn* + *Cu* + *Cr*)/20 + *Ni*/60 + *Mo*/15 + *V*/10 + 5*B*   (3.2)

The terminology is explained in chapter 5. The International formula is used for steels with high carbon levels and, nominally acceptable for all carbon levels, is generally considered applicable for carbon levels down to 0.15. For the low carbon steels and steels with C <0.15, the PCM formula is claimed to be more reliable as a predictor of steel behavior during welding. Usually the CE is limited to 0.43 for manual metal arc welding, though for modern steels and semi-automatic welding, this will be reduced to 0.34 or even 0.32. The PCM is usually specified as from 0.18 to 0.22. The distribution of alloying elements appears to follow a normal distribution and, as a result, the CE and PCM will also tend to be normally distributed between individual pipe joints.

The EEMUA recommend in their Document 166 that full-scale weldability trials be undertaken if certain element analysis limits are exceeded.[6] These limits are summarized in Table 3–5.

Table 3–5 *Composition Limits before Weldability Trials Required (EEMUA 166)*

Alloying Element	Limits of Compositions (%)	
	Seamless Pipe	Welded Pipe
C	0.16	0.10
Mn	1.10 to 1.40	1.05 to 1.55
Si	0.20 to 0.45	0.20 to 0.45
Ni	0.25	0.30
Cu	0.20	0.35
Cr	0.20	0.10
Mo	0.20	0.10
V	0.05	0.08
Nb	0.04	0.05
Ti	0.015	0.02
Al	Twice N to 0.05	
Ca	Twice S to 0.05	
Ce + Mo + Ni + Cu	0.80	
Nb + V	0.10	
Nb + V + Ti	0.12	

3.3.5 Grain refinement

Steel fabrication techniques were revolutionized in the 1950s by the realization that refinement of the crystalline ferrite grains resulted in increases in both strength and toughness and without sacrificing weldability.[9] The ability to alter grain size results from the iron metamorphoses at elevated temperature. When heated above the transition temperature the body-centered cubic ferrite (α) iron is transformed into face centered cubic austenite. The transition temperature varies with the extent and nature of the alloying elements. On transformation, the iron grains recrystallize and in reforming become smaller but then grow in size if the temperature is held constant. Recrystallizing the steel and quickly reducing the temperature freezes the smaller grains.

Alloying elements, however, can interfere. The ferrite structure has very limited voidage between the iron atoms compared to the austenite structure. At a high temperature, the iron can dissolve relatively large volumes of alloying elements; but these elements must partition out when the steel reverts back to the ferrite form. The form of carbon precipitates has a particular effect on the steel mechanical properties. Some of these are beneficial to the mechanical properties of a pipeline steel while others must be avoided, e.g., martensite.

The relationship of strength to grain size is represented by the following formula:

$$Yield\ Strength = Grain\ Friction + Dislocation\ Locking\ Constant \times \delta^{-0.5} \quad (3.3)$$

The grain friction factor is a measure of the forces that oppose slippage between the individual grains and the dislocation-locking constant is a measure of the reduction in dislocation movement within the grains. Both forces are fixed for a given chemistry of steel. Other models have been derived that relate yield and tensile strengths to steel chemistry and grain size, and these provide an insight into the effects of the alloying elements:

$$Yield\ Strength = 53.9 + 32.3\ Mn + 83.2\ Si + 354\ \%N_{free} + 17.4\ \delta^{-0.5} \quad (3.4)$$

$$Tensile\ Strength = 294 + 27.7\ Mn + 83.2\ Si + 3.85\ Pearlite + 7.7\ \delta^{-0.5} \quad (3.5)$$

where

Mn   = weight percent of manganese

Si   = weight percent of silicon

N_{free} = weight percent of free nitrogen

δ = average grain size

Grain size reduction can be achieved in several ways. The most common technique for pipeline steels is to use aluminum as a grain-refining alloy. The aluminum, about 0.03%, is added to the ladle steel. It combines with nitrogen to form aluminum nitride as a dispersion throughout the steel. During heat treatment, these particles lock the austenitic grains and prevent their growth. On cooling, the fine austenitic grains convert to fine ferrite grains. Addition of carbon and manganese depresses the transformation of austenite to ferrite, and this and more rapid cooling would have a similar effect on grain size. The combination of basic steel compositions and aluminum grain refining can produce steels with strengths up to X42. To produce the higher strength steels, other strengthening mechanisms need to be applied.

3.3.6 Precipitation strengthening

There is an upper limit to the strengthening that can be achieved with aluminum grain refining because too much aluminum (above 0.03%) has a deleterious effect on the fatigue performance of the steel. The mechanism of strengthening by the aluminum nitride precipitates can be exploited using other precipitation-hardening alloying elements. For the structural steels, the cost effective precipitation strengthening alloys are vanadium, niobium, and titanium. These materials are sparingly soluble in steel and have an affinity for combination with carbon and nitrogen. Typical additions are 0.1–0.2% V and 0.05–0.1% Nb. The microalloying elements dissolve in the molten steel and react to form carbides (-C), nitrides (-N), or carbo-nitrides (-CN) that precipitate at the austenite-ferrite boundaries as the steel transforms from austenite to ferrite. The grain growth of the ferrite is restricted. During normalization, the transformation to austenite and back results in additional grain size pinning; and a very fine grain steel is produced. The precipitates also increase the frictional forces between the grains, which have a direct effect on the steel strength. Of these micro-alloying agents, vanadium has the most powerful strengthening effect, 1,500 MPa per unit weight (% v).

3.3.7 Transformation strengthening

When the steel is heated above the austenite transition point, carbon and many other alloying elements dissolve into the austenite. On cooling and conversion to ferrite, excess carbon separates as an iron carbide secondary phase. The nature of this precipitated iron carbide depends on the rate of cooling. Slow cooling results in the formation of alternate layers of ferrite and iron carbide, termed *pearlite*. Fast cooling, with/without solid solution strengthening results in the formation of a less defined micro-structure: two forms of ferrite and iron carbide microstructures occur, termed *bainite* and *martensite*. Both these microstructures result in very high strength steel, but toughness and ductility are sacrificed. Martensite is particularly deleterious and welding would also be affected: the use of martensite is not a viable option for pipeline steels.

Ferritic and pearlitic steels have tensile strengths in the range 450–600 MPa (X65–X80). The same steel composition with a bainite structure would have strengths ranging from 600–1,000 MPa; and a martensitic structure would have a tensile strength of 1,000–1,200 MPa. The transformation microstructure that is used for the higher-grade steels is pearlite; however, the increase in the tensile strength is at the expense of a reduction in the toughness of the steel. Transformation strengthening can be of critical importance in obtaining adequate strength in the heat-affected zone adjacent to the girth welds and for hot-formed bends.

3.3.8 Dislocation strengthening

If steel is plastically deformed, i.e., stressed beyond the yield point, then the dislocations moving through the crystal lattice, which provide the ductility of the material, become locked in place. The material increases in strength but loses ductility. This process is termed *dislocation strengthening*; but when it occurs fortuitously, it is generally termed *work hardening*. This technique is used for improving the strength of pipeline steels in one of the thermo-mechanical control processing (TMCP) options for fabrication of the plate. There is also a degree of dislocation strengthening of the pipe fabricated by the U-O-E process during the expansion of the pipe subsequent to completion of the longitudinal weld.

Some work hardening may occur during installation of the pipe when it is strained over the upper bend during S-lay or during reeling of a pipe onto a reel ship. The strain imposed on the pipe during these procedures is highly controlled and within the elastic limit, though the total stresses imposed, including residual stresses, can be up to 95% of yield. Accidental work hardening can occur when the pipe is dropped, indented, or other-wise roughly handled. Such uncontrolled work hardening of the steel should be avoided, as the material usually must be re-evaluated for fitness for service if significant deformation has occurred. The pipeline material, and installation specifications need to include clauses detailing what steps need to be taken if inadvertent work hardening of the pipe does occur, particularly of pipes to be used for sour service.

3.3.9 Measurement of strength

Yield strength is used as a major criterion for design of a pipeline. The standard strengths of line pipe are given in Table 3–1. The value generally quoted is the specified minimum yield stress (SMYS). The most common method of measurement of tensile strength is by pulling shaped samples uniaxially. There are standardized specimen dimensions and test procedures. A typical relation between stress and strain is shown in Figure 3–2. Stress is defined as the ratio between the tensile force applied to the specimen and its initial cross-sectional area before the test began. Strain is elongation over a defined gauge length divided by the original gauge length.

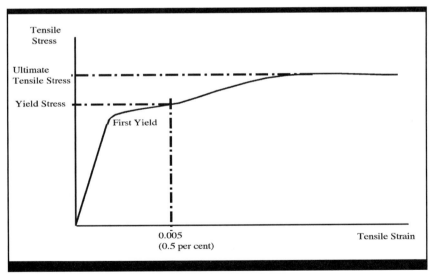

Fig. 3–2 *Relation between Stress and Strain in a Tensile Test*

The yield stress is conventionally defined as the stress at which the strain reaches 0.005 (0.5%). This is not the strain at which the material begins to deform inelastically, which is somewhat lower, about 0.002. Continued deformation after yield leads to an increased stress because of strain hardening. The ultimate tensile stress is reached at a much larger strain, typically about 0.20.

Tensile samples are cut from representative pieces of pipe and must be cold flattened to produce the tensile test specimens. The flattening process itself necessarily creates plastic deformation and, therefore, strain hardening—one manifestation of the Bauschinger effect, which in plasticity theory describes the effect on stress-strain relations of previous plastic strain. This effect may alter the apparent tensile strength of the material. The strength may be increased or decreased. In X70 pipe, for instance, the effect can reduce the apparent yield strength by 70–80 MPa (~15%). There are alternative methods of testing, such as inducing circumferential tension in a ring cut from the pipe. It can be argued that ring testing gives a more reliable indication of the stress/strain behavior. Ring tests take longer and are more expensive, and pipe fabricators are generally not geared up to do large numbers of these alternative tests.

Several possibly contentious issues can arise as a result of the Bauschinger effect. A suitable piece of pipe may be rejected because the tensile strength appears to be below specification. An unsuitable piece of pipe may be passed even though its true strength is inadequate. More usually the pipeline material passes the test, but the steel strength is considerably above that required and for which the pipe has been designed. This may create complications during S-laying or reeling of the pipe.

The magnitude of the Bauschinger effect depends on the microstructure of the steel, as this is a major parameter affecting the magnitude of change in yield stress. Steels with a ferrite-pearlite composition show the maximum loss of yield strength. Steels containing martensite show little or no effect, and those with low temperature transformation products may show an increase in strength.[8] The steels can be classified as:

(a) Steel showing a significant loss of strength

- High carbon ferritic-pearlitic steel with niobium or vanadium as microalloying agents
- Normalized high-carbon ferritic-pearlitic steels with niobium or vanadium as microalloying agents
- Low-carbon, controlled-rolled, pearlitic steels with niobium or niobium and vanadium as microalloying agents

(b) Steels showing little or no loss of strength

- Controlled-rolled, low-carbon steel with pearlite-reduced transformation products that contain vanadium, niobium, and chromium

(c) Steels showing an increase in strength

- Cold-worked, high-grade ferritic steel with reduced pearlite and with niobium or vanadium as microalloying agents
- Very low carbon, high-grade, controlled-rolled, two-phase steels with niobium or vanadium as microalloying agents

Submarine pipeline steel strength is generally restricted to grade X65, and the usual problem is an apparent reduction in strength.

3.3.10 Ductility

Ductility of steel inevitably reduces as the strength is increased. A correct blend of strengthening mechanisms should ensure that strength is achieved without sacrificing ductility. Ductility is measured in a tensile testing machine as the percentage elongation of a sample of the material at a defined tensile stress that is slightly above the nominal yield strength. The standard sample for pipeline testing is 50.8 mm (previously 2 in.). The elongation required ranges from 20% to 40% depending on the tensile strength of the steel. Longitudinal and transverse properties of the pipe have different strengths and ductility, and the specification must make clear the value and orientation that is to be tested. Simple bend tests are also used as practical guides to ductility, and these involve various degrees of flattening of the pipe and examination of the surface for cracks.

The stress range of the region of ductility is important. As the strength of the steel is increased, the leeway between yield and tensile strength is reduced, i.e., the ductile region between the elastic behavior of the pipe and tensile failure is narrowed. This can have important consequences during the installation and service of the pipeline. If the pipe is overstressed, e.g., as a result of heavy weather during installation, then it may fail by tensile tearing, rather than deforming and remaining intact. The level of stress imposed on the pipe is carefully controlled during installation, and pipe that is damaged would be replaced. However a tensile failure might allow the pipeline to fall to the seabed, and this action could result in an expensive recovery operation. Even within the controlled stress domain if there is an inadequate ductile range, the pipe may buckle, rather than bend. In this case the problem is related to the transverse strength, rather than to the longitudinal strength.

To allow an adequate window between yield and tensile strength, it is usual to specify a minimum ratio between the yield strength and the tensile strength. For example, a value of 0.93 would mean that the yield strength

is limited to 93% of the tensile strength or, alternatively, that the tensile strength is at least 7% above the yield strength. Typical ratios used in pipe specifications are 0.90 maximum transverse and 0.92 longitudinal for a sweet service pipeline and 0.93 transverse and 0.95 longitudinal for a sour service pipeline. Land pipelines usually require a larger difference, and a typical specification would be 0.85 transverse and 0.875 longitudinal. The API Specification 5L yield/tensile ratios based on minimum yield and tensile strengths are much lower than these values but may not be guaranteed in practice.

3.3.11　Overstrength

In the past, the steel makers struggled to produce steel of high grade, but nowadays it is increasingly common to find that the steel provided has strength above that required for the particular service, i.e., the steel is overstrength. For example, nominal X52 steel may actually be X60 steel. In most cases overstrength of the steel does not matter, provided it is supplied at no additional cost; but it is important that the pipeline engineer knows the actual mean strength of the steel, rather than assuming it merely meets the level specified.

Knowing the actual strength of the steel may save money as the behavior of the pipe in circumstances leading to upheaval buckling and in spans is affected by the strength. However, it is often the case that these calculations are fixed, and the advantages afforded by the higher strength cannot be realized. On the other hand, there may need to be restrictions on the strength of the steel, e.g., for reel ship laying. Generally, the disadvantages of overstrength are seen to outweigh the possible advantages, and increasingly specifications are written with restrictions on the upper level of strength that is acceptable. The upper limit might be set as an actual strength value, e.g., 450 MPa, -0, +140 MPa, or as a percentage, e.g., 125% YS.

DNV rules require a minimum safety factor of 80% that is achieved by setting minimum limits on the mean yield strength, tensile strength, and wall thickness. In practice, this rule sets the limits for the quoted SMYS, SMTS, and rolled pipe wall thickness as:

Mean YS - 2 ◊ standard deviation ≥ SMYS

Mean UTS - 3 ◊ standard deviation ≥ SMTS

Mean thickness + 2 ◊ standard deviation ≥ minimum wall thickness

When available, the statistical distribution data of strength of the pipe material should be collected and retained in the pipeline file. Most pipe fabricators collect this data from their test programs and can readily provide it in electronic, graphical, or tabular form. In the future, this distribution will be useful for evaluation of accidental damage and residual life assessments that take the statistical properties of the pipe and use these to calculate the risk of failure of the pipeline. Such information may be impossible to replace once the pipeline is installed, and everyone has moved on to the next project.

3.3.12 Toughness

Toughness is a measure of the resistance of the pipe material to impact loading. Tough materials yield on impact and fail in a ductile, progressive manner. Non-tough, brittle materials do not. Brittle fracture is sudden, and this may have catastrophic results. Glass is usually used as an example of a brittle material, but this is not a true comparison as glass has an amorphous structure. Perhaps more relevant as an example of a brittle metallic material would be old fashioned gray cast iron. Toughness is measured by impacting a sample of the steel with a heavy object moving at a defined velocity.

BCC crystalline materials such as steel are susceptible to a change in toughness as the temperature falls. When the steel temperature falls, the atoms in the crystal lattice become closer together as the metal shrinks. The close packing of the BCC lattice restricts the degree of shrinkage that is possible. Highly alloyed material can shrink less. A lower bound is reached beyond which further shrinkage is not possible. As a consequence, the material has no residual elasticity remaining to absorb impact energy and fails in a brittle manner. When impacted, the brittle material fails by cleavage between the crystal planes. The temperature at which the material changes from ductile to brittle behavior is loosely termed the *transition temperature*. A fracture surface, which has failed in a brittle manner, shows a glittering, crystalline appearance because of the flat facets produced by failure across the planes in the material.

A plot of toughness against temperature appears as two plateaus linked by a sloping ramp. The toughness of steel, measured as its resistance to sudden impact, is only slightly reduced with reducing temperature from ambient until the upper critical temperature is reached. At and below this temperature, the toughness of the steel declines. Below the lower critical temperature, the impact resistance of the steel remains relatively constant at the lower value. The non-ductile transition temperature is the tempera-ture roughly half way between the two critical temperatures. For high

carbon steels the transition ramp between the upper and lower *shelves* is steep while for low carbon steels the ramp is gradual. The plateaus are termed the upper and lower shelves. The broken surfaces are usually examined after the impact testing. Brittle failures show a crystalline appearance, and the ratio of this crystalline area to the complete fracture surface is used as a parameter in evaluating the point of change over from ductile to brittle behavior.

The point of transition from ductile to brittle is defined as the non-ductile transition temperature (NDTT). This is the temperature at which the fracture surface appears 50% ductile and 50% brittle. One major unresolved problem is that the different test methods and their interpretations give different critical temperatures. It is important that the test method to be used and the test result interpretation are agreed upon by the material specifier and pipeline provider.

The NDTT temperature is a function of metal composition and grain size, the item thickness, and the rate of impact. The slope of the ramp reflects the statistical variation of these properties within the steel. Thick section material has a higher transition temperature than thin metal because the thin material can yield, thus reducing the effective rate of transfer of stress. Slow impact loading allows the material to deform in a plastic manner, reducing the NDTT. Pipeline testing generally uses full thickness specimens of pipe though thinner samples can be used if the pipe wall thickness is very large, but this variation in testing is subject to negotiation between the purchaser and the pipe producer.

Whereas strength of materials is well understood, toughness is not. Toughness is increased by reducing alloying content and/or by decreasing the grain size. However, other factors are important, for example, the degree of segregation of phases within the steel, the presence of inclusions, a failure to fully kill the steel leaving gas vacuoles within the steel matrix, the presence of surface, and/or internal defects. Reduction in alloying improves toughness, but because reduction in alloying would lead to low strength steel, heat treatment to temper or normalize the steel is used to ensure toughness by reducing grain size and eliminating phase segregation and hard brittle material or areas within the steel. The effect of alloying additions and grain size, δ, on the yield, tensile strength, and non-ductile transition temperature can be deduced from equations used by steel makers to model these effects:

$$NDTT = -19 + 44\,Si + 700\,(Ni)^{0.5} + 2.2\,Pearlite - 11.5\,\delta^{0.5} \qquad (3.6)$$

where

Si = weight % silicon

Ni = weight% nickel

δ = average grain size

The toughness required and the acceptable transition temperature depend on the service and environment to which the pipe will be exposed. Gas and condensate pipelines in particular require significant toughness. There is a very high energy of gas compression in such pipelines; and, if a failure were to occur, the expansion of the gas would result in a rapid fall in the temperature of the gas and of the adjacent steel by Joule-Thompson cooling. The temperature fall could lead to a potential shift from ductile to brittle behavior of the steel. The initial failure may become a running crack in susceptible material. The crack runs at the speed of sound in the material and will continue until the driving energy of the gas falls below a critical level or the crack hits a sufficiently tough section of pipe.

Ensuring conformance with a defined toughness criterion is supplemental to API Specification 5L and will not normally be provided by the pipe fabricator unless included in the material specification under Supplemental Requirements SR 5A and SR 5B. SR 5A relates to the NDTT and SR 5B to the absorbed energy. Even if good impact resistance at low temperature is not essential, it is good practice to specify impact testing at the standard testing temperature of 0 °C. The standard Charpy V-notch test gives a good indication of the general quality of the steel, and the tests can be used for quality assurance/quality control (QA/QC), rather than as definitive guides to the behavior of the steel under a sudden imposed load. If the pipeline will be installed or operated at temperatures below freezing or is a gas or multiphase pipeline; then toughness at a lower temperature will be required, and the impact energy and test temperature must be defined clearly in the material specification under SR 5A and 5B.

A range of tests is available to ensure that the steel has adequate toughness. All the test methods employ suddenly applied forces. Drop weight testing, tear testing, and Charpy testing are typical test methods. The function of the tests is to ensure that the pipe material will meet two design values decided by the pipeline engineer: the NDTT and the absorbed energy. The NDTT specified should be slightly more severe than the worst-case temperature expected in service, either environmental or resulting from a loss of pipeline integrity. If the NDTT is set for installation conditions, for example, if installing a pipeline in a cold climate where the

temperature may fall below zero; the absorbed energy specified should be sufficient to tolerate the impacts likely to occur during installation. For example, the NDTT may be based on a piece of pipe at a temperature of –10 °C being dropped on the lay barge deck. If the NDTT is set for operational conditions, the absorbed energy should be determined by calculation of the expected forces resulting from in-service loading and/or forces resulting from a product leak. For the North Sea, typical values specified are 68 J at –10 °C with a minimum allowable value of 27 J; these values reflect installation, rather than service requirements.

The water temperature at the seabed generally will not be lower than 4 °C as below this temperature the density of water decreases and the water rises. This temperature limitation means that submarine pipelines do not usually require toughness at very low NDTTs during service. Even with gas cooling by expansion, the heat transfer from the seawater to the pipeline limits the temperature fall. Pipelines, however, may need to be installed in cold, windy conditions when the material temperature may fall well below zero because of chill factors. The riser, particularly the section above mean sea level, will generally be the section of the pipeline that may need to be considered separately.

Charpy testing is the most common requirement in pipe steel specifications. In this test, a 2 mm deep by 55 mm V-shaped notch is machined in the sample of pipe, and the sample cooled to the relevant test temperature. After cooling, the sample is clamped in a vice and impacted by a weighted hammer attached to a pendulum that is allowed to swing down from a defined height. The type of hammer used depends on the test code being followed (ISO or ASTM). The residual upswing of the pendulum after breaking the sample is used to calculate the energy absorbed in breaking the sample.

The Charpy test is rapid has a modest cost, and gives some indication of the resistance to initiation of a brittle fracture. Usually, the steel is required to accommodate a particular energy (e.g., 25–75 Joules) at a defined temperature (0 to –40 °C). Typically, the values for the larger diameter pipes for cold-water service depend on the pipe wall thickness and the microalloying used for precipitation strengthening. Pipe below 20 mm wall thickness would be specified as having a mean absorbed energy of 60 J at –20 °C while pipe with a wall thickness of 20–30 mm would be expected to have a mean absorbed energy of 60 J at –30 °C. The microalloying agent does not alter the absorbed impact energy but alters the transition temperature.

It is important to require toughness tests on the formed pipe, rather than on the plate from which the pipe is formed. Typically, the toughness of pipe will have an impact toughness of about 10–20 J less than the plate. The heat-affected zone (HAZ) of the longitudinal weld of U-O-E pipe would be typically 10 J less than the parent plate and would be unlikely always to pass this criterion at temperatures below –20 °C. The longitudinal weld composition is also important. Carbon alloying would only provide adequate toughness at 0 °C, and it is necessary to provide nickel and molybdenum to achieve toughness at –20 °C and titanium and boron to achieve lower transition temperatures.

There are large differences in the impact behavior of the steel between transverse and longitudinal directions because of the shape of the steel grains resulting from the rolling of the steel during pipe or plate fabrication. For typical pipeline steels, the longitudinal sections absorb at least twice the impact energy of transverse sections. The finer the grain structure, the smaller the difference; but there is always a difference because of the grain extension that results from the rolling. For in-service toughness, the Charpy tests should be specified on transverse sections, rather than on longitudinal sections because crack propagation is in the longitudinal direction and resistance to the crack is required perpendicular to the direction of crack propagation. For installation toughness, the most likely area of failure is at the girth welds; and Charpy testing of the heat-affected zone adjacent to the welds is a good practice.

The Charpy V-Notch Test does not give a true reflection of the tear resistance and additional and/or alternative testing may be required. The Battelle drop weight tear test (DWTT) is the test most widely used as an alternative and/or adjunct to the Charpy V-notch Test and should be in accord with API RP 5L3. In this test, a falling weight impacts a horizontally mounted sample cooled to the required test temperature. The 75 mm x 305 mm sample contains a machined notch—representing a flaw—at the point of impact. The complete energy of the impact is absorbed by the sample, and the length of the crack from the impact point to the point of crack arrest is used as a measure of the ability of the material to resist propagation of a brittle fracture. The DWTT is done on full-thickness, flattened samples of pipe taken from transverse sections of the pipe. It provides a more realistic evaluation of the likely behavior of the material compared to a Charpy V-Notch Test. The sample is examined for shear area of fracture, rather than for the absorbed energy; therefore, the test evaluates the resistance to fault propagation, rather than initiation as with the Charpy V-notch Test. Typical acceptance values for a sweet pipeline would be 85% at 0 °C and 75% at –10 °C and for a sour pipeline 80% at 0 °C and 70 % at –10 °C.

To obtain data for fracture mechanics from which the maximum acceptable flaw size can be calculated, the pipe material can be evaluated by crack tip opening displacement (CTOD) testing. A full thickness sample is notched across the full width in the area of interest, e.g., at the fusion line of the longitudinal weld; then the sample is pulled open and the displacement of the crack tip measured. Typical acceptance values are 0.15–0.2 mm at 0 or –10 °C.

There are other tests available that use similar techniques to evaluate the behavior of the material subject to impact, but most are of more relevance in evaluating plate to be used for tanks or ships than for pipelines. If Battelle DWTT and/or other tests are required, they should be specified under Supplemental Requirement SR 6 of the API Specification 5L.

Emphasis on the intrinsic toughness of the material is important, but attention must be paid to other factors that can markedly affect the toughness of the pipe sections or completed pipeline. Surface blemishes on the pipe and weld strikes need to be minimized as they can act as crack initiators. Weld morphology must be controlled to ensure that adverse features (cracks, lack of side wall fusion, excessive porosity) are avoided. Pitting corrosion at welds also must be suppressed during service.

3.4 Pipe Production

3.4.1 Steel-making for pipe fabrication

In the 1950s, the steel for pipe manufacture was produced by the acid open-hearth process. This was the most efficient method then available for producing moderately low carbon steels. The molten steel was treated with manganese to remove the bulk of the sulphur and oxygen, then killed with calcium, silicon, and aluminum to remove the residual oxygen and nitrogen before being poured into ingots. The ingots would be transferred to the shaping plant to be cut into billets for seamless pipe manufacture or hot rolled into plate for welded pipe manufacture. Steel is still made by this process, but it should be avoided for pipe production except for pipes of the lower strength grades in non-severe service.

Quality pipe is fabricated from slabs of steel continuously cast from molten metal produced by the basic oxygen process (Basic Oxygen Steel), illustrated in Figure 3–3, or by electric arc smelting. The production process must produce steel that is low in carbon and sulphur, phosphorous, oxygen, hydrogen and nitrogen (termed SPOHN), and manganese and microalloying elements need to be added. The sequence to produce the slabs involves primary steel production, secondary or ladle steel production, and casting. Intermediate steps include magnetic stirring and degassing.

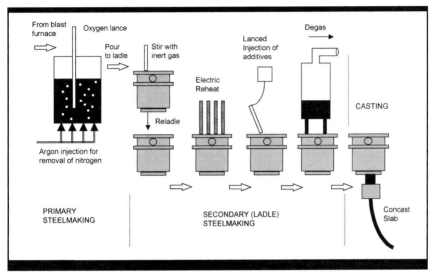

Fig. 3–3 *Basic Oxygen Steel-Making and Secondary Processing*

The primary furnace to produce the primary steel is lined with a basic material, usually dolomite. The charge of graded ore, pig iron, and scrap is treated in the furnace to remove excess carbon by blowing oxygen through the steel either using a lance inserted deep into the molten metal or by injecting the oxygen by bottom entry. The other additives are also introduced by lancing as this ensures a good dispersion. Sulphur and dissolved oxygen are removed by sequential additions of manganese, silicon, and aluminum.

During production, the steel can absorb various gases. These reduce the fatigue properties of the steel because the small voids in the steel increase localized stresses and may coalesce under cyclic loading to produce micro-voids. For pipelines the steel must be fully *"killed."* Oxygen and nitrogen dissolved in the molten steel may be removed by purging with argon, which also stirs the steel and aids in the decarburization and dephosphorization of the steel, aids flotation of the slag, and reduces the volume of nitrogen absorbed in the steel. Various additives are used to absorb the residual gases, e.g., silicon, aluminum, and calcium; and vacuum degassing is also routinely used.

The treated molten steel is poured into ladles for secondary steel making. The first pour is stirred magnetically to aid removal of the residual slag and is transferred to a second ladle usually from the bottom of the first ladle to reduce carry–over of floating slag. After electrical arc reheating whilst being magnetically stirred to ensure dispersion, the steel is injected with the relevant microalloying additives and the other major alloying elements to the appropriate level. The steel is then degassed and immediately and continuously cast into slabs.

Continuous slab casting is used because it provides a more consistent product and also helps to reduce the segregation of phases within the slab, which, when rolled, can produce plate with discontinuous mechanical properties. A typical continuous casting system is illustrated in Figure 3–4.

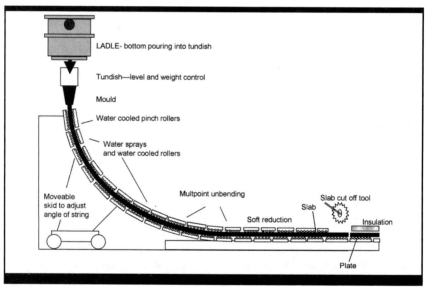

Fig. 3–4 *Continuous Casting of Steel Slab*

The slab caster comprises a tundish from which the steel is poured from the bottom at a controlled rate into a mold. The outer surface of the steel is chilled in the mold. Jets of water continuously cool the strand of liquid steel within the viscous case of solidifying steel. The strand then flows through a sequence of water-cooled rollers that direct the strand from the vertical to the horizontal where it is hot reduced by a sequence of rollers to the final slab geometry. Deleterious materials in the steel have low melting points and congregate at the center of the strand. To reduce the volume of these unwanted materials, the strand of steel from the mold is squeezed by pinch rollers, which retain the unwanted material as a molten, floating core within the liquid steel. The pinch results in a hot reduction of about 12 mm in the slab thickness. When the pour is completed, this plug of material is cut off the end of the final slab and recycled to the primary steel production unit.

3.4.2 Typical compositions of steel for line pipe

Alloying elements are added to pipeline steels to alter the mechanical properties of the steel but also have an impact on the corrosion behavior. Table 3–6 lists the typical alloying elements and their effect on steels. The API Specification 5L allows for a wide range of compositions, specifying only the maximum levels of a limited number of alloying elements. This tolerance allows the steel producer to achieve the specified strength, toughness, and weldability by a range of fabrication routes.

The wide range of combinations of chemistry and heat treatments able to achieve the required specification of steel highlights the complexity of the technical decision to be made by the pipeline engineer. Fortunately, it is usually not necessary to be overly specific about the steel-making route taken to produce the primary slabs. It is usually sufficient to restrict the steel to that produced by the basic oxygen process or to steel produced by electric arc furnace as both these routes produce low carbon, low SPOHN steel with fine grain and good low temperature impact toughness. The steel should also be specified to be killed (fully deoxygenated) and of fine grain. A usual range of compositions of typical pipeline steels is given in Table 3–2.

Modern steel for coiled plate for electric-resistance-welded (ERW) pipe is increasingly produced with very low carbon contents. The low carbon permits rapid welding while maintaining adequate toughness. However, when the pipeline is fabricated, the HAZ adjacent to the girth welds may have notably lower strength than the parent pipe and weldment. This results because there is insufficient carbon to strengthen the steel by transformation hardening. Also, grain refinement can be inadequate resulting in excessive grain growth. Usually this is of no consequence but may affect pipe to be used for reeling. When the pipe is reeled, there is some ovalization, and tearing of the pipe in the HAZ has been observed on a reeled pipeline in the Gulf of Mexico as a result of the disparity in ultimate tensile strength (UTS) between the parent pipe and HAZ.

Alloying Element		Effect on Steel Mechanical and Corrosion Properties
Carbon	C	Increases tensile strength and hardness but reduces toughness. Reduces weldability. Increases corrosion.
Manganese	Mn	Increases tensile strength, hardness, and abrasion resistance. Decreases porosity and cracking. Forms sulphides that may cause hydrogen-induced cracking.
Phosphorous	P	Increases brittleness and cracking. Increasingly being restricted to <0.025% for sweet and < 0.015% for sour service pipe.
Sulphur	S	Increases porosity, brittleness, cracking. Forms manganese sulphide that traps hydrogen leading to internal cracking. Surface emergent sulphides initiate pitting. Increasingly restricted to 0.01% sweet and <0.005% for sour.
Silicon	Si	Increases tensile strength but markedly reduces toughness. Added as a deoxidizer to kill the steel (remove gases). Restricted to 0.35–0.4%.
Aluminum	Al	Used to refine grain size. Increases hardness. Added as a deoxidizer to kill the steel. Aids weld toughness when added to 0.02 - 0.05%.
Copper	Cu	Improves sour cracking resistance for environments at pH >4.5. Affects corrosivity of weld HAZ. In conjunction with Ni claimed to stabilize corrosion films and reduce corrosion. Often used with Ni for pipe for bends and in thick section pipe.
Chromium	Cr	Increases tensile strength and hardness. Decreases weldability. Has major effect on corrosion resistance. Material becomes a stainless steel if Cr ≥ 12%.
Calcium	Ca	Deoxidizer and desulphurizer. Secondary addition used for inclusion shape control for sour service pipe steels.
Molybdenum	Mo	Increases tensile strength and corrosion resistance. Reduces pitting attack. Used in high-grade bends.
Titanium	Ti	Micro-alloying element. Increases tensile strength, hardenability, and wear resistance. Combines with carbon to form carbides that may reduce toughness.
Niobium	Nb	Micro-alloying element in C-steel and always added to steels above X42.
Vanadium	Vn	Increases tensile strength, hardenability, and wear resistance. Used as a micro-alloying element for thick pipeline material.
Nitrogen	N	Increases strength but reduces low temperature toughness. Restricted to 0.01%.
Nickel	Ni	Increases tensile strength and toughness at low temperature and improves corrosion resistance. Reduces susceptibility of weld corrosion and improves weld strength. Used in pipe for bends. Steel with > 1% nickel not permitted in sour service.

Table 3.6 *Effect of Alloying Elements on Steel Properties*

3.4.3 Production of plate and strip for welded pipe fabrication

The steel slabs must be rolled to produce plate of the correct dimensions from which the pipe can be fabricated. Flat plate, approximately of the dimensions of the eventual pipe, is used for the fabrication of U-O-E pipe that is longitudinally welded by submerged arc welding. Flat plate is formed in thickness up to 50 mm. After production, the flat plates are stacked while they are hot (~ 450 °C) as this maintains the temperature longer and encourages residual hydrogen to escape. Coil plate is used for production of ERW and spiral welded pipe and is formed as long sheets of plate by successive rolling of a slab of steel until the correct thickness is achieved. The upper limit of thickness for coil plate is about 25 mm. The plate is then coiled for storage and transport to the pipe fabrication plant. The coil is formed while the steel is hot and is stored in still air to maintain the coil hot for longer and to encourage the escape of residual hydrogen.

The initial slab should be cast to the size needed for production of the final plate, but some steel producers cut larger slabs down to the appropriate width. If this is done, an odd number of smaller slabs should be cut; and under certain circumstances, the central cut may be discarded for pipe fabrication. If the slabs for the pipe plate are produced from slabs cut in two, one side of the longitudinal weld of the pipe will be made from steel from the center of the slab where the highest levels of inclusions and tramp compounds are present. When formed into pipe, the inclusions would be adjacent to the longitudinal weld. This type of pipe may be unsuitable for sour service.

3.4.4 Thermo-mechanical controlled processes for plate production

In the past, hot rolling was used to roll the slab to the correct dimensions, and the slab was then heat-treated to obtain the final mechanical properties. In the traditional hot rolling process, the steel slabs are heated well into the austenite region, 1,200–1,250 °C, and held at that temperature (soaked), where re-crystallization produces coarse austenitic grains. The plate is hot rolled starting at this temperature, and the rolling is continued as the plate cools to about 1,000 °C. The rolling results in the formation of large, elongated grains. After rolling, the steel is allowed to cool in air. Recrystallization of the austenite grains occurs as well as some grain refining.

To refine the grains further, the steel is normalized by re-heating into the austenitic region to re-crystallize the grains and then cooled at a controlled rate to obtain a fine ferrite grain size. An alternative is to quench and temper (Q&T) the hot rolled steel plate. In this process, the steel is heated into the austenite region to re-crystallize the grains and then cooled rapidly to freeze the grains. By cooling rapidly, the non-equilibrium structure, martensite, is formed. The steel is fine grained and very strong but brittle and requires tempering that reduces the strength and improves the toughness. The Q&T procedure gives more control of the final properties than does normalizing. The final strength of the plate material produced by hot rolling and heat treatment is achieved by a relatively high alloying content, microalloying, and limited grain refining. Consequently, the upper limit of strength is around X60, though higher grades can be achieved with careful control of the heat treatment. The carbon equivalent of these steels is high, which has an impact on the weldability. Typically, these steels have carbon equivalents in the range 0.35–0.43. The blend of strength and toughness that can be produced by hot rolling is limited. The higher grade, higher quality plates for pipeline fabrication are increasingly being produced by the controlled rolling processes. There are several variants of the controlled rolling process, often termed thermo-mechanical control process (TMCP) that involve rolling the plate in several stages at carefully controlled temperatures. Figure 3–5 contrasts schematically the temperature and rolling histories for normalized, quenched and tempered, and TMCP pipe. Figure 3–6 describes the various TMCP procedures.

Option 1 of Figure 3–6 is the simplest TMCP and is the original controlled rolling process developed to produce large sheets of high strength steel for construction of ships and super tankers. The slab is heated to 1200–1250 °C, and the steel is rolled to the 'roughing' stage. The plate is allowed to cool to a temperature where it remains austenitic, but reworking the grains will not re-crystallize as a result of the work input during the second phase of rolling. The steel is cooled further until it is on the borderline of the transition zone of austenite to ferrite and is finished by re-rolling at this temperature to form elongated, austenite grains. On further cooling, the steel re-crystallizes into fine ferrite and pearlite grains. The energy input in working of the austenite grains at the transition temperature results in a relatively fine ferrite grain structure, though the steel retains a marked banded appearance, rather like the grain in wood. All the working of the plate occurs in the austenite region, and this option is termed TM(γ).

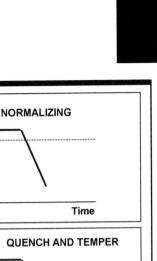

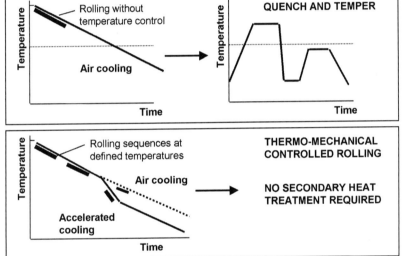

Fig. 3–5 *Hot Rolling Steel Plate and Strip Production Processes*

Option 2 is to roll the plate for the third time while it is in the austenite + ferrite zone before complete conversion to ferrite has occurred; this results in a very fine ferrite grain structure with residual structural banding. Some dislocation strengthening also occurs, as there can be no relaxation of the imposed stress by re-crystallization. Because the working of the steel occurs in both the austenite region and the ferrite region, this option is termed TM($\alpha+\gamma$).

Option 3 is to roll the plate as in Option 2 and to cool the steel using water-cooling immediately after the third rolling is completed. This action allows the conversion to ferrite grains but freezes the growth of the grains that would occur if the plate were allowed to cool slowly in the air. This option involves working the plate in the austenitic and ferritic regions and accelerated cooling and is termed TM($\alpha+\gamma$) + ACC. The rate

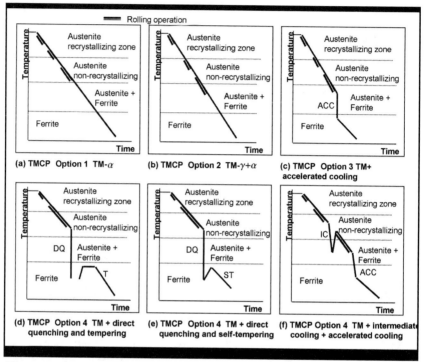

Fig. 3–6 *Alternative TMCP Processes*

of cooling is restricted to avoid quenching the structure, which would result in the formation of bainite and martensite, which would require subsequent tempering.

Option 4 is to roll the plate at the high temperature and then allow some cooling and re-rolling of the steel in the temperature region where re-crystallization of the worked austenite grains will not take place. The steel is then water cooled until the temperature is in the ferrite region; the steel is essentially quenched. This method produces a very strong material with a fine but disordered grain structure and requires tempering. The working occurs in the austenitic region and grain growth is restricted, this process is termed $TM(\alpha) + Q\&T$.

Option 5 is Option 4 with the steel quenched to a sufficiently high temperature so that it has adequate residual heat to self temper. This option avoids the need to reheat but can only be applied to thick material. This option is termed $TM(\alpha) + QST$.

A final option is to hot roll the steel as with the previous options, then allow the steel to cool to the zone where re-crystallization does not occur, and to re-roll and then cool the material into the ferrite zone. Ferrite formation starts to occur, and the material is re-heated slightly to raise it into the austenite zone where it is re-worked as it cools through the austenite-ferrite transformation temperature. The material is then acceleration cooled. This option is termed TM(γ)IC + (α)ACC.

The initial slab heating stage for hot rolling is important as it controls the microalloying process and the starting size of the grains. Each of the microalloying elements requires a different temperature for solubilization and precipitation. During the rolling process, the grains are deformed and, if this is at the re-heating temperature between the re-crystallization temperature and the transition temperature, or in the dual phase austenite-ferrite temperature band, then the grain size is refined by re-crystallization between rolling passes when imposed strain is above the critical deformation level. Grain growth is restricted by the precipitation strengthening precipitates formed at the grain boundaries; the grains cannot grow thermally and are refined by each recrystallization step.

Accelerated cooling is increasingly being used because the rate of cooling from the dual phase ($\alpha\gamma$) temperature band has a marked effect on steel strength. The more rapid the cooling, the greater the increase in the strength of the steel even with a low alloying content. Because a low alloying content reduces the carbon equivalent and hence the weldability, these steels provide very high strength and toughness with excellent weldability. The rate of cooling should avoid quenching the structure as this would result in the formation of bainite and martensite, and these would require subsequent tempering.

After forming, the plate will contain residual hydrogen that takes time to diffuse out. Most steel manufacturers stack plate to maintain a high temperature for longer, thereby encouraging hydrogen degassing. The pipe fabrication specification provided by the pipe producer may require checking that stacking is allowed for plate of 20 mm thickness or above and for plate that will be used for fabrication of pipe for sour service.

3.4.5 Inclusion shape control

In the 1960s, there was an increase in production of oil and gas and a concomitant demand for large-diameter pipe. Very wide steel sheets of strong, tough steel were needed to form the pipe. For example, producing a 48 in.-diameter pipe by the U-O-E process requires a sheet of steel about 4 m by 12.5 m. Much of the new production was also sour. It was known that pipelines able to tolerate sour product had to be constructed in a manner that avoided hard spots in the material at welds to avoid the risk of sulphide stress cracking (SCC) as defined in NACE MR-0175.[10] This requirement is best met by the use of steel with a low alloying content, but as discussed earlier, obtaining high strength steels with low alloy contents required steel with a fine grain size to provide both strength and toughness. The plate for these large diameter pipes was created using the controlled rolling process, which was restricted to Option 1 at that time.

When this early controlled, rolled steel was used for transmission of sour crude oil, the pipe failed due to internal cracking resulting from the formation of hydrogen gas within the steel. Some of the atomic hydrogen evolved during internal corrosion migrates into the steel and is reformed as hydrogen gas at manganese sulphide inclusion platelets in the steel. The build up of hydrogen gas develops large pressures that raise sufficient internal stress to cause internal cracking of the steel, a similar process to the blistering of tank steel at laminations. The rolling of the steel had resulted in a banded steel microstructure with the grains elongated in the rolling direction. Such steel has marked anisotropic properties between the longitudinal and transverse directions, e.g., differences in tensile strengths and a factor of 2–3 in absorbed impact energy. More importantly, for sour service the inclusions of manganese sulphide were also flattened and elongated and presented a large surface area for absorption of the atomic hydrogen migrating through the steel. These aspects are discussed in detail in the section on sour service. That section examines the methods used to alter inclusion morphology to reduce steel susceptibility to HIC and also to reduce the anisotropy.

The first attempt to produce steels resistant to HIC was the incorporation of copper into the steel. Subsequent to some corrosion, the copper forms an enriched copper sulphide layer on the inner surface, which catalyzes the combination of hydrogen atom to form molecular hydrogen. The quantity of atomic hydrogen entering the steel is reduced. However cracking still occurred when copper-containing pipe formed from controlled rolled steel was used for transmission of wet sour gas. Steel composition had to be reformulated again to reduce the sulphide inclusions and to alter their shape.

The primary method of improving resistance to HIC was to reduce residual sulphur in the steel. The small quantity of sulphide remaining, as manganese sulphide, could be further modified by shape inclusion. Primary calcium treatment of steel is used after the aluminum additions to reduce the concentration of sulphur. The calcium is added after the aluminum addition as calcium silicide or carbide into the molten steel and combines with sulphur and oxygen in the steel; the reaction products are removed in the floating slag. Very low levels of sulphur are achievable, typically in the range 0.001–0.003%. Magnesium and cerium have the same effect but are more expensive than calcium. A secondary addition of calcium was introduced at the ladle stage to shape the residual sulphides as spheroids. With calcium treatment, the residual sulphide is associated with the calcium principally as calcium sulphide, rather than as manganese sulphide. These particles tend to be fine and globular; and because they are hard, they retain their shape through the hot rolling process. The area presented to the atomic hydrogen is markedly reduced, which results in lower hydrogen accumulation and a marked improvement in resistance to HIC. The ratio of Ca to S should be restricted to a minimum of 1.5:1 to 2:1 at the ladle analysis. A lower calcium ratio provides inadequate calcium to form the spheroids while excess calcium forms unfavorable compounds that affect the mechanical properties in the HAZ. A typical composition for internal crack resistant steel is given in Table 3–2.

3.5 Pipe Fabrication

3.5.1 General comment

Pipe for the oil and gas industry is made by one of four fabrication routes: seamless, longitudinally welded by electrical resistance welding, helical or spiral welded, and longitudinally welded using submerged arc welding. Production of pipe by furnace butt-welding of hot plate, though permitted, cannot produce the large diameters required. Pipe formed with more than one longitudinal weld is also not used for submarine pipelines.

Nowadays, many pipe fabricators are independent and not part of an integrated steel company. Fabricators also tend to specialize in certain pipe fabrication techniques, and few produce pipe material over the full range of diameters and wall thicknesses. The *Oil and Gas Journal* makes a regular survey of pipe fabricators and their production capabilities. Independent pipe mills form pipe from plate or coiled plate that is bought from whatever source can provide suitable material at an acceptable price. For the production of a large quantity of pipe, the plate may be sourced from several steel suppliers. The mechanical properties of the pipe will be consistent with specification, but caution is necessary regarding the weld procedures for welding of the pipe on the lay barge, particularly if automatic welding processes are to be used as small variations in composition can affect the quality of the weld achieved.

3.5.2 Seamless pipe

Seamless pipe is formed by hot working steel to form a pipe without a welded seam. Several processes are available, illustrated schematically in Figure 3–7.

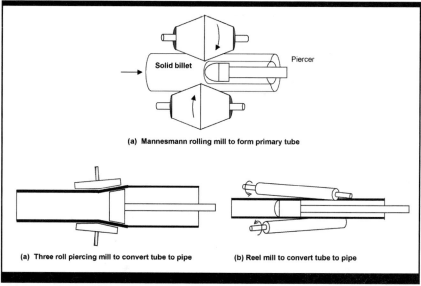

Fig. 3–7 *Seamless Pipe Manufacturing Processes*

The initially formed pipe may be subsequently cold worked to obtain the required diameter and wall thickness and heat treated to modify the mechanical properties. A solid bar of steel, termed a *billet*, is cut from a slab and is heated and formed by rollers around a piercer to produce a length of pipe. The Mannesmann mill is perhaps the best-known type of piercing mill. In this mill, the steel billet is driven between rotating, barrel-shaped rolls set at a slight angle to each other. The rolls rotate at about 100–150 rpm. The billet also rotates. The piercer is placed just beyond the point where the billet is squeezed by the rolls so that as the formed billet passes through the *pinch* zone between the two rollers, the reducing stress tend to open the metal over the piercer.

The piercing mill produces the primary tube that requires finishing to form the pipe. During finishing the wall thickness is further reduced. In a plug rolling mill, the pipe is driven over a long mandrel fitted with a plug of the correct internal diameter between rollers that extrude the tube to the required external diameter. Other methods use multiple conical or conventional horizontal rolls or offset rollers. Pipe is finished in a reeling process in which it is driven between slightly conical rollers, followed by passage through a sizing mill that ensures circularity.

An older process is the Pilger process. This process uses eccentric rolls to form the pipe in discrete stages. A mandrel is inserted into the partly formed pipe from the piercing mill. The assembly is driven into the open rolls and, as the rolls rotate back and forth, sequential sections of the pipe are drawn into the eccentric rolls; and the outer diameter formed to the required dimension set by the roller eccentricity. This process is also used for production of corrosion resistant alloy pipe.

After forming, the pipe may be delivered as produced or, more usual for the oil industry, as either normalized or quenched and tempered. These heat treatments homogenize the mechanical properties and further improve strength and toughness. After forming is completed, the pipe is inspected for internal laminations and pressure tested hydraulically.

This type of pipe is generally available in diameters up to 16 in. but can be obtained from a restricted number of suppliers in sizes up to a maximum of 28 in. with wall thickness to 2 in. The larger diameter pipe is made by hydraulically expanding smaller diameter pipe. Seamless pipe is the preferred material of several operators for small diameter pipelines. Its principal advantages are its good track record in service and the absence of welds in the pipe sections. However, the larger diameter, seamless pipe may be more expensive than pipe fabricated by the alternative processes. The disadvantages of seamless pipe are a fairly wide variation of wall thickness, typically +15%/−12.5% and out-of-roundness and straightness. The premier pipe fabrication mills can produce seamless pipe to closer tolerances.

The outer surface of the pipe may be highly distorted such that when it is grit-blasted, prior to coating, tiny slivers of steel rise up. These slivers can be a drawback when the pipe is to be coated with a thin anti-corrosion coating such as fusion bonded epoxy (FBE), and it is prudent for the pipeline engineer to check for such effects when pre-qualifying pipe suppliers.

3.5.3 Electrical resistance welded (ERW) pipe

ERW pipe is formed from coiled plate steel. The plate is uncoiled and sheared to the required width, flattened and the edges dressed. The plate is passed through a sequence of rolls to form the pipe. The rolls crimp the edges of the plate; then progressively bend the plate into a circular form ready for welding of the longitudinal seam. The longitudinal seam weld is made by electrical resistance welding (hence the name of the pipe). No welding consumable is used to produce this type of pipe. The process is illustrated schematically in Figures 3–8 and 3–9. When a new coil of plate is started, it is welded to the end of the previous coil to allow it to be pulled through the rolling mill. A pipe is formed with a joint in the middle (a jointer). Pipe of this type is generally not accepted for use for submarine pipelines.

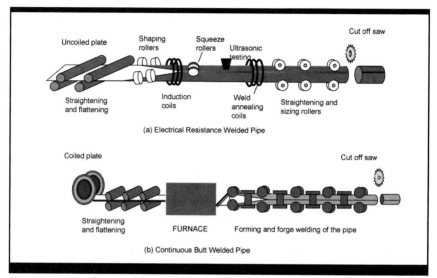

Fig. 3–8 *Manufacture of ERW and Continuous Butt-Welded Pipe*

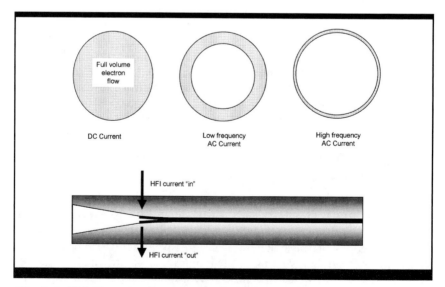

Fig. 3–9 *High-Frequency Concentration of Current at Surface*

An electrical current is passed across the interface to heat the steel pipe faces that are to be ERW welded. Once heated, the faces are pressed together to produce the longitudinal seam weld. The heating may be by low frequency AC current, typically 60–360 Hz, introduced directly into the pipe by rolling contacts, or induced into the steel with induction coils operating at high frequencies of above 400,000 Hz. The later process of producing the pipe is termed high frequency induced (HFI) ERW pipe.

Pipe used for submarine pipelines is almost exclusively produced using HFI welding. HFI is effective because of the skin heating effect that results from the use of very high frequency AC current. As the frequency of an AC current in a conductor increases, the electrons that carry the current tend to flow increasingly at the outer surface of the conductor (*skin effect*). At very high frequencies, the electron movement is exclusively in the outer 1 mm or less of the conductor. When applied to steel pipe, the result is a very high rate of heating of the steel faces sufficient to melt the metal in the weld region.

The pressure exerted on the faces during the weld forming process result in the molten metal at the faces being squeezed outward to form stubs of metal above and below the weld. Any debris or oxides on the steel faces is discharged in the stubs of metal. The stubs of metal are trimmed off, and the weld is inspected using ultrasonic probes. The weld is then locally heat-treated to temper the weld and heat-affected zone. Depending on the diameter, the pipe is re-heated and stretch reduced slightly or is passed through sizing and straightening rollers to obtain the required dimensions and circularity. The weld is re-inspected, and the pipe cut to length. The pipe lengths may be further heat-treated though this is rarely necessary. The final inspection is hydraulic testing. In the premier mills there will also be a full body inspection after the hydrotesting.

The original ERW pipe was fabricated with low frequency welding, and problems with this form of pipe have resulted in a poor track record. In response the HFI-ERW pipe fabricators provide a high quality inspection to restore confidence in the product. Conventional, low frequency ERW does need a high level of inspection because the welds are prone to a form of failure termed *stitching*. Dirt or oxide on the weld seam surfaces, arcing or insufficient pressure during the welding causes a lack of through-wall-thickness welding.

The weld is extremely fine as the bulk of the molten metal is squeezed out. It is not possible to detect the weld by eye, and often a line is painted to mark the weld line. Samples of weld are cut from the ends of pipe for metallographic inspection, analysis, tensile, ductile, and toughness testing.

Because the weld formed is minute, failures of the heat treatment may leave the weld area vulnerable to enhanced corrosion and affect the toughness values of the HAZ. Most of these problems have been overcome in the HFI production process; but because of problems with the early ERW pipe, there is reluctance to use this type of pipe for high-pressure submarine pipelines. The material is specified routinely for land pipelines and has been specified for less critical subsea projects.

ERW pipe is the main competitor to seamless pipe. It is cheaper than seamless, and it can have considerably tighter tolerances on wall thickness. Pipe lengths are typically standard length ±50 mm, and the pipe can be produced in lengths up to 27 m. Though API Specification 5L permits a wide tolerance on wall thickness, +19.5/-8%, typical modern wall thickness tolerances are ±5 %. It is also claimed that pipe wall thickness can be specified to 0.1 mm, and non-API Specification 5L sizes are available. These tight tolerances have a cost benefit. The smaller tolerance in wall thickness and circularity permits a more rapid set up on the lay barge and a lower number of welding complications. The resulting increase in productivity on the lay barge may have a significant benefit to the overall project cost. Savings of up to 20% on the combination of material cost and installation have been estimated to be possible.

3.5.4 U-O-E (or SAW) longitudinally welded pipe

U-O-E pipe is formed from individual plates of steel by first forming a plate into a U, then into a tube (O). After longitudinally welding, the pipe is then expanded (E) to ensure circularity (Fig. 3–10). Because the longitudinal weld is produced using the submerged arc welding process, the pipe is sometimes termed *SAW* pipe. The submerged arc welding procedure is described in chapter 5 on welding.

The plate is cut to exact size, and the edges dressed. The edges are then crimped, and the complete plate is progressively bent into a U-shape and then into a tube in presses. The O-press leaves a residual 0.2–0.4 % compression in the pipe; a higher compression may be provided for pipe for sour service. Tab plates attached to the butting edges are tack welded to prevent movement during the main welding. The butted edges of the tube are then welded using submerged arc welding with multiple head welding devices. At least two welding passes are made: the internal weld is formed first; then the pipe is rotated through 180° and the second, external, weld pass is made. Tab plates are provided to allow the weld to start and finish beyond the end of the pipe to ensure a quality weld at the start and end of the pipe. Some pipe fabricators use gas welding to produce a complete first weld pass to joint the two edges of the plate. The weld is then completed using SAW.

The weld is inspected by ultrasonic examination and X-ray radiography. If the weld is acceptable, then the pipe is expanded using a hydraulic die that is inserted at one end of the pipe and, after each expansion, is moved incrementally along the pipe. At each halt, the die expands to form the tube into an exactly circular pipe. Typically the cold expansion is limited to below 1–2%, and the expander is shaped to avoid damage to the weld. The pipe is hydrostatically tested and then re-inspected using automatic ultrasonic testing and radiography. The weld may also be inspected by magnetic particle inspection. Finally, the ends of the pipe are squared or beveled, fitted with end protectors, and the pipe moved to the storage racks.

U-O-E pipe is used for the larger diameter pipelines. It is competitive with seamless pipe for the intermediate diameters (14-in. to 28-in.). For the smaller diameter pipes, the pipe fabricator may use cut down plate because producing narrow plate is less economical. Because inclusions and segregation tend to concentrate at the center of the plate, a pipe formed from a split plate may have inclusions and segregation adjacent to the weld. Such pipe may be unsuitable for sour service. If split plates are to be used, then an odd number should be cut to avoid the center line of the original plate abutting the longitudinal weld.

U-O-E pipe is the premier pipe material for large diameter, high-pressure pipelines. It has good wall thickness and out–of–roundness tolerance; a typical wall thickness tolerance is +12%, −10%, and ovality ±1%. Many pipe fabricators can produce pipe to tighter tolerances, typically ±5% and ovality ±½%. Specifications may be based on percentages or dimension as mm. Usually, the percentage route is better for the larger diameters and the dimension better for small diameter pipe. Press/bent pipe is used for very heavy wall pipe and is formed similarly to U-O-E by forming a cylinder from plate and welding longitudinally. The bending uses three rollers, one offset and moved to form the required diameter.

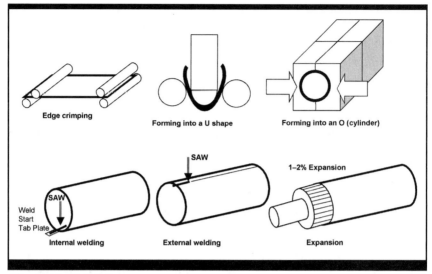

Fig. 3–10 *U-O-E Process*

3.5.5 Helical-welded (spiral-welded) pipe

A coil of hot-coiled plate is uncoiled, straightened, flattened, and the edges dressed. The plate is then helical wound to form a pipe. The width of the strip and the angle of coiling determine the pipe diameter. As the pipe is formed, the helical internal seam is welded using inert gas welding or SAW first, and as the seam rotates to the top position, the external weld is made. A continuous length of pipe is produced. After forming, the pipe is passed through a sequence of rollers to enhance circularity.

The pipe weld is tested using radiography or ultrasonic testing, and the pipe is then cut to the required lengths. The pipe joints are hydrostatically tested before being re-inspected. If the pipe passes inspection, it is end-faced or beveled, fitted with the end protector caps, and transported to the pipe racks. A schematic of the process is given in Figure 3–11.

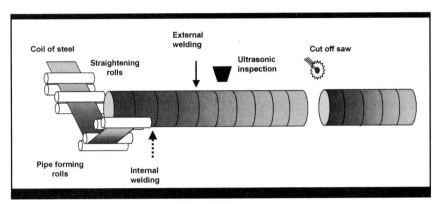

Fig. 3–11 *Manufacture of Helically-welded Pipe*

The end of the coiled plate is welded to the start of the next plate coil, resulting in a weld perpendicular to the helical weld forming the pipe. This weld joint should not be less than 300 mm from the end of the pipe. This weld may not receive the same degree of scrutiny as the helical welds, so the pipe specification may need to call for an additional inspection of these welds after the hydrotesting stage.

Helical-welded pipe can be made in a wider range of diameters and wall thickness than nominal API Specification 5L sizes. It can also be produced in long lengths above the normal double random length of 12 m. Because the pipe is formed from rolled strip, the wall thickness tolerances are good, being similar to U-O-E pipe, though there may be higher out–of–roundness tolerance. Helical-welded pipe has been used for large diameter submarine pipelines, both crude oil and gas, but is generally considered a less reliable material than U-O-E-formed pipe. It is cheaper than U-O-E and is widely used for caissons, sleeves, outer pipe for bundles, low-pressure hydrocarbon service, dried gas service, and for water transportation where the service is moderate. The large length of weldment and the need for some of this weld to be at the bottom of the pipe where the worst case corrosion occurs is seen as potential weaknesses that tend to preclude this type of pipe from general high pressure corrosive service. The helical welding may reduce the data gained in intelligence pig surveys because some types of sensors skip at the welds, resulting in blind spots just downstream of the welds.

3.6 Pipe Variables

3.6.1 Selection of pipe wall thickness

The diameter of a pipeline is selected based on hydraulic studies as described in chapter 9. In some cases, the wall thickness is determined by installation stresses, but usually the wall thickness of the pipe is selected dependent on the service pressure, the design factor, the corrosion allowance, and the fabrication tolerance (wall thickness variation). These factors are described in chapter 10. It is then common practice to select the nearest higher API Specification 5L/ISO 3183 standard thickness pipe. The larger negative tolerance in wall thickness for seamless pipe as compared to SAW pipe means that for a given diameter, a seamless pipe will need to be thicker on average than an ERW or SAW pipe. As wall thickness increases, the complexity of welding may increase. However, many modern pipe fabricators can produce pipe to tighter tolerances, and this option may need to be reviewed using cost benefit analysis. Many companies insist on smaller tolerances than those given in API Specification 5L and also specify exactly the wall thickness they require. However for small batches of pipe, this may not be economical, as standard API pipe tends to be cheaper than non-standard pipe.

The corrosion allowance is determined by calculation, and a method is given in chapter 7. In non-corrosive environments, there is no need for a corrosion allowance. In a mildly corrosive environment, a corrosion allowance alone may be adequate, though it is usual to use a corrosion inhibitor to suppress pitting corrosion. More aggressive environments may demand a corrosion allowance and continuous corrosion inhibition. For very corrosive environments, carbon steel may not be a suitable material; and an alternative corrosion-resistant material may have to be used. See chapter 4 for a detailed discussion of this topic.

There are no easily applicable criteria for these changeovers. The design life expectancy of the pipeline, the pressure and flow profiles over the design life, and the expectations of changes in production fluid composition, all need to be taken into account. Other relevant factors include the criticality of the line, the ease of inspection, the availability of inhibitor, the level of skill of operatives, and, increasingly, the environmental aspects.

3.6.2 Pipeline joint lengths

In land pipeline construction, the welders move along the pipeline, welding the pipe joints to the end of the growing pipeline. This allows considerable flexibility in the lengths of individual joints. On a lay barge, the weld positions are fixed; and as the line is laid and the barge advances in steps, the pipe moves through a sequence of welding stations. The welding stations are relatively fixed in position, and so, for submarine pipelines, it is important to specify a constant length of joint: the standard length is 40 ft (12.19 m). Line pipe for submarine pipelines is typically specified as 12.2 m ±0.3 m. As the complexity of the welding procedures increases, so the need to reduce the tolerance on pipe length generally increases; manual welding remains the most flexible welding procedure. Overspecification should be avoided, and the lay barge contractor requested to advise an acceptable tolerance.

3.6.3 Bends

Few bends are required for submarine pipelines; and it may be possible to select the pipe to be formed into bends from the general pipe joints, as there are usually some over-thick pipe joints that can be used. However, it is important to check that the material can be hot formed while retaining its strength. The hot bend fabrication process may excessively reduce the strength of very low carbon steels. If this is the case, special steel joints may be required from which the bends can be fabricated.

If a pipeline is to be constructed to allow inspection by magnetic flux leakage pigging tools, there are restrictions on the acuteness of the bends. The bend size decreases with increasing pipeline diameter. For 6- and 8-in. pipes, the pigs may require 5-D bends. For 24-in. pipes, the more advanced pigs are designed for $1\frac{1}{2}$-D or 2-D bends.

3.6.4 Straightness

For offshore service, the importance of pipe straightness depends on the welding procedure used and the depth of water. Automatic welding demands a lower tolerance than semi-automatic and manual welding. Usually the out of straightness of the pipe follows a log-normal distribution with diameter; and, as would be expected, the larger the diameter of the pipe, the straighter is the pipe. API Specification 5L allows 24 mm in a 12 m joint. Usually, it is possible to obtain pipe to 12 mm in 12 m. To reduce total cost, overspecification of straightness should be avoided.

Pipelines that are to be used at high operating temperature may be fabricated using alternate pipe joints curved by 10° so that the installed pipe has a zigzag form. This construction provides the pipe with a longitudinal expansion capability, thereby eliminating the risk of upheaval buckling, which is described in chapter 14.

3.6.5 Internal cleaning

It is common practice for gas pipelines to be internally cleaned if there is excessive oxide (termed *millscale*) on the inner surface of the pipe. In service, this millscale will detach; and a fine black dust is blown thorough the pipeline blocking filters, valve seats, and process equipment. Seamless pipe that has been quenched and tempered is particularly problematical because the double heat treatment forms thick oxide layers. The pipe is cleaned by internal grit blasting. For water and oil service, it is not usual to require the pipe to be internally cleaned.

3.6.6 Residual magnetism

The final step in pipe fabrication is to bevel the ends of the pipe to ensure that they are square to the longitudinal axis of the pipe and are ready for welding. The bevels may be inspected for laminations using magnetic particle inspection (MPI). After MPI, the pipe should be de-magnetized (typically to 30 gauss) because residual magnetism can cause problems during welding. Control of the plasma arc is lost because the plasma is an electrical current that is distorted by the residual magnetic field. Note that problems with magnetism may also arise when attempting to repair a pipeline that has been inspected using a magnetic flux inspection pig.

3.7 Specification Checklist

- Type of pipe
- Steel-making route
- Identification of pipeline service conditions
- Limitations on pipe steel chemistry
- Limitations on longitudinal weld chemistry
- Minimum yield strength (SMYS)
- Upper strength limitation

- Non-ductile transition temperature (NDTT)
- Minimum acceptable average and minimum toughness
- Toughness test temperature
- Non-destructive testing, acceptance criteria, procedures for out-of-specification product
- Hydrostatic testing, acceptance criteria, procedures for out-of-specification product
- Wall thickness tolerance
- Length tolerance
- Ovality
- Straightness
- End finish: square, standard, or special bevels
- Internal cleanliness of the pipe
- Provision of caps at pipe ends
- Avoidance of jointers, split skelps, residual magnetism limitations
- Inspection requirements, frequencies, reporting procedures
- Performance tests required, special tests, acceptance criteria
- Track record of fabricator
- Warranties
- Pre-qualification requirements, mill visits, complaints procedures

References

1 Craig, B. D. (1991) *Practical Oilfield Metallurgy*. Tulsa: PennWell.

2 API Specification 5L: "Line Pipe (42nd ed.)" American Petroleum Institute.

3 ISO 3183: "Petroleum and Natural Gas Industries—Steel Pipe for Pipelines." International Organization for Standardization. 1996.

4 Publication No. 166: "Specification for Line Pipe for Offshore Pipelines (Seamless or Submerged Arc Welded Pipe)." Engineering Equipment and Materials Users Association. 1991.

5 Courteny, T. H. (1990) *Mechanical Behaviour of Materials*. New York: McGraw Hill.

6 Takeuchi, F., Yamamoto, A. and Okaguchi, S. (2002) "Prospect of High Grade Steel Pipes for Gas Pipelines," 185–203. Proceedings of Pipe Dreamer's Conference. Pacifico, Yokohama, Japan.

7 Palmer, A.C. (1998) "Innovation in Pipeline Engineering: Problems and Solutions in Search of Each Other." *Pipes and Pipelines International*, 43, 5–11.

8 Llewellyn, D.T., (1992) *Steels: Metallurgy and Application*. Oxford: Butterworth-Heinemann.

9 Higgins, R. A. (1993) *Engineering Metallurgy*. London: Arnold.

10 NACE Materials Recommendation MR-0175: "Sulfide Stress Cracking Resistant Metallic Materials for Oilfield Equipment." National Association of Corrosion Engineers.

4 Increasing Corrosion Resistance

4.1 Introduction

Carbon-manganese steels are the cheapest pipeline material and are used wherever possible, often in combination with corrosion inhibition. However for some services, this combination has inadequate resistance to internal corrosion by the transported fluids; therefore, corrosion resistant materials must be used.

Pipelines for corrosive service may be fabricated in solid corrosion-resistant alloy (CRA) or in carbon steel clad, lined with CRA, or made as flexible pipes. In this chapter the metallic solid and clad rigid pipeline options are discussed. Flexible pipe, composite pipe, and composite-lined steel pipe are discussed in chapter 6. Solid CRA pipe or pipe internally clad with CRA are usually selected only if conventional steel is unsuitable for the service.[1] Selection is clear if the fluid to be transported is too corrosive for carbon manganese steel even with inhibition. Borderline cases arise where carbon steel with inhibition is theoretically feasible, but the use of corrosion inhibition may not be economically or technically feasible. Examples of scenarios that may arise are unmanned satellite platforms, production from subsea installations, and production in remote areas. Another reason might be the need to minimize risk of environmental damage. Under these circumstances, it may be cost effective to use a corrosion resistant alloy for the pipeline, rather than carbon steel plus inhibition.

In conjunction with the technical reason for selection of the CRA, there will need to be a justification of the particular CRA selected on strictly economic grounds.[2, 3, 4] In all cases, the design life is of critical importance. If the cost-benefit analysis leads to a choice between a CRA and carbon steel, it is important to include all the hidden costs of operating a carbon steel pipeline, e.g., the regular monitoring and inspection, transport and application of the inhibitor, and the higher risk of failure of the pipeline. If the analysis is to distinguish between different CRAs, then the costs associated with installation and potential failure risk tend to be dominant. Several reported studies indicate that the more expensive material option can be the cheapest overall for modest diameter (to 10-in.), short length (to 15 km) pipelines with design lives of 15 or more years.[5,6] Outside this diameter-length domain, the overall operation of the field may need to be re-evaluated to alter the corrosiveness of the produced fluids. In this context, the use of mixed metallurgy for long pipelines is being considered: CRA pipe at the inlet of the pipeline and carbon steel when the temperature and corrosiveness have reduced sufficiently.

Many CRAs are available, but only a limited number can be used as solid material for pipelines that are to be fabricated by welding. Once the CRA is welded, part of the heat-affected zone of the weldment is in the solution-annealed condition. Many CRAs have lower yield strength in this condition compared to their strength in the wrought condition. Regaining strength by cold working, heat treatment, or ageing cannot be applied to the fabricated pipeline. Overcoming the low strength would necessitate thick wall pipe and consequent severe economic impact. The range of materials that can be used for internal cladding is wider than the range suitable for use as solid CRA material as the strength of the material is not important, but there are limitations on the melting point of the cladding material. Studies have been made on the use of screwed connectors as an alternative to welding for small diameter pipelines. That connection method would permit a much wider range of materials to be used for both solid and clad pipelines.

As a consequence of these limitations, the most commonly used solid CRA at present for submarine pipelines is duplex stainless steel. The duplex stainless steels have yield strengths of X60 or above and an operational temperature range from -50 to +300°C. However, the upper temperature would be limited by stress corrosion cracking considerations.

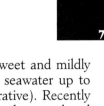

Duplex stainless steel has good corrosion resistance to sweet and mildly soured fluids and can also tolerate external exposure to seawater up to 150 °C (in case the cathodic protection system is inoperative). Recently developed alloys that are intermediate in price between carbon steel and duplex steels are the low nickel 12–13% chromium (Cr) alloys. These are weldable and suitable for high temperature sweet and sour service though limited in their tolerance to the chloride content of the product.

There are limitations on the materials for internally metallic cladding or lining carbon manganese steel pipe because of the heat treatment of the cladding material required after fabrication of the pipe. The cladding does not contribute to the pressure retention strength of the pipeline, and it is usual to optimize its corrosion resistance by using the cladding in the solution-annealed condition. To achieve this, the cladding needs to be heat treated, but the heat treatment must not exceed about 1000 °C. Above this temperature, recrystallization and grain growth of the carbon steel will occur and this would affect the mechanical strength of the steel pipe. This feature restricts the cladding materials to austenitic stainless steels, high nickel alloys, type 825 and type 625, which can be annealed at 950–980 °C. Compositions of these materials and others mentioned in this chapter are given in Table 4–1.

Table 4–1 *Compositions of Corrosion-Resistant Alloys*

(a) Martensitic Stainless Steels

Alloy	UNS Code	Cr	C	Mn	Si	Mo	Ni	Fe
410	S41000	11.5–13.5	<0.15	1	1			Residual
420	S42000	12–14	>0.15	1	1			Residual
WMSS	—	12–14		1	1	1.5	4–5	Residual

(b) Austenitic Stainless Steels

Alloy	UNS Code	Cr	Ni	C	Mo	Mn	Ti	Si	Fe
304	S30400	18–20	8–10	0.08	---	2	---	1	Residual
316L	S31603	16–18	10–14	0.03	2–3	2	---	1	Residual
321	S32100	17–19	9–12	0.08		2	5x C	1	Residual

(c) High Molybdenum Chromium Nickel Stainless Steels

Alloy	UNS Code	Cr	C	Ni	Cu	Mo	Mn	Fe
904L	N08904	19–23	0.02	23–28	1–2	4–5	2	Residual
25-6MO	N08925	19–21	0.02	24–26	0.05–1.5	6–7	1	Residual

Table 4–1 *Compositions of Corrosion-Resistant Alloys (cont'd)*

(d) Duplex Stainless Steels

Alloy	UNS Code	Cr	C	Ni	N	Mn	Mo	Si	Fe
2205	S31803	21–23	0.03	4.5–6.5	0.08–0.2	2	3	1	Residual
2507	S32750	25	0.03	7	0.1–0.25	1.5	3–4	1	Residual
Ferralium 255	S32550	24–27	0.04	4.5–6.5	0.1–0.25	1.5	2–4	0.03	Residual Cu 1.5-2.5

(e) High Nickel Alloys

Alloy	UNS Code	Cr	C	Ni	Mn		Al	Cu	Mo	Fe
825	N08825	19.5–23.5	0.02	38–46	1	Ti 1	0.2	1.5–3	2.5–3.5	22
625	N08925	20–23	0.02	Residual	2	Nb 4	0.4	0.05–1.5	8–10	5

Other alternatives that may be considered for solid pipelines or as cladding are the high molybdenum super austenitic stainless steels, e.g., modified type 904L or 25-6MO, which have improved corrosion resistance and are mid-way in price between the common austenitics and the high nickel alloys. However, these materials are on the limit of the heat treatment range and may have to be used in a less than optimum condition.

4.2 Methods of Improving Corrosion Resistance

4.2.1 Nickel and chromium addition

The corrosion resistance of carbon steel is markedly improved by the addition of nickel and/or chromium. The separate effects of nickel and chromium are shown in Figure 4–1. Other elements, for example molybdenum, are also added in small quantities to improve resistance to specific types of corrosion attack. Corrosion in sweet systems (fluids containing carbon dioxide) is suppressed by relatively small additions of chromium and reaches a minimum when the chromium content equals or exceeds 12%; a similar effect is observed with nickel additions, with the minimum corrosion reached at a nickel content of 9% or greater. Using a single alloying element would minimize material cost. Chromium alone would be the preferred alloying element because it is cheaper than nickel, but the simple chromium alloys have poor weldability, showing delayed cracking and a marked reduction in the mechanical properties of the HAZ. To overcome these limitations, blends of chromium, nickel, and iron are used.

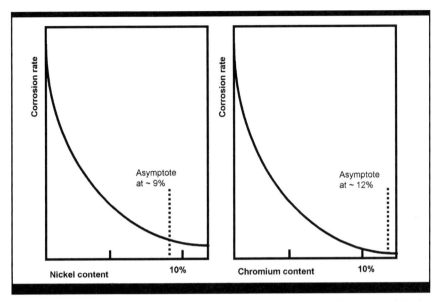

Fig. 4–1 *Effect of Chromium and Nickel Content on the Corrosion Resistance of Steel*

Unfortunately, though general corrosion is reduced, other corrosion mechanisms become dominant, for example, chloride stress corrosion cracking, crevice corrosion, and pitting. Because the pipeline is exposed in seawater, the external corrosion issues must also be addressed. The selection of the CRA is largely determined by the need to avoid these potential internal and external failure mechanisms. In addition to nickel and chromium, other alloying elements may be used to improve the material resistance to pitting and crevice attack. Molybdenum and nitrogen are the two elements that have the most significant effect.

As noted earlier, other internal and external corrosion mechanisms such as sulphide or chloride cracking need to be considered. A nickel content in the range 1–9% is perceived to reduce the resistance of the material to sulphide stress corrosion cracking. Because the addition of alloying elements alters the weldability, sometimes adversely, the additional costs associated with special weld procedures should also be taken into account in the cost-benefit analysis of pipe material selection.

4.2.2 Passivity

Chromium-nickel steels are corrosion resistant because after a small amount of corrosion, the alloys form a continuous and tenacious surface film enriched in the corrosion resistant element. This is particularly the case with chromium where the chromium forms a chromium oxide film.

This film is normally invisible, but it can be thickened by anodizing and/or oxidizing and can become sufficiently thick to appear tinted by light diffraction. The film has a high electrochemical potential compared to the base material and is less likely to corrode. The corrosion resistance resulting from the presence of the surface film is termed *passivity*, and the film is termed the *passive film*. The passive film is not a static entity and is constantly being reformed but at such a low rate that the material appears to be uncorroded.

If conditions are such that the surface film is preserved and can reform correctly, the material will give long service. High levels of chloride and the other halides can prevent local reformation of the film causing corrosion to occur at these defective areas. Because the normal chromium oxide film is cathodic to the anodic (corroding) areas, a galvanic couple is established. The rate of corrosion at the anode depends on the ratio of the cathodic and anodic areas. If this area ratio is large, then the continued corrosion may appear as pitting attack. The effective area ratio of cathode to anode depends on the local conductivity of the water.

The factors that most influence the behavior of the stainless steels, in rough order of importance to corrosion, are as follows:

- Presence of oxidizing species, which aid reformation of the oxide film
- Chloride ion concentration because chloride hinders oxide film repair
- Conductivity of the electrolyte, which affects the cathode/anode ratio
- Crevices that can initiate corrosion
- Sediments that prevent reformation of the oxide film
- Scales and deposits that prevent reformation of the oxide film
- Biological activity that can produce slimes and deposits that prevent reformation of the oxide film or cellular metabolites that initiate oxide film breakdown
- Chlorinating practice that alters the chlorine content of the environment
- Surface condition of the stainless steel
- pH (if below 5) that increases the cathodic reactions
- Temperature that alters the relative rates of oxide film breakdown, corrosion processes, and oxide film reformation rate

4.2.3 Pitting resistance number (PRE_N)

The pitting resistance number, sometimes called the *pitting index,* is a useful factor in evaluating the corrosion resistance of a CRA. Examples of PRE numbers are given in Table 4–2. The number has no fully developed theoretical basis but is empirically proven. The PRE_N is calculated from the following equations:

$$PRE_N = Cr + 3.3\,Mo \quad or \qquad (4.1)$$

$$PRE_N = Cr + 3.3\,Mo + 16\,N \qquad (4.2)$$

Table 4–2 *Mechanical and Corrosion Properties of Corrosion Resistant Alloys*

Material	Yield (min) MPa	Tensile (min) MPa	Hardness	PRE_N	Specific Gravity
410	205	415	95 BHN	11.5–13.5	7.7
420	760	965	302–352 BHN	12–14	7.72
304	205	515		18–20	7.94
316L	170	485		22.6–27.9	7.94
321	205	515		32.2–36.1	7.94
2205	450	620	30.5 HRC	32.2–36.1	7.85
2507	550	760	31.5 HRC	32.2–44.2	7.85
Ferralium 255	550	760	31.5 HRC	32.2–44.2	7.81
904L	220	490		29.2–39.5	8.0
25-6MO	300	600		38.8–44.1	8.1
825	172	517		27.8–35.1	8.14
625	414	827		46.4–56	8.44

Specifications for these steels should take into account the relative importance of these elements and ensure that minimum compositional levels are specified. It is often possible to obtain markedly improved corrosion resistance within the common specification by ensuring that the upper concentrations of alloying are observed. Materials with a PRE_N above 40 are often referred to as *super-alloys,* either super-austenitic or super-duplex.

4.2.4 Pitting and crevice corrosion

The surface film may be damaged in use, but it rapidly reforms. However, if the environmental conditions are such that the passive film cannot reform or is incorrectly restructured, then corrosion will occur at the area of the blemish in the film. Chlorides and the other halides are the most common elements that damage the passive film, and pitting can initiate at these areas.[7] There is also substantial evidence that many of the pits initiate at non-metallic inclusions in the steel surface.

Because corrosion is suppressed where the chromium film is intact, the pits in stainless steels tend to be narrow-mouthed and deep. Very high rates of pitting attack can occur, and thin wall material may fail by pinholing. Where pits form on the sides of pipes in low flow velocity conditions, in most cases the damage spreads downwards, i.e., with gravity, because the corrosion product from the pit initiates further pitting. The corrosion products remain on the metal surface, slowing down the diffusion of passivating ions. The corrosion process is postulated to be a sequence of steps:

1. Metal into solution   (anodic reaction)

$$M = M^+ + electron(s) \qquad (4.3)$$

2. Cathodic reaction

$$O_2 + 2H_2O + electrons = 4OH^- \qquad (4.4)$$

3. Concentration of cations (M^+) builds up in the pit causing chlorine ions to migrate inwards to equalize the electrical charge.

4. Formation of soluble metal chlorides

$$M^+ + Cl^- + H_2O = MOH + H^+Cl^- \qquad (4.5)$$

If the metal surface is covered with a non-porous material, e.g., a gasket, then the tiny gap between this material and the metal surface can form a crevice. The narrower the gap, the more likely it is that crevice corrosion will occur. Crevice corrosion occurs under less severe conditions than does pitting corrosion; and hence if the design can obviate the risk of crevice attack, then pitting is unlikely.

The aqueous fluid (the electrolyte) in the crevice reacts with the metal, and its ionic concentration changes faster than can be maintained by diffusion from the bulk electrolyte. The usual initiators are chloride ions. They migrate into the crevice faster than the other ions. An imbalance develops, as described earlier, with the pH and chloride content of the crevice becoming markedly different from the bulk fluid outside the crevice. Under these conditions, the metal continues to corrode until either the crevice blocks up with corrosion product or the gap becomes sufficiently large for the diffusion processes to catch up and repassivate the metal.

Pitting and the critical crevice gap for initiation of crevice corrosion are related to the PRE_N described in section 4.4. Figure 4–2 shows pitting and crevice corrosion rates for several CRAs as a function of PRE_N.

Type 304 stainless steel will suffer crevice corrosion if the crevice gap is below 1 mm while for type 316L crevice attack occurs if the gap is below 0.35 mm. For type 904L, crevice attack will occur if the gap is below 0.25 mm; and for alloy 625, crevice attack may initiate if the gap is below 0.1 mm. Gaskets are an obvious crevice area, but inadvertent crevices can be formed by weld spatter, hard layers of debris or exfoliated coatings, or for some materials, the oxide layers that are formed during welding.

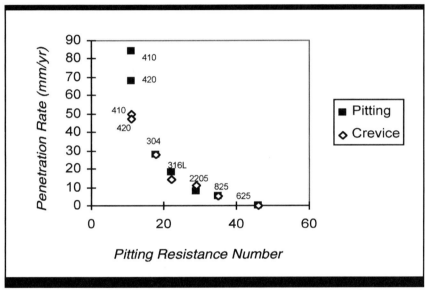

Fig. 4–2 *Pitting and Crevice Corrosion Rates as a Function of PRE Number*

The pH and chloride concentration in the crevice at which corrosion starts is also linearly related to the PRE_N. For example the 410/420 chromium steels have a critical pH of about 2.9 and chloride concentration of 35,000 ppm. Type 316L has a pH of 1.6 and chloride of 150,000; alloy 625, a pH of 0 and chlorides of above 200,000 ppm. Alloying with molybdenum has a major effect. For example, 25-6Mo stainless steel has a critical pH of below 1 and a chloride concentration of above 200,000 ppm. The rate of corrosion within a crevice is related to the crevice dimensions, typically being around 0.1 mm/year at the critical crevice size; but corrosion rate within the crevice may increase exponentially to 3–4 mm/year as the crevice dimension decreases.

The susceptibility of materials to pitting and crevice corrosion is tested by immersion of the material in an acidic ferric chloride solution to ASTM G48 and G78.[8,9] This is a somewhat artificial test but useful for ranking alloys and for QA/QC. Either an elastomeric band around the test specimen or attachment of an inert fluorocarbon polymer washer forms

the crevice. Alternative tests use electrochemical techniques that sweep the electrochemical potential from noble to base and back to noble with a record kept of the current flux.[10] The results are plotted as potential vs. log current flux, and the susceptibility of the material is estimated from the area (termed *hysteresis*) between the upward and downward sweeps. Other parameters used to rank the CRAs are critical pitting temperatures and potentials, and the pitting behavior can be shown as ISO-potentials in log chloride-potential plots. A listing of the crevice temperatures for corrosion resistant alloys is given in Table 4–3.

Table 4–3 *Critical Crevice Corrosion Temperatures: Tests in 10% FeCl3.6H2O at pH 1 for 24–Hour Exposure*

Material	UNS Code	Crevice Temperature °C
304 stainless steel	S30400	< -2.5
316 stainless steel	S31600	0
904L Cr-Ni alloy	N08904	0
2205 duplex steel	S31803	17.5
Ferralium 255	S32550	22.5
254 SMO	S31254	32.5

4.2.6 SCC

SCC is a corrosion process involving the conjoint action of stress and a specific environment on a susceptible material, as illustrated in Figure 4–3. For the austenitic and duplex stainless steels, chloride stress corrosion cracking (CSCC) and SSC are the prevalent forms. The risk to a pipeline by CSCC is essentially an external issue but could occur internally in an operational pipeline if it were exposed to aerated seawater while the temperature was high. SSC is an internal cracking phenomenon. Martensitic stainless steels suffer sulphide stress cracking in sulphide environments but are immune to CSCC. Duplex stainless steels are significantly more resistant to CSSC than the austenitic stainless steels. Figure 4–4 illustrates the difference between type 316L austenitic stainless steel and type 2205 duplex stainless steel. The high nickel alloys, which contain over 40% nickel, are resistant to both forms of cracking.

For CSCC the specific environment is the presence of oxygen, chlorides, and a high temperature. Even very low levels of oxygen are sufficient to cause CSCC. For example, stressed type 316L will crack in environments containing above 50 ppm at a chloride at a temperature above 50°C. The classic crack morphology of CSCC is highly branching cracks, rather like tree roots that spread outward from the origin. The cracking may be either trans- or inter-granular, a situation that can lead to difficulty in defining the mechanism.

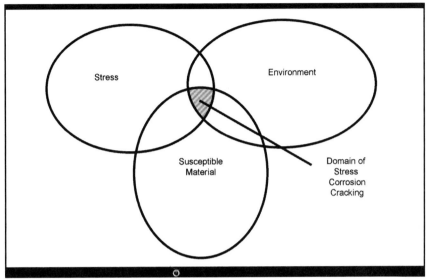

Fig. 4–3 *Stress Corrosion Cracking*

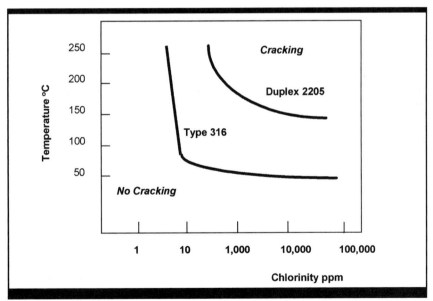

Fig. 4–4 *Comparison of Stress Corrosion Cracking Susceptibility of Steels*

The cracking initiates at pits and crevices or other surface blemishes. There are two hypotheses to explain propagation of cracking. The first hypothesis relates to differential corrosion. If the pit is sufficiently deep, the pit behaves as a tiny crevice; and corrosion proceeds similarly to crevice corrosion. Once the pit is over a certain depth, there is a stress amplification which results in a plastic yield at the crack tip and this continuously damages the oxide film. Therefore the material at the crack tip does not repassivate while repassivation occurs on the walls of the crack. The anodic area is small, and the cathodic surface large so that the crack proceeds by severe anodic dissolution at the crack tip. The alternative hypothesis relates to hydrogen embrittlement of the material at the crack tip leading to material fracture. The hydrogen migrating through the crack tip area collects at the area where there is a concentration of dislocations (resulting from the yield). This is a similar mechanism to that postulated for sulphide stress cracking.

Sulphide stress corrosion results from localized embrittlement of the alloy by atomic hydrogen that migrates from the surface through the steel. The atomic hydrogen is formed by corrosion. At areas of yield, there is high concentration of dislocations that act as traps for the atomic hydrogen. If the local stress level exceeds the embrittled strength of the material, the crack will propagate into the ductile material. Single, unbranched, transgranular cracks are typical. Because corrosion rates are lower than carbon steel and permeability of hydrogen through the stainless steel is low, these materials have a markedly better resistance to SSC than do the carbon steels. However this resistance is sensitive to chloride concentration. Conjoint high sulphide and chloride can cause SSC; and, as with the carbon steels, the material strength (evaluated as hardness) must be restricted. NACE MR-0175 gives acceptable values for several CRAs.[11]

Cathodic protection is effective in preventing external pitting and crevice corrosion and CSCC of these steels. Typically, the cathodic protection current required is 10–20% of that required for carbon steel.[12] Consequently, CP of stainless steels is both effective and efficient. There is always, however, a risk that a pipeline CP system may not be operative; and in such a case, a susceptible pipeline material could fail. Consequently, solid austenitic stainless steel is not used for submarine pipelines. Though the CRA may require a lower current density for CP because the steels are almost inevitably linked to carbon steel, it is usual to design for an overall protection potential of -800 mV SSCE and to consider the stainless steel as if it were carbon steel.

4.3 Available Corrosion Resistant Alloys

4.3.1 Metallurgical structure

The structure of a stainless steel can be determined from the Schaeffler diagram (Fig. 4–5). Some alloying elements promote the formation of ferrite (e.g., chromium) while others promote the formation of austenite (e.g., nickel). The diagram in Figure 4–5 maps the complete range of stainless steels. The formulae used to determine where an alloy is located on the Schaeffler diagram are listed here:

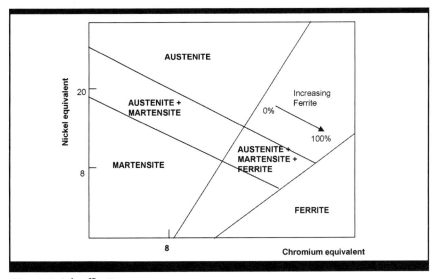

Fig. 4–5 *Schaeffler Diagram*

Chromium equivalent (ferrite forming) =

$$Cr + 2\,Si + 1.5\,Mo + 5\,V + 5.5\,Al + 1.75\,Nb + 1.5\,Ti + 0.75\,W \quad (4.6)$$

Nickel equivalent (austenizing) =

$$Ni + Co + 0.5\,Mn + 0.3\,Cu + 25\,N + 30\,C \quad (4.7)$$

4.3.2 Low-chrome carbon steels

Weldable carbon steels containing up to 0.75% chromium have been used for a limited number of pipelines. The modest chromium content does not affect the welding procedure. These are not corrosion-resistant alloys but show improved sweet corrosion resistance compared to conventional carbon manganese steels. This reduced corrosion rate allows a reduced corrosion allowance. Conventional corrosion inhibitors are often, but not always, suitable for these steels. However the corrosion resistance of these materials is very sensitive to chloride content, pH, and flow rate. When considering use of these steels, it is most important to define clearly the expected service conditions and to evaluate the steel in simulation tests. Proposed corrosion inhibitors should also be screened, and the testing should include a weldment to check that the inhibitor protects the weld and HAZ as well as the parent plate.

4.3.3 Martensitic stainless steels

Martensitic stainless steels are commonly used for OCTG and valve components. These materials are essentially a carbon manganese steel enhanced with 12–13% chromium, the minimum amount to ensure good resistance to sweet corrosion. (See Figure 4–1.) Typical types used are AISI 410 and AISI 420. AISI 420 has a higher carbon and manganese content and significantly higher strength and slightly better corrosion resistance. The steels have superior erosion-corrosion resistance compared to carbon steel, permitting about double the critical erosion velocities. Martensitic steels have poor toughness at low temperature, which may restrict the range of installation conditions. It is common practice to use a double heat treatment—Q&T twice—to improve the toughness. This practice also improves the resistance to SSC but reduces the general corrosion resistance because of the growth of the grains during the second tempering stage. A single heat treatment is recommended whenever possible to maintain the higher corrosion resistance. Martensitic stainless steels are approximately three times the cost of carbon steel. A comparison of costs of CRAs is given in Table 4–4.

Unfortunately the martensitic stainless steels have poor weldability, as expected from the high carbon equivalent, and also suffer delayed hydrogen cracking after welding. However a pipeline internally clad with 12% Cr and fabricated by centricasting was installed in Indonesia and has operated successfully for more than 20 years. To circumvent the poor weldability, the 13% chromium materials have been reviewed as candidates for use in pipelines joined by screwed connections.

Table 4-4 *Relative Costs of Solid Corrosion Resistant Alloys*

Material	UNS Code	Relative Yield Strength	Price Factor	Relative Price Factor
316L	S31603	1.00	1.0	1.00
2205	S31803	2.64	1.1	0.42
317L	S31703	1.21	1.2	0.99
904L	N08904	1.29	2.0	1.55
25-6MO	N08925	1.76	2.7	1.53
Alloy 825	N08825	1.01	3.2	3.17
Alloy 625	N06625	2.44	3.3	1.38
Titanium	R50250	2.82	8.0	2.84

To overcome the welding limitation, new alloys of low carbon, 13% chromium have been developed. These alloys contain nickel, manganese, and molybdenum. The alloying element concentration is enhanced for sour service. Corrosion resistance is as good as the 13% chromium steel and weldability is sufficiently improved for these materials to be used for welded pipelines. The materials are commonly welded using procedures and consumables used for duplex stainless steels. These new materials are a halfway house between carbon steel and duplex stainless steel.

There are limits to the corrosion resistance of the martensitic chrome steels because the passive surface film, which provides the resistance, is susceptible to breakdown by chlorides. The upper limit of use is around 120 °C and 10% brine for a partial pressure of carbon dioxide of 20 bar. Corrosion inhibition using inhibitors suitable for carbon steel is possible for these materials, but this could defeat the object, which is to minimize the operating expenditure (OPEX).

The conventional 13% chrome martensitic steels are susceptible to SSC. If these materials are to be used, then NACE MR-0175 should be consulted for the latest tolerance factor of hydrogen sulphide if there is any likelihood of souring of the production in service. An upper limit for use would be a partial pressure of 0.1 psia of hydrogen sulphide equating to a concentration of 200–300 ppm for a typical gas pipeline.[13]

4.3.5 Austenitic stainless steels

The austenitic stainless steels contain both chromium and nickel. The nickel content must be sufficient to extend the austenitic range so that the steel is austenitic at all temperatures; and, hence, the nickel content must vary depending on the chromium content. For chromium contents of 18%, the minimum nickel required is around 8%. This is the basic stainless steel:

type 304. The austenitic stainless steels are non-magnetic (they can be magnetic when they have been severely cold worked) and are widely used in chemical process plant, refineries and gas plant, and for cutlery, decorative trim such as lift doors and shop fronts. There are several hundred formulations ranging from the minimum material: 18% chromium plus 8% nickel (type 304) through to the super austenitics containing over 50% alloying elements, for example, 27% Cr, 30% Ni, 3% Mo. Molybdenum is used to improve resistance to pitting and crevice corrosion, and the PRE_N increases by 3.3 times the molybdenum content. The molybdenum stabilizes the structure of the chromium oxide film, reducing the damaging effect of chloride. Molybdenum content of 2.5% is the minimum required for stabilization of the oxide film in seawater (2.5–3.0% chloride). Types 316 and 316L, which contain 2.5% molybdenum, are the most widely available of the austenitic stainless steels.

The austenitic stainless steels have high general corrosion resistance but can suffer pitting, crevice corrosion, and CSCC. It should be recalled that while the austenitic stainless steels are resistant to corrosion by the produced fluids, they are at risk of corrosion during seawater hydrotesting and of CSCC in service if aerated seawater were to enter the pipeline during service at high temperature. CSCC of austenitic stainless steels occurs if there is the conjoint action of stress and the presence of chloride on the hot steel in an oxygen-containing environment. The stress is the combination of both the residual and applied stress. The critical chloride concentration for type 316 steels is around 50 ppm, and the critical temperature is 50 °C. The level of oxygen to prevent CSCC must be very low, as cracking problems have been reported at 5-10 ppb of oxygen. There is no method of predicting the initiation of CSCC; but once the cracks are initiated, they can propagate rapidly.

Because of the risk of external CSCC in the aerated seawater if the cathodic protection system were to become inoperative, the austenitic stainless steels are not used in the solid form for submarine oil or gas pipelines. The low yield strength of these steels also makes them unattractive for use as solid CRA pipe. Stainless steel type 316L, however, has been used extensively as a liner in carbon manganese steel pipe for gas and gas-condensate service and is also used for vessels, valve trim, and minor components. A 3166 clad carbon steel pipeline would be approximately from 4 to 5 times the cost of a carbon steel pipeline.

Austenitic stainless steels have a good resistance to SCC compared to carbon steels, provided they are not work-hardened. However, SSC can occur in the presence of a high partial pressure of hydrogen sulphide together with chloride. The material strength (evaluated as hardness) must be restricted for service in such environments.[11]

The stainless steels are readily welded, though for some compositions caution is needed to avoid weld sensitization.[14] This occurs in the heat-affected zone (the areas adjacent to the weld itself) because the chromium reacts with carbon to form carbide, which reduces the availability of the chromium to protect the steel from corrosion. To avoid this problem, either a very low carbon content is used (below 0.03% C) or stabilizing alloying elements are added. Titanium and niobium are stabilizers. They combine preferentially with the carbon so that the chromium level remains unaffected. Titanium-stabilized steels are not recommended for submarine pipelines because the non-metallic carbides that are formed can initiate pitting. Materials for pipelines are low carbon and such problems do not occur; however, if used as cladding, the welding of the carbon steel pipe subsequent to welding of the cladding needs careful attention. Low carbon content steels are identified by the suffix L attached to the generic identification number, e.g., type 316L.

A problem more likely to occur with austenitic stainless steel cladding materials is solidification cracking, which occurs when the molten weld material is just about to solidify. The austenitic steels have high thermal expansions; and at the point of solidification, the shrinkage of the parent material can pull the solidifying crystals apart while they are surrounded by liquid metal. Careful control of the root gap is important, and the weld procedure usually allows for the formation of a small fraction of delta-ferrite material in the weld material. Composition of consumables can also be balanced to prevent solidification cracking.

4.3.6 Duplex stainless steels

Duplex stainless steels have been used for several submarine pipelines for the transport of hot, wet, sweet fluids. (See Table 4–5.) Solid duplex is about four to five times the cost of an equivalent carbon steel pipe. The composition is similar to the austenitic stainless steels, but the nickel content is lower, around 5–7%, insufficient to ensure that the material is wholly austenitic. As a consequence, the material has a double, or duplex, structure with intimately intertwined alternate regions of ferrite and austenite. A 50:50 phase balance between ferrite and austenite is the optimum for mechanical properties and corrosion resistance.[15]

Duplex steels are strong having about twice the yield strength of an austenitic material while retaining good ductility. They have slightly lower corrosion resistance than the austenitic stainless steels but markedly better resistance to chloride stress corrosion cracking and also maintain good toughness at low temperature.[16] Both these latter qualities result from the composite structure that prevents crack propagation. While the austenitic

material remains susceptible to CSCC cracking, a crack does not propagate through the CSCC resistant ferrite phase. At low temperature, the ferrite material shows reduced toughness, but the austenitic phase retains high ductility and can absorb high impact. Provided the material has equal austenite and ferrite phases free of inter-metallic and non-metallic compounds, the duplex steels show a gradual ductile-brittle transition. They are suitable for arctic conditions (e.g., they have been used in Alaska) though they are not suitable for cryogenic service. The lowest acceptable service temperature is about -50°C.

Table 4–5 *Examples of Submarine Duplex Pipelines*

Location	Service	Diameter (in.)	Length (km)
Netherlands	Gas	6	5.9
Netherlands	Gas	10	3.8
Netherlands	Gas	12	9
Netherlands	Multiphase	20	4
Netherlands	Gas	8	10
Netherlands	Multiphase	10	8
North Sea	Gas	12	4.7
North Sea	Gas	20	4.7
North Sea	Multiphase	Twin 6	7
North Sea	Gas	Twin 9	11.5
North Sea	Gas	6	11.5
North Sea	Multiphase	6	14
North Sea	Multiphase	Triple 6	16
North Sea	Multiphase	14	5.5

The tolerance of high-chloride environments is largely due to the higher molybdenum content of these materials. The more commonly available third-generation duplex stainless steels also include copper and nitrogen to 0.15–0.2% to aid in increasing both the yield strength and the PRE_N. Initially, the materials replaced austenitic stainless steels in chemical plants (in particular for heat exchangers) because of the enhanced resistance to CSCC, crevice, and pitting attack and because duplex stainless steels have slightly higher thermal conductivity. Though having high resistance to CSCC, these materials are not immune. The standard test for SCC is boiling 42% MgCl in which all the duplex materials will eventually crack.[17] A more realistic test for the oil industry is the boiling 25% sodium chloride test and the *Wick Test*.[18]

Heat treatment of the material as occurs when hot forming bends may affect the balance of the two phases; it is this effect that also complicates welding. When held at high temperature, the material converts to austenite. This is a useful effect during the fabrication of plate steel because the austenite has a lower strength. However as the temperature drops, the material converts to mixed austenite-ferrite. The difference in high temperature strengths can result in internal cracking during the cooling phase.

It is also important to avoid the formation of sigma and/or chi phase materials within the duplex material. These iron-chromium-molybdenum phases are very brittle and are formed if the material is held in the range 700–955°C. These inter-metallic phases reduce both the corrosion resistance and the toughness of the steel. The rate of formation of sigma and chi phase materials is greatest over the range 800–850°C, and the risk of production of sigma phase increases by increasing chromium and molybdenum content but reduces by increasing the nitrogen content. The material composition is, therefore, a compromise between fabrication requirements and the eventual mechanical and corrosion properties.

Because of the ease of transformation of the material to ferrite and the risk of formation of inter-metallic phases, during welding it is necessary to maintain low heat inputs which results in relatively slow welding. As the chromium and molybdenum content increases, the heat input restrictions become more severe. For girth welding in the field, TIG welding is normally used.[19] During welding, the inner bore of the pipe must be purged with inert gas to prevent the formation of thermal oxide films around the weldments, which could lead to pitting/crevice corrosion. Welding speed of duplex stainless steels is about a third the rate of welding of a geometrically similar carbon steel pipeline. The slow welding can have a major impact on the overall installation cost and, hence, fabrication of the pipeline onshore for installation in a bundle or by reel lay may have economic advantage over conventional S-lay.

The welding specification will usually include a limit on the ferrite content of the welds, but, despite care, the weldments do tend to have a higher ferrite content than the parent pipe. As the ferrite content rises the material loses toughness and corrosion resistance. The critical value is about 75% ferrite when the material properties become similar to ferritic material. Presently ASTM specifications do not include control of ferrite and inter-metallic phases. This is presumably because there are no simple tests that will provide the required information. Metallographic testing is not reproducible, and impact and corrosion testing are not readily applicable as standard tests. The pipeline engineer will undoubtedly add the requirement for supplementary testing and compositional requirements to the ASTM specification, e.g., controlled nitrogen level,

metallographic tests to substantiate phase balances and the absence of inter-metallic (sigma and chi phases), toughness testing at specified temperature, and corrosion resistance in ferric chloride or 25% NaCl.

A portion of the yield strength of the duplex stainless steels results from dislocation strengthening, which occurs during the manufacturing process of the pipe. At a high operating temperature, there is a gradual relaxation of the dislocation strengthening and a consequent reduction in yield strength. Consequently, steels with the same chemical specification may not have the same yield strength if sourced from different manufacturers. Depending on the source of the steel, this different degree of strength loss of the steel complicates design, as it is not possible to define the required wall thickness. One approach is to require the steel supplier to guarantee an SMYS at the design temperature and then to use this value to determine the optimum wall thickness. The supplier can then bid against the wall thickness relevant to his guaranteed SMYS.

Because of the loss of tensile strength with increase in temperature, there will be an upper limit on the temperature at which a solid duplex material is applicable. Because the wall thickness would have to be increased, there are consequent economic penalties for the cost of the material itself, for welding. Also there is the subsequent risk of upheaval buckling when the pipeline is operational. Some advantage may be gained by design of the pipelines to DnV Codes, using strain related design rather than designing to adhere to API Codes that use stress-related design formulae. This issue is discussed in chapter 11. The present trend is toward the use of higher strength materials, e.g., type 2507 compared to type 2205.

Duplex steels in the wrought form are not immune to SSC. Resistance is improved by heat treatment, but this action reduces the material strength. Not all duplex stainless steels are presently incorporated into NACE MR-0175. The behavior of the newer formulations is unclear at present and subject to research. European studies indicate that the resistance of duplex steels to SSC is markedly affected by the concentration of chlorides in the aqueous phase and that NACE MR-0175 may not be conservative in high chloride concentration brines at elevated temperatures. It may be necessary to test the materials in simulated service conditions; and it is, therefore, important that the water composition is available when evaluating these materials.[20, 21, 22]

4.4 Manufacture of Corrosion Resist Alloy Pipe

4.4.1 Solid CRA pipe

API Specification 5LC is the basic specification for CRA pipe.[23] Duplex stainless steel pipe is fabricated by a variety of routes including extrusion, mandrel and plug mills up to 16-in. as used for carbon steel seamless pipe. The length of individual pipe joints is limited by the billet size; and full length (12.19 m) seamless pipe to API 5LC may include a factory girth weld. Larger diameter pipe is produced by longitudinal welding after forming the pipe by a U-O-E process. The welded route is generally favored for pipe with a diameter above 10-in. Pipe fabricated from coiled strip can be supplied in standard lengths.[24]

4.4.2 Internally clad pipe

Clad pipelines are formed from a carbon manganese steel outer pipe lined internally with a thin layer (2–3 mm thick) of corrosion resistant material. The internal cladding is not considered to act as a pressure retention item. The clad pipe may be formed with the lining either metallurgically bonded to the steel pipe or with a tight fit into the steel pipe. The clad pipe represents an economic solution of mechanical and corrosion requirements.[25] The clad pipe confers the high strength of carbon steel (typically API 5L X65 or X70) with the corrosion resistance of the liner.

Commonly used CRAs are austenitic stainless steel type 316L and high nickel alloys type 825 and 625. 13 Cr material has also been used, and there is interest in using super grade stainless steels, e.g., nickel alloy 904L or 25-6MO. Clad pipe and its usage have been extensively reviewed.[26] Examples of clad pipelines, both subsea and land, and risers are given in Table 4–6.

A variety of methods of forming clad pipes have been developed. Bonded pipe is formed by hot extrusion of a pipe within a pipe in a process similar to the fabrication of seamless pipe. Clad pipes also may be formed from plate that has had the lining material bonded to the steel plate. Explosive bonding, as used for fabrication of composite materials for vessels, is under investigation. Cast clad pipe is also available. Bonded pipe formed by extrusion processes or from roll-bonded plate is the most commonly used type of pipe.

Table 4–6 *Examples of Internally Clad Pipelines and Risers*

Location	Service	Liner	Diameter (in.)	Length (km)
Indonesia	Gas	13 Cr	10	12
Netherlands	Gas	316	18	5
Netherlands	Gas	316	4	3
Netherlands	Multiphase	316	8	1
S. North Sea	Gas	625	12	Riser
S. North Sea	Gas	316	8	12
S. North Sea	Oil	316	30	Riser
S. North Sea	Gas	625	12	0.7
North Sea	Oil	625	6	1
Algeria	Gas	410	8	---
Netherlands	Sour oil	316	10	1.6
Netherlands	Gas	316L	12	1.5
Saudi Arabia	Oil	316L	8 to 24	1.1
Netherlands	Sour gas	825	36	Slug catcher
Netherlands	Sour gas	825	12	13
Netherlands	Sour gas	825	8	3.3
Netherlands	Sour gas	825	8	5
Netherlands	Sour gas	625	6	1
Indian Ocean	Sour gas	825	24	6.5
Indian Ocean	Sour gas	825	20	4.4
Netherlands	Sour gas	825	6	3
New Zealand	Gas	316L	20	16
Mobile Bay, USA	Sour gas	825	5	5.8
Mobile Bay, USA	Sour gas	825	6	1.1
Alabama Gulf, USA	Sour gas	825	6 to 12	0.4
North Sea	Oil	316L	36	Riser
Alaska	Oil	825	6	0.3
USA	---	825	6	10.6
North Sea	Gas	825	3 to 12	Risers

4.4.3 Hot-formed bonded seamless pipe

A tubular liner is inserted into a seamless pipe; then the liner and pipe are co-extruded in a manner similar to the fabrication of a conventional seamless steel pipe. See Figure 4–6. This method can be used to form pipe with a 2–16 in. diameter and wall thickness up to 25 mm. The length that can be formed is limited by the weight of the billet and the extrusion equipment so that thick wall pipe may be in short lengths of 6 m. Conventional 12-m lengths are available in the more modest wall thickness. Standard lengths are formed by factory welding of the shorter lengths.

A variation of this process is the production of a short, thick-wall steel pipe internally lined with CRA applied as a weld overlay. Then the pipe is hot extruded to the required dimensions.

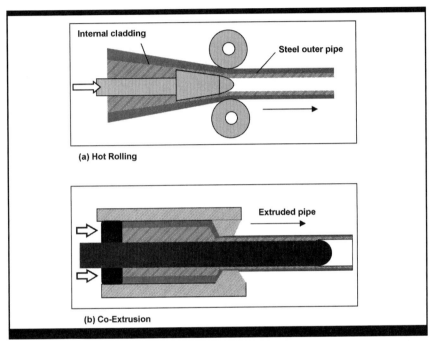

Fig. 4–6 *Fabrication of Hot-Bonded, Internally Clad Seamless Pipe*

4.4.4 Hot isostatic pressing (HIP) clad seamless

For HIP clad seamless pipe, the liner is a sintered material bonded to the carrier pipe by high pressure at high temperature. This method can also be used to clad fittings and valves. A *blind* HIP cladding process to form solid alloy fittings (tees and Ys) is also available. The minimum cladding thickness is 2 mm, and the process can be applied to pipe with a diameter from 10–30 in., though only short lengths, around 2–4 m, can be formed. Again, to produce standard lengths, the pipes are factory-welded.

4.4.5 Roll-bonded pipe

To form plate steel covered with the CRA, liner sheet is hot rolled to bond the cladding to the steel.[27] Usually two plates are co-rolled with the CRA liners facing inward separated by a non-bonding interleaf. This produces two clad plates in one rolling operation and avoids damage and contamination of the cladding. After rolling, the plates are separated, trimmed, and formed into cylinders by a U-O-E or press-bend technique, and completed with a double longitudinal weld. The process is shown schematically in Figure 4–7. When possible, the cladding is welded from the inside of the pipe, but this procedure is limited to a minimum diameter of 8 in. Smaller diameter pipe is welded completely from the outside. The process is illustrated in Figure 4–8. A maximum total wall thickness of up to 30 mm is available. Most fabricators produce pipe of 8 m length although some mills can produce 12 m lengths in some diameters.

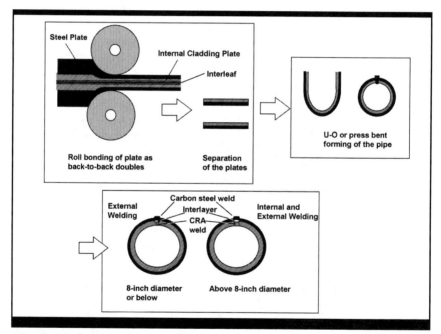

Fig. 4–7 *Fabrication of Hot-Bonded, Internally Clad U-O-E*

4.4.6 Explosively bonded plate

The steel plate is covered with the sheet cladding, which is separated slightly from the steel plate using plastic spacers scattered over the steel plate. The face of the cladding is coated with explosive that is ignited at one end of the cladding. As the explosive front runs across the cladding, the cladding is pressed onto the pipe and metallurgically bonded to the plate with a characteristic wavy bond between liner and steel plate.[28] The pressure of the explosive bonding is sufficiently high to remove debris from the surface of the steel plate, which reduces the preparation costs compared to roll bonding of plate. The size of plate that can be clad is limited by the logistics of maintaining a constant separation between the steel plate and cladding sheet, though the process can bond almost any material to any other. Pipe lengths tend to be short, but wall thickness can be substantial.

4.4.7 Explosively bonded seamless

For explosively bonded seamless pipe, the liner is bonded to the carrier pipe by a controlled explosion of the cladding to the outer pipe. Typically, pipe with around 10-in. diameter and 10–20 mm wall thickness is fabricated in this manner. Pipe lengths are short, typically 5 m. Conventional lengths are formed by shop welding of the short lengths.

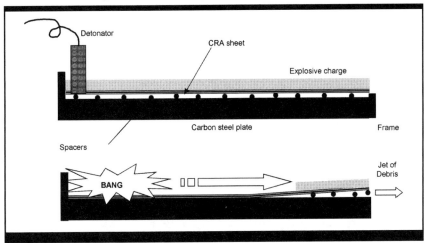

Fig. 4–8 *Explosive Cladding of Steel Plate for Internally Clad, U-O-E Pipe*

4.4.8 Centricast seamless

The outer carbon steel pipe is formed by centrifugal casting, and the liner is introduced while the carbon steel is still hot.[29] Some intermixing and diffusion occurs, creating a very strong bond between carrier and cladding. The internal cladding may be martensitic or austenitic stainless steels or a high nickel material. After casting, the inner surface of the pipe is bored internally to remove the slightly porous inner surface. The exact dimensions of the centricast pipe permit fit up for welding to fine tolerances. Pipe can be cast up to 90 mm thick in diameters from 4 to 16 in. Lengths are short, however, around 6 m. Standard pipe lengths are produced by shop welding. At present, this form of pipe is unobtainable as Kubota has closed the production line.

4.4.9 Non-bonded, tight–fit pipe (TFP)

There are two processes used to produce non-bonded TFP. In the first process, the outer steel pipe is heated; and the cladding, in the form of a cylinder, is inserted and expanded into the outer pipe. On cooling, the outer pipe shrinks on to the liner. The second process relies on cold hydraulic expansion of the liner into the outer pipe. The liner is stretched, but the steel pipe remains in the elastic region. See Figure 4–9. In both cases, the CRA liner is welded to the steel pipe at each end of the pipe, and the welding bevel cut through this welded section. The welding ensures that the cladding resists any movement of the liners during installation.

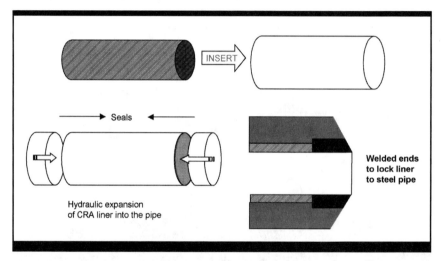

Fig. 4–9 *Fabrication of Internally Clad TFP*

The technique produces pipe with wall thickness up to 24 mm with a CRA liner thickness up to 20 mm in diameters from 2–16-in. The pipe length available is dependent on the diameter but is a minimum of 9 m. At present, this form of pipe is commonly used for larger diameter line pipe because of the lower cost compared to bonded pipe.

The lining is bonded to the external pipe for mechanical and corrosion reasons. Most CRA materials have low yield strength compared to carbon steel and deform plastically under conditions where the carbon steel behavior remains elastic. This can arise if the pipe is subject to a strain reversal, and the stress is sufficient to yield the cladding. Such a case could occur during reeling and unreeling or as a result of acute sag and over–bending. In this case the bending can cause wrinkling of the liner in pipe that is not metallurgically bonded.

During the corrosion process, atomic hydrogen is formed at cathodic sites. The bulk of this atomic hydrogen combines to form hydrogen gas that dissolves into the fluid streams. Some of the atomic hydrogen can migrate through the cladding and reform as molecular hydrogen on the other face. Concern arises with non-metallurgically bonded pipe as a result of hydrogen build up in the annular space between the cladding and the carbon steel pipe. If sufficient hydrogen gas is formed in the annulus over time, there may be a risk of collapse of the liner if and when the internal operating pressure in the pipeline decreases. Hydrogen could also be produced by excessive external cathodic protection of the carbon steel pipe in which case atomic hydrogen would migrate inward to the annular space. Hydrogen permeation studies on tight-fit pipe showed that hydrogen build up in the annulus is very limited: (a) corrosion rates are low and little atomic hydrogen is produced, and (b) the diffusion rate of atomic hydrogen through most CRAs is low compared to the diffusion rate through carbon steel.[30] Consequently, the time to produce sufficient hydrogen to risk an exfoliation of a liner off the outer pipe would be beyond a normal pipeline design life. However, some evaluation of these rates would be required to confirm this for a specific pipeline scenario.

4.4.10 Nickel alloys

Nickel alloys are expensive materials because of the high nickel and other alloy content. Alloy compositions are given in Table 4–1. The high nickel alloys are unlikely to be used as solid CRA materials for pipelines. If they are used, consideration should be given to the use of alloy 625 compared to alloy 825. Though alloy 625 is more expensive, it has much higher yield strength than alloy 825, about $2\frac{1}{2}$ times higher, allowing a considerably thinner wall, and is more readily weldable than alloy 825. Alloy 625 also has significantly better corrosion resistance than alloy 825. When used as a cladding, the higher strength is of no economic benefit; and economics favor the use of alloy 825.

The high nickel materials have extremely good corrosion resistance and are immune to stress corrosion cracking; they also have good resistance to crevice and pitting corrosion. They are likely to be used for hot, sour, acidic production. Pipelines clad with a high nickel alloy (type 825) have been used for submarine lines, the largest to date being in India and Mobile Bay in the United States. Flow loops and manifolds clad with these alloys have been used in the North Sea.

Price guides are not readily available for these materials, but a factor of 6–9 times the cost of carbon steel is probable based on published comparative studies. The cost of the pipe is markedly affected by the method of fabrication and the pipeline diameter. Alloy 825 clad pipe is straightforward to weld and has the high tensile strength of the carrier pipe (usually X70 grade) at elevated temperatures. Because there is an intermediate inspection stage, the overall welding time is two to three times that required for a similar geometry carbon steel pipeline. The overall project cost, therefore, may not be excessive when compared to use of thick wall solid duplex for hot, high-pressure service and the welding implications are taken into account. Further details are given below in the detailed discussion of welding clad pipeline materials.

4.4.11 Possible future materials

The super-austenitic materials are possible materials for use either as solid pipelines or, more likely, as a cladding material. The lower nickel content would reduce the cost of the steels to mid-way between the type 300 austenitic steels and the high nickel alloys. Super-austenitic steels have high PRE_N values and hence good resistance to pitting, crevicing, and stress corrosion cracking. These materials are also readily weldable compared to the duplex stainless steels.

Candidates are 904L and 25-6MO. The 904L requires modification by the addition of more molybdenum to a minimum content of 6.1%. The multiplier of 3.3 for Mo moves the PRE_N for the 904L into the super-austenitic domain. 25-6MO is a generic type, but is presently only available from a single source.

4.5 Welding of Internally Clad Pipe

Welding of internally clad pipe must generally be done in two steps: welding of the CRA liner followed by welding of the outer steel pipe. A major factor in preparing the CRA weld is the fitting up of the pipe because the liner is only 2 – 3 mm thick. The CRA weld is made with a material that is compatible with the liner. The internal corrosion behavior of the pipe is that of the liner, and particular attention is needed at the welds to avoid any dilution of the material by the subsequent carbon steel weld, which would reduce the corrosion resistance of the lining.

Welding of the carbon steel directly onto this CRA pass would dilute the corrosion resistant liner weld and, therefore, one of two procedures is used. In the first process, a pure iron weld is made as an intermediate layer followed by conventional carbon steel welding of the carrier pipe. In the second process the carbon steel weld is made with a corrosion resistant material, e.g., for a type 825 alloy cladding, the liner weld and carbon steel weld could be with alloy 625 consumables. The use of a pure iron interlay is cheaper, but there remains the additional complexity of multiple welding consumables, and there have been reports of cracking in the interlayer

The liner must also be X-rayed before the carbon steel weld is made. This step delays the completion of the weld. Consequently, welding is slow, typically at rates equal to those for solid duplex stainless steel. However, weld rates have significantly improved as experience has been gained.

4.6 Evaluating Corrosion Resistance

As discussed earlier, the resistance of a chromium alloy to pitting and crevice corrosion can be qualified from the pitting resistance number, PRE_N. It has been empirically observed that for duplex steels, a minimum PRE_N value of 33 is necessary to minimize corrosion problems with submarine pipelines. Super duplex stainless steels are now available with PRE_N values above 40. Crevice propensities are based on standard tests using exposure to hot ferric chloride solutions or, alternatively, using critical pitting temperature (CPT) testing. In this latter test, the temperature at which crevicing or pitting commences is taken as an indication of the susceptibility of the material to crevice or pitting corrosion.

Standard sulphide stress cracking tests are four-point bend tests to NACE TM-0177 in the NACE solution (5% NaCl + 0.5% acetic acid) saturated with hydrogen sulphide at 25°C for sulphide stress cracking; ASTM G35, which considers polythionic acids; and ASTM G36 (boiling 45% magnesium chloride) for chloride stress corrosion cracking.[17, 31] There has been an increase in the use of slow strain rate testing of these materials for comparisons and QC. In these tests, the stress level in a sample of the material is continuously increased at a constant strain rate while the sample is exposed to the test solution. The samples are strained to failure, and the failure point used as an indicator of the resistance to crack propagation.

The variations in composition, microstructure, and residual stress in the CRAs indicates that the types of qualification tests, which also serve to rank competitive materials, should include:

- Mechanical tests, perhaps at ambient, service, and design temperature
- Critical pitting potential (CPT)
- Critical crevice temperature (CCT)
- Stress corrosion cracking
- Static C-ring testing at 80% of yield of fatigue pre-cracked specimens
- Sulphide stress cracking tests in autoclaves with simulated service environments
- Hydrogen embrittlement testing (if protected by impressed current CP or if the pipeline is to be electrically bonded to a steel structure)

4.7 External Protection

Martensitic, austenitic, and duplex stainless steels require greater care in preparation for coating compared to carbon steel. Iron-free grit must be used to clean and profile the pipeline surface. Any residual iron introduced into the steel surface would corrode to produce ferric chloride that could initiate pitting of the stainless steel. To avoid iron contamination, aluminum-based abrasive is used although garnet is now becoming more widely used. Garnet is heavier than aluminum oxide and, therefore, more effective; and it is also environmentally neutral.

The use of conventional coatings such as reinforced coal tar or asphalt are unsuitable as they may induce crevice attack. FBE coatings detach when damaged thereby reducing the risk of a crevice being formed. Hence, FBE remains the most commonly used coating though it is increasingly used as a primer coat underneath extruded polymer coatings or elastomeric coatings for elevated temperature pipelines.

Laboratory testing indicates that the duplex steels are susceptible to hydrogen charging by cathodic protection. At a potential more negative than -950 mv silver-silver chloride electrode, the materials exhibited brittle fracture during slow strain rate testing. In these laboratory tests, protection was provided by a potentiostatic impressed current CP system. If a CRA pipeline is to be protected by an impressed current CP system or is to be electrically connected to a structure that is protected by an impressed current CP system, it may be necessary to evaluate the resistance of the material to hydrogen charging. Hydrogen evolution is less likely to occur in a seawater environment when cathodic protection is provided by sacrificial anodes, so this feature of duplex steels is unlikely to be of concern for the majority of submarine pipelines.

For internally clad pipe, the carbon steel carrier pipe can be protected by coatings and cathodic protection as applied to conventional carbon steel pipe.

4.8 Cost Comparisons

Cost comparisons are difficult to provide because so many details are project-specific. The materials cost alone is not adequate to analyze overall project cost. From the operator's viewpoint, it is also necessary to take into account the savings over the life cycle of the pipeline, e.g., reduced inspection, corrosion monitoring and chemical inhibition. A typical analysis is given in Table 4–7.

Table 4–7 *Cost Comparison of Pipeline Material Options*

Cost Item	Units	Carbon Steel	Duplex SS	825 Clad CS
Material				
6 & 12 in.	$/ton	800	7,000	13,500
36-in.	$/ton	800	7,000	8,000
6-in.	$/mile	40,500	356,400	950,200
12-in.	$/mile	111,700	984,300	2,417,800
36-in.	$/mile	401,400	3,536,700	4,944,400
Labor & Welding				
6-in.	$/mile	80,000	113,150	200,000
12-in.	$/mile	100,000	132,000	330,000
36-in.	$/mile	250,000	198,000	395,000
Welding Rates				
6-in.	Welds/day	120	40	24
12-in.	Welds/day	100	27	22
36-in.	Welds/day	80	25	17
Filler Metal Costs				
6-in.	$/mile	N/A	5,500	11,700
12-in.	$/mile	N/A	11,000	24,000
36-in.	$/mile	N/A	33,000	70,500
Corrosion Inhibition				
Labor	$	200,000	---	---
Equipment	$	40,000	---	---
Inspection				
Smart Pig	$/mile	30,000	30,000	30,000
Frequency	Years	3–5	6–10	6–10

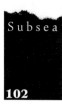
References

1   Peters, P. A. et al. (1986) "Line Pipe for the Transportation of Highly
    Corrosive Media." International Conference on Duplex Stainless
    Steel. The Hague, October.

2   Schofield, M. J. (1991) "Corrosion Resistant Alloys for Oilfield
    Applications—Alternatives and Selection Criteria." International
    Conference on Duplex Stainless Steels. Beaune, France.

3   Swales, G. L. & B. Todd. (1989) "Alloy-containing Alloy Piping for
    Offshore Oil and Gas Production." NiDi Technical Series (initially
    presented at 28th Annual Conference of Metallurgists) Canadian
    Institute Mining and Metallurgy. Halifax, Nova Scotia, August.

4   Smith, L. M. (1992) "Weighing the Higher Cost of Alternative
    Pipeline Materials vs. Their Potential for Greater Corrosion
    Protection." Corrosion Protection of Offshore Pipelines Conference,
    IIR, London, November.

5   Smith, L. M. (1992) "Cost Effective Applications of Corrosion
    Resistant Alloys in Offshore Operations." Corrosion Asia Conference,
    Singapore, September.

6   Craig, B. D. (1992) "Economics of the Application of CRA Clad
    Steel Pipe Versus Carbon Steel with Inhibitors—Factors Involved and
    Examples." International Seminar on Clad Engineering. NiDi & I
    Corr ST, Aberdeen.

7   Fielder, J.W. & D. R. Johns. (1989) "Pitting Corrosion Engineering
    Diagrams for Stainless Steels." UK Corrosion '89. ICorrST-NACE,
    April.

8   ASTM G48: "Pitting and Crevice Corrosion Resistance of Stainless
    Steels and Related Alloys by the use of Ferric Chloride Solution."
    American Society for Testing and Materials.

9   ASTM G78: "Crevice Corrosion Testing of Iron-Base and Nickel-Base
    Stainless Alloys in Seawater and Other Chloride Containing Aqueous
    Environments." American Society for Testing and Materials.

10 ASTM G61: "Cyclic Potentiodynamic Polarisation Measurements for
    Localized Corrosion Susceptibility of Iron-, Nickel-, or Cobalt-Based
    Alloys." American Society for Testing and Materials.

[11] NACE Materials Requirement MR 0175: "Sulfide Stress Cracking Resistant Metallic Materials for Oilfield Equipment." National Association of Corrosion Engineers.

[12] King, R.A. (1992) "On the Cathodic Protection of Stainless Steels." Conference on Redefining International Standards and Practices for the Oil and Gas Industry, IIR, London, March.

[13] Details extracted from product literature provided by NSC. Sumitomo; Kubota; Mannesmann; British Steel Corporation.

[14] ASTM A262: "Detecting Susceptibility to Intergranular Attack in Stainless Steels." American Society for Testing and Materials.

[15] Redmond, J. D. (1986) "Selecting Second-Generation Duplex Stainless Steels." *Chemical Engineering*, October, 152.

[16] Watts, M. R. (1989) "Material Development to Meet Today's Demands." *Anti-Corrosion*, February, 4.

[17] ASTM G36: "Stress-Corrosion Cracking Tests in Boiling Magnesium Chloride Solution." Pennsylvania: American Society for Testing and Materials.

[18] Schofield, M. J. & R. D. Kane. (1991) "Defining Safe Use Limits for Duplex Stainless Steels." International Conference on Duplex Stainless Steels. Beaune, France.

[19] Heikoop, G. G. & D. E. Milliams. (1984) "Girth Weld Corrosion in Duplex Stainless Steel Pipeline." Offshore Northern Seas Conference, Stavanger, Norway..

[20] Eriksson, H. & S. Bernhardsson (1990) "The Applicability of Duplex Stainless Steels in Sour Environments." (Paper 64) Corrosion 1990 Conference. National Association of Corrosion Engineers, Las Vegas, NV.

[21] Fujita, S., Y. Kobayashi, & M. Sugar. (1990) "Factors Affecting Corrosion Resistance of Duplex Stainless Steel Line Pipes in H_2S Containing Environments." (Paper 62) Corrosion 1990 Conference. National Association of Corrosion Engineers, Las Vegas, NV.

[22] Place, M. C., R. D. Mack, & P. R. Rhodes. (1991) "Qualification of Corrosion-Resistant Alloys for Sour Service." (Paper 6603) Offshore Technology Conference. Houston, TX.

23 API 5LC: "Specification for CRA Line Pipe." American Petroleum Institute.

24 Parlane, A. J. A. & J. R. Still. (1998) "An Overview of Pipelines for Subsea Oil and Gas Transmission." *Mat Science and Tech.*, April, 314.

25 Smith, L. M. (1992) "Engineering with Clad Steel." Proceedings of the International Seminar on Clad Engineering, NiDi & I Corr ST, Aberdeen.

26 NACE TPC T1F21D: "Review of CRA Usage" [Report] National Association of Corrosion Engineers, 1993.

27 Charles, J., D. Jobard, F. Du Poiron, & D. Catelin. (1989) "Clad Plates: An Economical Solution for Severe Corrosive Environments." *Material Performance*, April, 70.

28 Nowell, D. "Bang on for Clad Plating." *Process Engineering*, August 1987, 47.

29 Swales, G. L. (1989) "Application of Centrifugally-Cast Alloy Piping and Pipe Fittings in Onshore and Offshore Oil and Gas Production." (Paper) NiDi Proceedings of the 28th Annual Conference of Metallurgists. Canadian Institute Mining and Metallurgy, Halifax, Nova Scotia, August.

30 Kane, R. D. & S. M. Wilheld. (1990) "Evaluation of Bimetallic Pipe for Oilfield Flowline Service Involving H_2S and CO_2." (Paper 557) Corrosion 1990, National Association of Corrosion Engineers, Las Vegas, NV.

31 NACE Test Method TM-0177: "Laboratory Testing of Metals for Resistance to Sulfide Stress Cracking in H_2S Environments." National Association of Corrosion Engineers.

5 Welding

5.1 Introduction

The methods of production of individual lengths of pipe to API 5L and ISO 3183 by forming plate or strip steel into cylinders that are completed by some form of welding are described in chapter 3. The welding procedures used are briefly described in this chapter. Submarine pipelines are constructed from lengths of pipe joined together by manual, semi-automatic or automatic fusion welding.[1,2] The principal codes for these welding processes are API 1104 and BS 4515.[3,4] Most of this chapter is concerned with these processes, but new welding processes are in development and are discussed in section 5.18.

The selection of the welding method is determined by the contractor's capability; the pipe diameter; wall thickness; and, to a lesser extent, fabrication location. Pipelines constructed on land for installation by reeling or in a bundle as well as pull outs and small diameter S-lay lines are almost always welded by manual welding. However, for offshore fabrication by S-lay, the larger diameter lines are more economically welded using semi-automatic or fully automatic welding.[5] This division of techniques arises because of the cost associated with use of the lay barge. Welding is the critical step in pipelaying because it dictates the length of time to construct the pipe and, for this reason, has a major impact on the cost of the project.[6]

Pipeline contractors are constantly researching for faster methods of welding to speed up the laying process and reduce the time the lay barge is required. Advances in welding procedures have been considerable.[7] The installation of the original Forties 32-in. pipeline in the North Sea required two barges working for two lay seasons, while the replacement 36-in. line laid 15 years later was laid by one barge well within one season. However, further large increases in the speed of conventional welding appear unlikely. Major advances will occur through the introduction of newer welding procedures such as flash butt welding, which was developed in the Ukraine and was widely used for land pipelines in the former Soviet Union; homopolar welding; electron-beam; plasma arc; and friction welding. The requirement for a faster welding procedure results from the limitation in J-laying used for installation of pipelines in deep water. In J-lay there is only one welding station; and, though multiple jointed pipes are used (up to six joints), the single weld station is the limiting factor as the complete weld must be produced at this one location.

5.2 Welding Processes

Welding joins metal by inducing coalescence of the material, by heating to a suitable temperature with or without pressure and with or without the addition of filler metal. Coalescence is the growing together of the grain structure of the metals being welded. There are three critical parameters:

- Heat input: Sufficient energy must be provided to melt the metal and consumable (W/m^2).
- Heat input rate: The rate of energy input controls the rate of welding ($W/m^2/m/s$).
- Shielding from the atmosphere: prevents oxidation of the molten melt that would produce a weak weldment.

The heat for melting the metal may be provided by laser, burning acetylene gas with oxygen, or by an electrical process.[9] Laser welding is not practicable at present for welding thick wall pipeline material, but in the long term, future pipe joining using laser heating coupled with pressure is a possibility. Acetylene gas welding is not used for pipeline welding but may be used for cutting, though plasma-arc cutting is a more common procedure nowadays. The electrical methods produce heat by either resistance heating or by production of an arc between the welding torch and the pipe. Electric resistance welding (ERW) and flash butt welding are examples of resistance heating.

Manufacture of welded pipe requires rapid welding procedures, and two processes are used: submerged arc welding and electrical resistance welding. The weld of pipe produced by submerged arc welding is generally made in two weld passes, one internal and one external. Electrical resistance welding is a single pass operation used for fabrication of modest diameter pipe. These rapid techniques cannot be used for production of field girth welds. At present the girth welds are made exclusively by a sequence of arc welding processes in which the arc is a plasma discharge of high temperature, and typically four to seven passes are required. The electrical process to generate the plasma arc depends on the welding process. Direct current (DC) welding machines with drooping character-istics are used for manual arc welding with cellulosic-coated electrodes, and the voltages used are between 80–100 V, whereas gas metal arc welding and tungsten inert gas welding generally use pulsed alternating current (AC).

Welding technology uses acronyms as shorthand for the processes.[10] The terms commonly encountered in pipelining include:

- SAW—Submerged Arc Welding is used for producing longitudinally welded pipe formed by the U-O-E process and also for producing double or triple jointed pipe (24–36-m lengths). The metal is joined by fusing with an electric arc or arcs struck between a bare metal wire electrode or electrodes and the pipe. A blanket of granular, fusible material spread in a deep layer over the weld area shields the arc and molten metal. The process is illustrated in Figure 5–1.

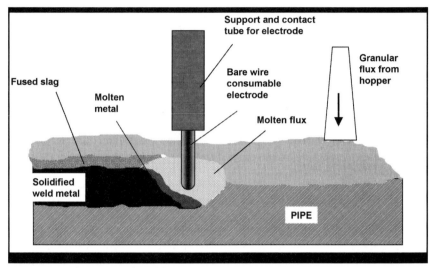

Fig. 5–1 *Submerged Arch Welding*

- SMAW—Shielded Metal Arc Welding is the conventional, manual arc welding process in which the heat for welding is supplied by an electrical arc struck between a consumable electrode and the pipe. The electrode, or stick, is covered with a basic or cellulosic coating that burns during use releasing carbon dioxide that shields the molten weld metal. The process is illustrated in Figure 5–2. The cellulosic cover is a blend of organic fibers. Cellulosic electrodes are sensitive to temperature changes, which alters their moisture content. This type of electrode is supplied in sealed cans, which are baked to ensure that they are dry. The electrodes are kept warm in *quivers*. Re-baking of electrodes that have been allowed to cool should not be permitted.

- GMAW—Gas Metal Arc Welding is a weld produced by heating with an arc struck between a bare metal electrode wire and the work. The electrode wire is fed continuously through the welding head. Shielding of the molten metal is provided by gas introduced through an annulus around the welding wire in the welding head. If the gas is inert, the procedure may be termed *metal inert gas welding*, MIG; and if the gas is active, the procedure may be termed *metal active gas welding*, MAG. For pipe welding, a mixture of argon and carbon dioxide is generally used. The process is illustrated in Figure 5–3.

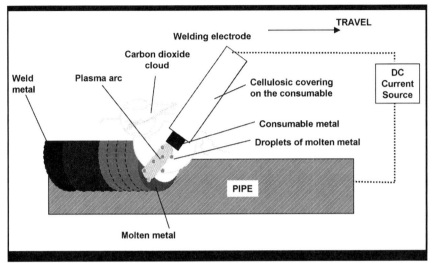

Fig. 5–2 *Manual Metal Arc Welding*

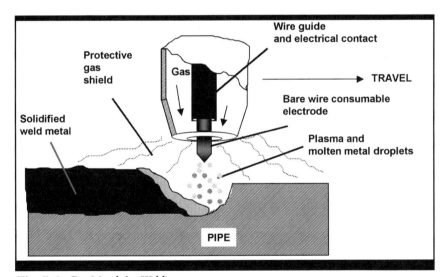

Fig. 5–3 *Gas Metal Arc Welding*

- GTAW—In Gas Tungsten Arc Welding, also termed tungsten inert gas (TIG) welding, the arc is struck between an inert, non-consumable electrode fabricated of tungsten, and the filler metal is introduced as a wire consumable fed into the molten metal pool. The molten metal is shielded by an inert or active gas introduced through an annulus around the tungsten electrode in the welding head. In the past, helium was used as the shielding gas; and the

process was termed *heli-arc welding*. GTAW is used for root passes and also for welding of corrosion-resistant alloys such as the duplex stainless steel with an argon shielding gas. TIG welding is slow because the heat input rate is limited. Welding using a hot wire feed into the weld pool is about 20% more rapid but is not a suitably robust process for offshore production welding. The procedure is illustrated in Figure 5–4.

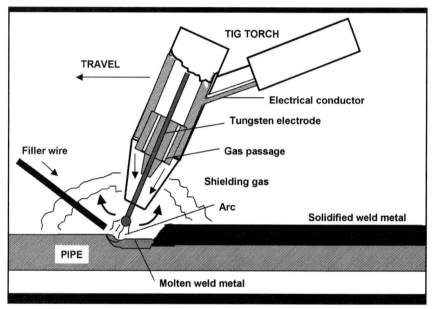

Fig. 5–4 *Tungsten Inert Gas Welding*

5.3 Preparation for Welding Operations

5.3.1 Pipe preparation

Small- and medium-diameter pipe is transported to the lay barge in racks. Large diameter pipe may be loaded as individual lengths. Before welding, the pipe is inspected to ensure that the weight coating is essentially intact. Concrete weight coating cannot be repaired offshore, and pipe with severe weight coating loss should be rejected. Gas service pipes will have been grit blasted internally to SA 2½ to remove mill scale, and all pipes are capped with plastic lids at each end to prevent ingress of moisture and debris. It is important that any desiccant bags (which should be attached to the end closure caps) are removed from the pipe when the closures are removed. On the lay barge, the pipes may need to be air blasted to ensure removal of any residual dust. Immediately prior to welding, the pipe joint ends must be dressed. Grinding or skimming can repair minor damage to bevels, but otherwise new bevels will need to be cut.

5.3.2 Welding bevel

The pipes to be welded must be prepared with a bevel at the ends of each pipe. The classic bevel preparation for most pipe is a straight angle cut of about 30° with a residual 1.5–2 mm land left for the formation of the root pass. The bevel is necessary to ensure that the first weld completely fuses the inside ends of the pipe.

The 30° angle was developed in the early days when all welding was with *sticks*, i.e., SMAW. The typical stick electrode is quite thick and this means that a large space has to be provided to allow the welder to reach into the joint and also to allow the shielding gas fumes to escape. The large bevel has to be filled subsequently with weld metal, and this takes time. Whenever wall thickness permits, it is advantageous to provide a slightly more acute bevel. With the advent of gas metal arc welding using thin continuous metal wire of small diameter, the requirement for a large bevel on thick pipes was reduced and narrower bevels were devised. Examples of bevels are shown in Figures 5–5 and 5–6. The narrow bevels, however, do increase the risk of lack of side-wall penetration, creating a particular problem with automatic welding if the set traverse of the wire is inadequate. To avoid this problem, the gas mixture can be modified to be a blend of argon and carbon dioxide used. The addition of about 5% carbon dioxide enhances lateral spread of the plasma arc, thereby increasing side wall penetration. The heat-input rate also needs careful control.

If the root pass is to be made from the inside of the pipe, which is common for larger diameter pipelines, then the bevel must be more complex with angle cuts from both sides. In all cases, the weld gap must be accurately set to ensure full root penetration.

Prior to moving the pipe into the firing line for welding, the beveled ends of each joint of pipe are thoroughly cleaned for about 40 mm at each end and inspected. Any pipe showing laminations requires cutting back and re-beveling and re-inspecting. Usually to avoid wasting time, ultrasonic testing for laminations within the 40-mm zone is used. It is also usual for the pipe to be inspected by magnetic particle inspection to ensure that all the laminations have been removed from the new bevel. Laminations result in weak welds and a high risk of cracking and may not show up with X-ray examination.

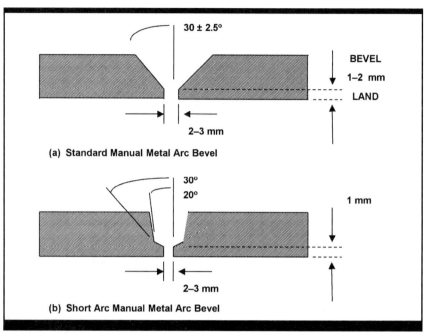

Fig. 5–5 Bevels for Manual Arc Welding

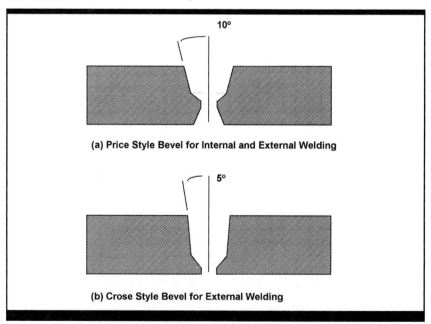

Fig. 5–6 Bevels for Semi-Automatic and Automatic Welding

5.3.3 Line-up for welding

The pipe lengths are pulled together and aligned with mechanical or hydraulic clamps. For pipes of 16 in. diameter and above, the clamps are usually fitted inside the pipe. Internal clamps usually must be used when the root pass is produced by an internal welding head. The internal clamps are pulled through the growing pipeline with cables. The pipes must be accurately lined up to ensure that the internal profile is as even as possible. Typically, alignment needs to be to less than 1.5 mm, and internal clamps almost always use a cooled copper or ceramic backing ring that fits behind the weld root gap to prevent excess protrusion of the root pass weld into the pipe.

5.3.4 Preheat

The weld area may need to be preheated. Offshore pipe often requires some heating before welding whatever the preheat requirement to ensure that the pipe is dry in order to reduce the risk of hydrogen generation which can lead to cold cracking. Hydrogen, produced by water dissociation, is quite soluble in molten steel and austenitic iron but is hardly soluble in ferritic iron. If the dissolved hydrogen has inadequate time to diffuse out of the steel before the transition from the face-centered cubic to body-centered cubic structure, then cracking can occur when the steel transforms to ferrite.

In thick wall pipe, the root pass weld will cool very rapidly and, depending on the carbon equivalent of the steel, this may result in the formation of martensitic material in the weld and heat-affected zones. Martensitic material is very susceptible to cracking by hydrogen. Preheating the pipe ensures that the weld will cool more slowly, giving time for the hot pass to be placed over the root pass or stringer bead which provides a longer period for diffusion of hydrogen. A rough guide to the need for preheating is given in Figure 5–7. The degree of preheat required depends on the carbon equivalent and is in the range 150–200 °C. Narrow bevels in thick pipes present a very small amount of weld metal in a large heat sink, and pre-heat becomes more critical to ensure correct tempering of the weldment by the successive passes. Pipe for sour service requires particular attention.[11]

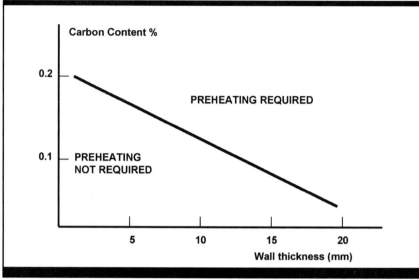

Fig. 5–7 *Requirement for Preheating for Cellulosic Electrodes*

5.4 The Welding Sequence

5.4.1 Root pass

The root pass is the initial and most critical weld. Because this weld is laid down in a straight line without weaving of the weld bead, the root pass is sometimes termed the *stringer bead*. The sequence of welds is illustrated in Figure 5–8.

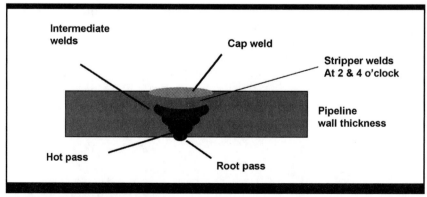

Fig. 5–8 *Girth Weld Fabrication*

The operation of a typical lay barge is shown schematically in Figure 5–9. Conventionally the root pass weld is started at the top of the pipe at the 12 o'clock position and is run down to the bottom, 6 o'clock position, in a straight line with two welders working one on each side of the pipe. This is *downhand* welding and is the fastest procedure. When welding thick pipe, the weld may be formed from the bottom up by *uphand* welding. Uphand welding is slightly slower, which reduces the risk of hydrogen cracking.

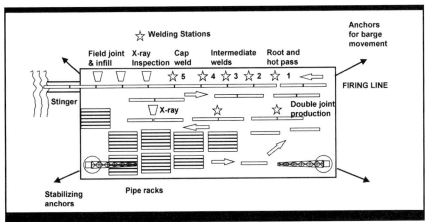

Fig. 5–9 *S-Lay Barge Welding Practice*

For pipes of diameter above 8 in., the welders start and finish together as this keeps the welding stresses in balance. Up to four welding positions may be needed on very large diameter pipes both for speed and stress balance. It is vital that this weld completely fuses the inner faces of the pipe joints without leaving unfused areas or excessive weld metal protruding into the pipe, termed *icicles*. These metal protrusions may initiate corrosion and will damage pigs. For large diameter pipes, it is possible to place the root pass from the inside of the pipe as this avoids formation of the icicles. However, this option requires complex equipment and is only cost effective for large diameter pipe and long-length pipelines. An alternative and common procedure to ensure that the root pass does not penetrate into the pipe bore is to use a backing plate across the face of the pipe joint, usually installed as part of the clamps. The water-cooling of the backing plate or ring is carefully controlled to avoid contamination and/or overchilling of the weld metal. During the forming of the root pass, it is important that the pipe is not strained as this may result in mechanical failure of the weld and the consequent need to part the pipe, rebevel, and repeat the root pass.

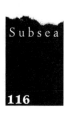
Semi-automatic equipment is commonly used for the root pass though fully automatic equipment is available and is used when the root pass is formed from the inside of the pipe. After the stringer bead or root pass is completed, the internal clamp is released and moves through the pipe to position the next joint as the pipe is advanced.

5.4.2 Hot pass

To obviate any risk of cold cracking of the root pass and the HAZ by hydrogen, a second weld is applied as soon as possible over the root pass. The hot pass remelts the first weld slightly and heat treats the HAZ. As a general rule, to ensure that hydrogen migration is effective, the temperature of the root pass should not be allowed to fall below 100°C ±25 °C if cellulosic electrodes are used. As the pipeline strength increases the interpass temperature needs to increase, so for welding grade X65 and above, the temperature required would be about 150 °C ±25 °C. The hot pass is usually placed within 4–5 minutes of completing the root pass. The maximum allowable delay to avoid falling below the minimum interpass temperature is about 10 min. During the cooling period, the root pass must be cleaned to bare metal to remove lateral slag which induces *wagon tracks*, named because of their appearance in the X-ray film. If the root pass is allowed to cool, then it may have a strength of 130 MPa above that of the finished joint. If there is an uncontrolled movement of the pipe, then the weld can crack. This situation would necessitate parting the pipe, re-beveling, inspecting and re-welding.

The hot pass should tidy up minor deficiencies in the root pass. The heat input of the hot pass must be sufficient to melt out the root pass and prevent wagon tracks, which are the external undercuts on the sides of the root pass. The hot pass will also melt out any slag inclusions that have been caught in the root pass because of the rapid solidification of the root weld. Upon completion, the hot pass is cleaned to bare metal with all slag removed and the subsequent welds applied.

5.4.3 Filler passes

The filler weld passes are less critical than the root and hot pass, and automatic and often semi-automatic welding machines that can lay down rapid volumes of weld metal are used. The filler passes need to be made with a slight weave—the movement of the molten filler metal from side to side. Weaving helps to ensure complete fusion of the bevel walls. Modern automatic and semi-automatic machines can now mimic this action. Between each pass, the welds must be scrupulously cleaned to bare metal.

The welding procedure can leave variations in the thickness of the weldment, so stripper passes at 2 and 4 o'clock may be needed to even up the thickness of the weldment before the cap pass is applied.

5.4.4 Cap/cover pass

The cap, or cover, pass(es) is the final weld(s). The cap pass is run around the pipe to fill the residual groove, leaving the weld 1–1.5 mm above the pipe surface and with an overlap on the outside surface of the pipe of around 1–2 mm. If manual arc welding is used, a typical electrode size is 5 mm. Slightly lower amperage is used to reduce porosity that can occur from overheating of the weld deposits or from excessive weaving. Care is also required to ensure that the overlap of the cap pass is fully fused with the parent pipe.

5.5 Manual, Semi-Automatic, and Automatic Welding

A welding procedure is manual if it is completely done by hand. Shielded metal arc welding (SMAW) using stick electrodes is the most common form of pipeline fabrication technique for smaller diameter pipes. The level of skill required for this type of welding is high because the stick electrode begins long and finishes short, and the weld must move smoothly from flat (1G) through vertical (3G) to underhand (5G). This geometrical complexity involves a constant assessment and readjustment by the welder. There is a limit on the practical length of the welding rod, and this may also result in short runs of weld metal before the welding rod must be replaced. The frequent changes of welding rod are a possible source of weld defects. The need to stop and start slows down the time to complete the weld: typically a 500-mm length weld will take around 4 minutes.

Semi-automatic methods are GMAW or GTAW. The filler metal is automatically fed into the weld by the machine, leaving the welder free to direct the arc or *spark*, The welding head is at a constant distance from the weld area, and the length of weld is not restricted. The absence of intermediate starts and stops means that there is less risk of weld craters and slag inclusions in the weld. Typically a 500-mm, semi-automatic weld can be completed in 2 min., representing an increase in welding rate of double that of the manual process. Semi-automatic welding methods include GMAW, Flux-Cored Arc Welding, SAW, GTAW, and Cold-Hot Wire Feed.

Automatic joining processes do not require constant adjustment by the welder. The machine is set up to complete the welds on a regular basis with only occasional re-adjustment. Automated joining is similar to automatic welding except that the equipment is usually less constrained in application. Most computer-controlled, robotic welding stations are automated. To date, the offshore pipeline industry mostly uses manual and semi-automatic welding procedures. Equipment hire cost and the long set-up time required means that fully automated systems would only be used for long, large diameter pipelines. A rule of thumb is that a pipeline's diameter is above 20 in. and its length is over 20 km before fully automated welding would be considered, but different contractors adopt different strategies. Many factors enter the decision, and technology changes. Most major contractors have bespoken fully automatic welding systems.

5.6 Weld Composition

A weld is a potential weak point in the pipe. A slightly greater thickness and an enhanced alloy composition compensate for the lower strength of the weld. The weld is generally 1–2 mm thicker than the pipe by virtue of the extra metal laid down by the capping weld. The composition is carefully selected to ensure that the weld has adequate strength. *Under matching* means that the strength of the weld metal is lower than the strength of the pipe metal. It should be avoided because imposed strains are concentrated in weaker material, which then has lower tolerance to defects. *Over matching* means that the strength of the weld metal is greater than the strength of the pipe metal and ensures that the imposed strains occur in the parent pipe, which as a wrought material should contain smaller production defects.

Usually, a weld is slightly stronger than the parent pipe, but excessively high strength should be avoided. Particular attention to weld strength is necessary for reeled pipe because excessive overstrength at the weld can lead to cracking of the HAZ during the ovalization of the pipe during reeling. The exact method of welding—consumables, preheats, heat input, and weld rate—are selected to ensure that the weld metallurgy can provide the required strength and that the HAZ is not adversely affected. Thus, it is vital that the approved weld specification and procedure are fulfilled. The welders must also demonstrate that they are competent with the selected weld procedure.

The weld represents a very small area of material in comparison to the parent pipes. If the weld itself or the area adjacent to the weld is anodic (more corrodible) to the parent metal, then galvanic corrosion is possible between the weld and the cathodic parent pipe; and rapid corrosion will occur because of the adverse ratio of anodic to cathodic areas. To avoid this, it is usual to enhance the composition of the weld to ensure that it is cathodic to the parent pipe. This is generally done by increasing the alloying content (e.g., nickel and copper) of the weld.

5.7 Weld Strengthening Mechanisms

Incorporating alloying elements into the steel matrix can strengthen the weld.[12] The alloying element atoms are a different size from the iron atoms, and their presence prevents slippage of the atoms in the crystal lattice by filling vacancies in the lattice (interstitial elements), or, if they are large, by causing distortion of the crystal lattice (substitution elements). The distortion prevents the slippage of the metal atoms relative to each other. Both interstitial and substitution alloying increase the strength of the steel though reducing the ductility.

The formation of transformation products is also used to increase weld and HAZ material strength. The ratio and form of the ferrite (pure iron with some dissolved carbon) and cementite (an iron-carbon compound) phases have a marked effect on the strength of the weld. Careful temperature change and heat treatment provides a balanced distribution of these phases in the material.

Temperature also alters the metal grain size. However this is a one way trip in that a large grain cannot be made small without a recrystallization process, i.e., heating the steel above the transition temperature into the austenite region. During welding, the grain size of material in the HAZ will be affected; this is discussed later in this chapter. Because the weld forms from molten steel, it will have a dendritic structure; but this will be modified by the tempering effect of the subsequent welds.

An additional strengthening method is to add micro-alloying elements, at concentrations of 0.1–0.2%, that dissolve in the weld while the weld metal is still molten but which come out of solution as the temperature falls. The micro-alloying elements concentrate at the grain boundaries and increase the friction between the grains reducing slippage and ductility of the metal by *stitching* the grains together. Small additions of titanium, niobium, and vanadium are used for this precipitation strengthening.

5.8 Heat-Affected Zone

During welding, fusion of the parent pipe occurs. At the weld, the metal temperature is around 1550 °C while the parent plate, some 300 mm away, is at close to ambient temperature. The temperature gradient produces a wide range of metallurgical features with consequent alteration to mechanical properties between the cast weld and the parent pipe. There is a region called the heat-affected zone (HAZ) on each side of the weld, as illustrated in Figure 5–10. The size of this zone depends on the thickness of the pipe, the amount of pipe preheat, and the rate of lay down of the weld metal, which is related to the weld heat input. Adjacent to the weld is a zone where a section of the pipe wall has been heated into the austenitic region and then cooled fairly rapidly, which results in a refinement in the grain size of the steel. These re-crystallized grains will be equiaxed. The repeated filler passes will raise the temperature of the grains resulting in some grain growth. Adjacent to this zone, the subsequent temperature excursions will be too low to cause grain growth and a zone of fine grains will persist. Adjacent to this zone, there is an area where the temperature will have been too low for re-crystallization but will have been high enough to allow grain growth and stress relaxation. Outside the HAZ, the elongated grains of the wrought parent pipe will persist. These changes in grain size alter the mechanical properties, strength, and toughness of the steel in the HAZ. The welding procedure should ensure that these properties remain adequate for service, and the welding specification must include suitable test procedures to ensure this.

Because of the changes in metal structure in the HAZ, this area of the metal may be very sensitive to corrosion. Though the material will have a composition similar to the parent plate, electrochemical studies indicate that there may be potential differences between the HAZ and the bulk pipe. A galvanic cell can be established. Corrosion inhibitors have a close potential range of effectiveness and, consequently, may not always protect the material in the HAZ as effectively as the parent pipe. If a new weld procedure or combination of pipe material and welding procedures are to be used, it may be prudent to test that this form of corrosion will not occur. There are two forms of testing. One test procedure involves sectioning of a weld into weld, HAZ and parent pipe, and the measurement of potentials and current flows between these elements when they are immersed in a suitable test solution and re-coupled through electronic test equipment. The alternative test procedure involves immersion of the complete weldment in the test solution and scanning of the weldment using a resonating reference electrode, which provides information on potentials and current flux between the different weld domains.

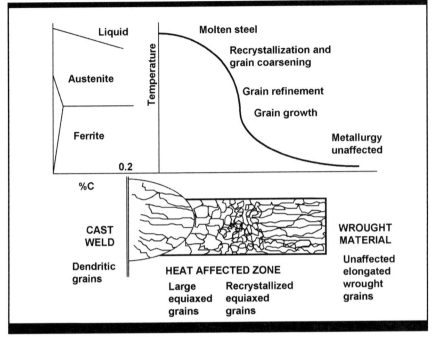

Fig. 5–10 *The Heat Affected Zone (HAZ)*

5.9 Weld Imperfections

5.9.1 Inspection

The completed welds are examined by radiography to ensure that they comply with workmanship standards or are fit for purpose.[13] Ultrasonic testing may also be used either as the primary inspection technique or as a support technique.

5.9.2 Porosity

Welds can suffer from porosity caused by gas or dirt being trapped in the cast metal. The porosity reduces the fatigue strength of the weldment. Most gases are highly soluble in liquid metal but hardly soluble when the metal solidifies. The gas forms isolated pores or clusters of pores or vacuoles in the weld. Single pores are usually not important, depending on both the size and the location. Porosity on the surface of the cap pass is caused by excessive and uncontrolled weaving, too high amperage causing

overheating of the weldment, or by excessive moisture content of a cellulosic electrode coating. Internal porosity is usually a delayed oxidation process caused by inadequate cleaning of the previous weld. Porosity accounts for about 50% of weld repairs. Engineering assessment studies have indicated that the level of porosity that can be safely tolerated is higher than specified in many standards though each pipeline would require individual assessment.

Nitrogen can cause porosity. Nitrogen has a high solubility in stainless steels so that it does not cause porosity in such material. It can increase the strength of some stainless steels and is deliberately added to duplex stainless steel welds by incorporation into the shielding gas.

Oxygen is a contaminant and must be avoided. The oxygen itself does not cause porosity, but it reacts with silicon and manganese and produces inclusions. This said, in some welding procedures, a small amount of oxygen (~1%) is added if argon is used as the shielding gas; the oxygen stabilizes the arc. For pipeline welding, carbon dioxide would be used instead of oxygen. Excessive oxygen can be drawn into the shielding gas if the gas flow is insufficient or if there is a strong wind.

For land pipeline welding, flux-cored wire, termed *self-shielded wire*, may be used in modified GMAW equipment to overcome excessive air movement. The gas produced is drawn into the molten weld metal, resulting in a gas blanket being formed at the metal surface, rather than being formed over it. On lay barges, the weld stations are enclosed; and the use of the more expensive flux cored wire is avoided.

Carbon dioxide is used as a cheap shielding gas but is not used alone for pipeline welding because the carbon dioxide reacts with the alloying elements in steel, as does oxygen. However it is less reactive. Typically, 5–8% carbon dioxide is mixed with the argon. The carbon dioxide stabilizes the arc and also increases the penetration of the arc.

5.9.3 Cold cracking

Welding codes prohibit all crack and/or crack-like defects. Hydrogen is the principal cause of cold cracking of steel. Cold cracking usually occurs in the root pass and HAZ where grain enlargement has occurred. Hydrogen is very soluble in molten steel and austenite but sparingly soluble in ferrite. See Table 5–1 and Table 5–2. If the weld cools rapidly, there may be inadequate time for the hydrogen to diffuse out of the steel before the steel transforms from face-centered cubic (austenitic) to body-centered cubic (ferritic). See Table 5–3. The trapped hydrogen causes cracks in the ferrite.

To reduce the risk of hydrogen cracking, the welding electrodes should be dry, the pipe preheated to ensure there is no moisture on the surfaces (to ~80 °C), and the use of light oil or silicon compound sprays to lubricate the filler wire in GMAW equipment should be prohibited.

Table 5–1 *Solubility of Hydrogen in Iron at Atmospheric Pressure*

Iron State	Temperature °C	Solubility of Hydrogen (cm^3 per 100 g iron)
Pure Iron BBC	20	<<3
Pure Iron BBC	900	~3
Austenite FCC	920	~5
Solid Iron	1535	~13
Molten Iron	1535	~27

Table 5–2 *Diffusion Coefficients for Hydrogen in Iron, Steel, and Stainless Steels*

Alloy	Crystal Lattice	Diffusion Coefficient at 25 °C
Pure Iron	BCC	1.6×10^{-5}
Pure Iron	FCC	5.4×10^{-10}
Steel	BCC + cementite	3×10^{-7}
27% chrome ferritic steel	BCC	6.7×10^{-8}
18Cr-9Ni austenitic steel	FCC	3.5×10^{-12}

Table 5–3 *Diffusivity of Hydrogen*

Temperature (°C)	Diffusivity ($cm^2/s \times 10^{-6}$)	Permeability (cm^3/hr)
20	15	---
100	35	0.00026
200	67	0.0045
300	100	0.029
400	138	0.11
600	204	0.59
800	269	1.00

Basic, low hydrogen electrodes are an alternative to cellulosic electrodes. The basic coating is essentially bonded calcium carbonate. Basic electrodes are preferred for thick wall pipe because they have a lower equilibrium moisture content. All coated electrodes must be heated (baked) before use and kept hot in quivers until used.

Some cracks occur because of movement of the pipe before the weld has set and/or there has been an excessive delay between root and hot passes. At areas of high stress, the embrittled metal will crack. The cracks occur in the HAZs where the grains have been enlarged and/or where brittle martensite and bainite occur. These cracks are not emergent on the outer surface of the pipe and are detected by radiography. The presence of hard material in and adjacent to the weld must be avoided if the pipeline will be in sour service. The welding must conform to the requirements of NACE MR-0175.[14]

One problem is that sometimes cracks do not appear immediately after welding but may take time to develop. The time to cracking is related to the quantity of hydrogen in the weld and the residual stress levels. If problems are anticipated, the weld procedure may require post-weld heat treatment to bake out the hydrogen. The high temperature increases the diffusivity of the hydrogen, allowing it to escape.

Typical bake-out temperatures are 175–205 °C, but higher temperatures can be used up to 300 °C. One side effect of delayed hydrogen diffusion is the hydrogen blistering of FBE coatings which are field applied at the field joints subsequent to the induction heating of the pipe to ~260 °C. Though unsightly, this blistering is insignificant; it represents a very small defect area that is adequately protected by the cathodic protection system.

5.9.4 Hot cracking

Hot cracking, or sulphur cracking, can occur if the pipe metal has a high sulphur content. This form of cracking is restricted to low grade seamless pipe and some forgings. The iron sulphides have a low melting point; and during the solidification of the liquid weld metal, they concentrate in the center of the weld—the area where the weld solidifies last. They are a weak material, and as the weld cools and shrinks they crack under the residual stress. A reduction in sulphur and phosphorous and an increase in manganese content prevent hot cracking.

5.9.5 Slag

Slag inclusion results from poor welding procedures; usually low amperage or inadequate cleaning before and between welding passes. Slag inclusion is more likely in narrow bevels. The narrow bevels also increase the likelihood of lack of fusion with the side walls and needs to be compensated for by increasing the amperage and attention to removal of oxides on the bevel faces. Heavy mill scale can be incorporated into the weld if the pipe has not been cleaned. Particular care is needed with internally scaled pipes as the hydraulic clamps may pull scale loose into the pipe joint at the 6 o'clock position.

5.9.6 Other defects

Other welding problems result from lack of welding parameter control. These faults include:

- Lack of sidewall fusion because of inadequate heat input or incorrect gas mixture and can be associated with narrow bevels
- Cold lapping between filler welds because of inadequate heat input
- Lack of root fusion due to incorrect arc placement or faulty filler wire feed, or if there is residual magnetism in a pipe
- Incompletely formed stringer bead, resulting from the root gap being too small or if persistent, due to a fault in the aluminum content of the steel
- Sagging root pass beads resulting from too large a root gap though too high an amperage may have a similar effect
- Incomplete root penetration because of misalignment of the joints or incorrect preparation of bevels or because the gas mixture is incorrect
- Rollover or overlap resulting from excess weld deposit from the capping weld because of an incorrect arc current
- Undercut in the root and cap passes resulting from incorrect gas mixtures, too high an amperage, or an incorrect weld weave pattern
- Excess reinforcement resulting from incorrect electrode manipulation that can be remedied by grinding, which is acceptable for onshore pipe fabrication but not feasible during an S-lay installation
- Striking marks that occur because the weld has been struck adjacent to the weld, causing hard spots that should be prohibited as grinding cannot always remove them without excessive thinning of the pipe wall area
- Crater cracking arising from inadequate cleaning of the weld at stop-start areas or over rapid cut-off of the arc
- Lamellar tearing occurring because of pre-existing faults in the pipe and can be avoided by testing the bevel area for cracks and laminations prior to welding

5.10 Weldability of Pipe Steels

Weldability describes the ease with which a metal can be welded to satisfactory standards. Poor weldability implies that the processes that can be used are limited and that considerable welding skill is needed. Good weldability means that a wide range of processes can be used and that only a moderate level of control and skill is required. Carbon steels generally have good weldability, and stainless steels have fair weldability.

For steel, the carbon content largely defines the weldability. To define weldability, the other elements in the steel are converted into a carbon equivalent; and the sum is used as a guide to the weldability of the steel. There are several formulae available to calculate the carbon equivalent of a steel; but for pipelining, the most commonly used is the international formula as given in API Specification 5L:

$$CE_{IIR} = C + \frac{Mn}{6} + \frac{Cr + Mo + V}{5} + \frac{Ni + Cu}{15} \qquad (5.1)$$

where

C	=	weight % carbon
Mn	=	weight % manganese
Cr	=	weight % chromium
Mo	=	weight % molybdenum
V	=	weight % vanadium
Cu	=	weight % copper
Ni	=	weight % nickel

For pipeline steels, it is usual to specify a maximum carbon equivalent of 0.32–0.39. For forgings and flanges, the carbon equivalent may be higher at 0.45. These values are easily achievable with modern pipeline steels. As the grade of steel increases, the carbon content generally has to fall. Consequently, grades X65 and above will have low carbon contents, below 0.1% C, and the international formula is less sensitive as a method of defining weldability.

The second most commonly used formula is the parameter for crack measurement, P_{CM}, which is calculated as follows:

$$P_{CM} = C + \frac{V}{10} + \frac{Mo}{15} + \frac{Mn + Cu + Cr}{20} + \frac{Si}{30} + \frac{Ni}{60} + 5B \qquad (5.2)$$

where the symbols are represented the same as those in formula 5.1 and in addition

Si = weight % silicon

B = weight % boron.

The P_{CM} value is generally specified as a maximum of 0.18–0.2. The P_{CM} formula is increasingly used for the modern low alloy pipeline steels where the carbon content is below 0.1% while the international formula used for steels conforming closer to API Specification 5L where the carbon content will be in the range 0.15–0.2%

5.11 Consumable Composition and Coating

The weld metal itself will be a casting and, therefore, may have a lower strength than a wrought pipe material of an identical composition. To overcome this, it must have a higher strength that is achieved by providing a higher alloying content. A typical root pass weld consumable will be 0.1% C, 0.15% Si, and 0.5% Mn. The weldment will have a yield strength of 380–450 MPa and a tensile strength of 450–520 MPa. The subsequent weld passes will be produced using a consumable with a higher carbon content, typically 0.1–0.15% C, 0.1–0.15% Si, 0.4–1% Mn, and ~0.5% Mo providing a yield strengths of 415–480 MPa and a tensile strength of 520–550 MPa. For cases where very high impact resistance are required, a typical composition will include 0.5–1% Ni. Copper is sometimes included to reduce preferential weld corrosion, but there are conflicting reports of the success of these weldments. See section 5.15 for a detailed discussion.

Manual metal arc consumables are coated with either cellulosic, basic, or acidic coatings. Cellulosic coated electrodes are cheaper and are widely used for girth welds on pipes up to API 5L X70. The cellulosic coatings produce a very large quantity of carbon dioxide gas, which allows larger electrode diameters and, hence, high amperages and rapid welding in the downhand mode. The impact properties of welds made with cellulosic coated electrodes are good, typically 40 J at -40 °C. However, the welds can contain a high hydrogen content; and this may be unacceptable for sour service pipelines and/or thick wall pipe.

Welds formed with consumables with basic coatings have a low hydrogen content because the coating has a hydrogen content of less than 5 ml/100 g, and the welds formed from basic electrodes also have very good mechanical properties. Consequently, the basic coated electrodes are used for thick pipe and high strength pipeline steel, e.g., API 5L X80, and for cases where high impact strengths are needed. Typically, a basic electrode will produce a weld that is twice as tough (Charpy impact energy) as a weldment formed with a cellulosic-coated electrode. The basic coatings are more expensive and often a mixed set of consumables will be used: cellulosic-coated electrodes for the root, and hot pass and basic electrodes for the remainder.

5.12 Welding of Duplex Stainless Steels

Duplex steels are a form of weldable stainless steel that combine good corrosion resistance, typical of the austenitic stainless steels, with improved resistance to chloride stress corrosion cracking, crevice attack, and pitting.[15] The mixed structure arises because the nickel content is insufficient to fully austenitize the steel, and as a result, the steel is an intimate mixture of the ferrite and austenite phases. This structure results from careful selection of alloy composition and very controlled heat treatments. The optimum corrosion and mechanical properties arise when the phase balance is 50:50. The pipe is usually supplied cold worked or in the solution annealed condition. The length of the duplex pipe available depends on the size of billet that can be handled; therefore, thick wall pipe may only be available in short lengths (typically 6 m). The pipes are jointed to standard length before the coating is applied, then double jointed before shipment offshore.

Welding of the duplex stainless steels is more complex than welding carbon steel.[16] The welding process must not overly alter the phase balance in the parent plate because doing so could result in a weld that is anodic to the parent plate, a situation that would result in preferential corrosion. The heat input must also be restricted to avoid formation of sigma and chi phases, which are hard intermetallic compounds and which markedly reduce the toughness of the steel. Ensuring the ferrite-duplex phases are in balance requires a low heat input rate and, therefore, the welding process is relatively slow, typically some ~30% the rate of carbon steel welding. The low heat input rate was initially achieved using tungsten arc welding, though developments have allowed other forms of welding to be used for parts of the weld process.[17, 18]

In view of these facts, there is an economic advantage for duplex pipelines to be fabricated onshore and to be installed in bundles or by the reel method, rather than by offshore fabrication by S-lay. For offshore construction, it is usual to double joint the pipes onshore to reduce offshore welding.

Corrosion resistance is markedly affected by slag or oxide inclusions or by any surface porosity. Oxygen must be rigorously excluded during the welding as the oxide films formed adjacent to the weld may result in pitting later in service. To avoid the formation of oxide films, the bore of the pipe is purged with argon. Moveable stopples are inserted into the line on both sides of the weld to minimize the quantity of argon used and to speed up the rate of de-oxygenation. A small percentage of nitrogen is included in the argon to ensure some nitrogen uptake by the weld as the nitrogen increases the strength of the weldment and the weldment pitting resistance (PRE_N). Corrosion of the parent pipe adjacent to the weld can occur under weld spatter. The TIG process produces little or no weld spatter.

The corrosion resistance of the duplex stainless steels is markedly affected by the surface condition. The presence of iron and certain iron salts on the surface can trigger pitting which may not passivate. Iron metal can be easily introduced into the surface of the softer stainless steel from steel implements and tools. The iron particles later corrode to form pits. Iron salts that oxidize to ferric ions can cause similar problems. It is important that the pipe surface preparation, beveling, and inter-weld cleaning are done without creating iron contamination. The brushes and grinding wheels have to be selected with care. All steel handling equipment must also be rubber covered to avoid iron contamination.

5.13 Welding of Clad Pipe

The liner of an internally clad pipe is only 2–3 mm thick, and so the fitting of the pipes is critical. To minimize fitting problems, there has been a notable reduction in the out-of-roundness and diameter tolerance of clad pipe, and most clad pipe manufacturers apply some expansion method to the ends of the pipes to ensure that they are within a tight tolerance. Closer tolerances will undoubtedly help to reduce the welding time. Consequently, additional cost of a higher specification of pipe and the offset gain from the improvement in welding rate should be taken into account when evaluating competitive bids from pipe manufacturers.

The high alloy liner in the clad pipe must be welded with a consumable that maintains the corrosion resistant properties of the liner. Weld consumables are usually of higher alloy content than the liner as some dilution occurs with subsequent passes from the carbon steel welds. First, the liner is usually welded either by an internally applied weld if the pipe diameter is large enough or by a single root pass from the outside. The hot pass may be a repeat weld or a pure iron pass to avoid overdilution of the root pass. Subsequent welds are with conventional carbon steel consumables. For some liner/steel combinations, such as type 825 and 625, it is possible to complete the full weld with the highly alloyed consumables.

Some contractors prefer to minimize the number of consumables (to avoid errors), and this approach favors a dual pass with the high alloy consumables followed by conventional steel welding. There is also some evidence that a pure iron pass may lead to internal cracking.

In almost all cases, the CRA liner is welded by TIG welding to reduce weld spatter that might result in pitting and crevice corrosion under the spatter. Saipem has recently developed a GMAW procedure for type 316 and alloy 825 linings using an internal clamp that incorporates a ceramic channel that prevents weld spatter onto the liner. This GMAW procedure is approximately twice the speed of TIG welding.

The welding of the cladding must be inspected by radiography before the carbon steel welding commences. This additional inspection step reduces the overall speed of pipe fabrication. The weld rate is relatively slow, being similar to that for solid duplex stainless steel, around ~30% of the rate of welding of carbon steel.

5.14 Weld Inspection

5.14.1 Radiography

Pipelines are pressure vessels, and all high-pressure oil and gas pipeline welds normally have to be 100% inspected by radiography.[19] The radiography station is situated after the weld stations but before the final stinger tensioners with sufficient space/time for repair welding. The radiographic technique involves high energy radiation either generated from a high voltage electrical source or provided from a radioactive isotope. The high–energy radiation can pass through steel, the percentage transmission depending on the thickness and density of the steel and the energy level of the radiation—termed *radiation hardness*. A photographic film placed on the opposite side of the pipe detects the transmission of the radiation. For pipeline weld inspection, the radiation hardness and the exposure time are selected to give the maximum contrast between pipe and weld.

The radiation source may be on one side of the pipe, and the film placed on the other side. The radiation must pass through the pipe wall twice to reach the film. This action is termed *double wall*. It is usual for large diameter and submarine pipelines to have the radiation source placed centrally inside the pipe and the film wrapped around the outside of the weld. Long lengths of film are involved for large diameter pipelines. Electrically generated X-rays are used whenever possible because the energy level can be tuned to give the best discrimination of defects on the film. Gamma radiography using a radionucleide source is used in remote locations and for small-diameter and modest thickness pipelines.

In both cases, a permanent record is provided. The process is rapid, and the method highly discriminatory. The location of the defect inside the metal is not always clear from the X-ray and may have to be found by subsequent ultrasonic examination. Some defects, for example longitudinal laminations within the steel, are not detected. Radiation is a health hazard, which implies that care is needed at the radiograph site. The evaluation of the films is also stressful and tiring, particularly if the lay rate is high because the films must be interpreted while they are still wet from the processing. New real-time radiography systems are available that present the X-ray on a TV monitor. The hard copy is recorded on videotape. This methodology has obvious attraction as the image can be digitized, allowing computer assessment to be used as an adjunct to conventional inspection.

5.14.2 Ultrasonic examination

Radiography does not give complete information on the three-dimensional location of a defect in a weld, only its two-dimensional location. Ultrasonic examination is used to give information on the spatial orientation of the defect within the metal. The technique pulses high frequency sound into the steel, the frequency being of the order of 1–6 MHz. The sound is reflected back from the areas where there is a density difference such as the inner face of the pipe or by internal defects. The signal and return sound pulses are viewed on a cathode ray tube.

By moving the probe around a suspected defect, the extent and orientation of the defect and its relative position in the weld can be determined.

The technique is portable and is sensitive to planar defects but is slow in comparison to radiography as only defects in the close vicinity of the probe are detected. The surface condition of the pipe must be good, and a couplant is needed to transmit the sound from pipe to probe and back. Coatings and the weld profile may interfere.

Advanced techniques have been developed, many of which are semi- or fully automatic. For example, ZipScan is an ultrasonic technique that uses separate transmitter and receiver heads. The two heads are mechanically moved around the pipe, one on each side of the weld, and the sound signal passes from the transmitter to the receiver at a fixed angle. Flaws in the metal result in a diffraction pattern of the sound that is analyzed by a microcomputer. The sound pattern is converted into a shaded pattern that identifies any flaws in the weld. The technique is rapid and can scan about 1 m of weld per minute.

The Dutch regulation NEN 3650 requires welds made by GMAW to be inspected using ultrasonic testing. API 1104 and BS 4515 do not require this.

5.14.3 Magnetic particle inspection

Magnetic particle inspection (MPI) is used to detect surface emergent flaws of a size undetectable by eye.[20] Typically, these would be cracks or laps and laminations at the pipe bevel. The technique uses a strong magnetic field to induce magnetism in the pipe surface. The magnetic field can be induced by electrical probes or permanent magnets. The area under examination is painted with a fluid containing ferromagnetic particles that line up at the areas of disturbance in the magnetic flux, typically at the discontinuities in the metal surface, which are at an angle to the magnetic flux.[21] The collection of particles makes the defect highly visible. An alternative is to gently blow dry iron particles across the metal surface. The technique needs a clean metal surface, and it is important to alter the direction of the magnetic flux as cracks and defects in line with the magnetic field may not be detected.

MPI is a slow technique, and it is only used over limited areas or if there has been a sequence of problems needing resolution. Typical use is for pre-inspection of forgings and castings prior to welding to line pipe and inspection of bevels prior to welding. The technique can be used only for magnetic materials. Demagnetization may be needed after testing as the magnetism may cause the welding arc to wander.

5.14.4 Dye penetrant inspection (DPI)

DPI is used for detection of cracks, similarly to MPI, but it is applicable to magnetic and non-magnetic materials. A colored or fluorescent dye is used to delineate the surface emergent imperfections. The surface to be inspected is cleaned and covered with a highly penetrating indicator fluid. The fluid is given time to permeate into surface defects by capillary action. After a set time, the surface is cleaned again and a developer coated onto the surface. The developer is a fine chalk dust that acts like blotting paper and pulls the indicator fluid out of the cracks and surface faults. If a fluorescent indicator is used, then an ultraviolet light is used to detect the dye. Cracks, porosity, lack of fusion, and laps are detected. A highly portable method, the technique can be used on any surface, and is cheap but slow. Surface cleanliness is important, and any surface coating must be removed.

5.14.5 Eddy current testing

An oscillating magnetic field is induced in the metal surface, and the resulting flow of current is measured. Surface-emergent faults are easily detected by alterations in the current flux. The technique can also be used to check heat treatment variations and alloy compositions though this is rarely required for pipeline inspection. The technique is similar in sensitivity to DPI and MPI but is faster. The technique is under study for use on pipelines as the simple geometry of the pipe allows an automatic method to be used. It would be a suitable technique for inspection of clad pipe.

5.15 Preferential Weld Corrosion

Preferential weld corrosion has been observed in water injection pipelines and some crude oil pipelines. The corrosion may be of the weld itself or of the HAZ. The variables implicated appear to include:

- Corrosion resistance of the weld metal
- HAZ composition and microstructure
- Hard transformed microstructures
- Use of basic coated welding consumables
- Low arc energy levels

Weld procedures themselves do not appear to have a noticeable effect. Post-weld heat treatment was found to reduce HAZ attack, hence, temper-bead welding would be advantageous. Corrosion enlarges with increasing partial pressure of carbon dioxide but does not appear to be sensitive to flow. Corrosion inhibitors are effective though it may be prudent to check the effectiveness of the inhibitor against a weldment. The addition of 0.6% nickel and 0.4% copper or 1% nickel or 1% chromium appears to be a suitable method of reducing the problem though not in all cases. Note that 1% nickel in welds is acceptable in sour service; the comment in NACE MR-0175 regarding nickel refers to wrought material.

5.16 Downstream Corrosion at Welds

When water separates from crude oil, the water can persist downstream of protruding root welds and initiate corrosion. Similar problems arise with mismatched pipe because of the large manufacturing tolerance of up to ±12.5%. In sweet systems once the corrosion initiates, it will spread downstream as the turbulent flow conditions at the downstream lip prevents the formation of the passive carbonate films. This corrosion is not preferential weld attack though it may appear as such initially.

Microbiological corrosion is also associated with protuberant welds. Presumably the bacterial colonies take advantage of the shelter and persistent water afforded by the weld.

5.17 Incomplete Penetration

Incomplete filling of the root pass leaves behind crevices that may increase weld corrosion and also act as stress raisers. However, lack of penetration should be detected by X-ray examination. In some cases, the corrosion occurs over half the weld. Usually the parent metal is not affected. This form of damage is excluded from analysis in ASME B31G because corrosion at welds results in sharp edged defects. The defects need to be analyzed using fracture mechanics techniques or the pipe prequalified by hydrotesting.

5.18 Possible Future Welding Techniques

Production in deep waters increasingly requires installation of the pipeline using J-lay to reduce the supported pipe weight and pipe tension. The rate of welding is critical as generally only one welding station is available. At present, multiple joint pipe is prepared prior to being lifted into position for welding at the single weld station. Novel methods are being reviewed to identify an alternative weld procedure that will allow faster welding to overcome this bottleneck. The options that have been explored include:

- Friction welding
- Flash butt welding
- Homopolar welding
- Magnetic impelled arc butt welding
- Explosive welding
- Shielded active gas forge welding
- Deep penetration welding

In friction welding one component is rotated at high speed, then progressively pressed against the stationary component. The frictional energy melts the contact faces, and the weld is formed by subsequent pressure. Three types of friction weld have been investigated. The simplest case is rotation of a complete pipe joint, but this involves considerable energy input to rotate the heavy pipe. A more attractive option is to use a rotating pup piece inserted between the pipeline and a stationary pipe joint. When up to speed, the pipe joint is pressed to squeeze the pup onto the pipeline. Considerably less energy is required, but two welds are produced for each added joint; therefore, two inspections are required. The third approach is to use a small specially formed ring that is treated in the same way as the pup piece. Low energy input is used because of the small weight of the joint, and the inspection of the two welds can be made at a single inspection.

Flash butt welding of pipelines was developed in the USSR in the 1950s and has been widely used there. Some 30,000 km of oil and gas pipes have been formed using this technique for diameters ranging from 4–20 in. The process has been further developed in the Unites States by McDermott for offshore welding on S-lay barges but has not yet been used in practice, in part because of the extreme conservatism of the industry.

The ends of the pipe are dressed square, and the internal face of the pipe cleaned. Internal shoes are inserted to contact the cleaned section of pipe and to pass a high current into the pipe while the new pipe joint is pressed against the pipeline. The asperities on the pipe faces melt, and the current rises to melting point over the complete pipe wall at which time the new joint is pressed against the pipeline to forge the weld. The flash produced is removed by skimming the internal and external faces of the weld.

Homopolar welding is similar to flash butt welding except that the high current is produced by a current pulse developed from the magnetic braking of a large rotor. The rotor is brought up to a specific speed while the pipes are prepared. The pipe faces must be prepared similarly to the flash butt welding. When ready, a magnetic field is applied to the rotor and the back electromotive force (e.m.f.) produces approximately 2 MW, which is directed to the pipe ends allowing a weld to be produced in about 3 seconds.

In magnetic impelled arc butt (MIAB) welding the pipes are set up similarly as for conventional welding. This is a nonconsumable welding procedure, and the weld gap is extremely small. An arc initiated between the weld faces and the arc is driven around the pipe circumference by a pulsating magnetic field.

Explosive welding is in principle an attractive procedure because it is a low energy, cold process although work hardening of the pipe at the weld area can occur. Two processes have been evaluated. In the first process, the new joint end is formed as a bell and placed over the normal end of the pipeline. Explosive charges are placed internally and externally and ignited. The explosion squeezes the faces together, and the final weld has the conventional wave-like interface of explosion-bonded plate. In the second process, a shaped ring, internally coated with explosive, is placed across the zone to be welded. An external anvil is situated on the outside of the pipe to contain the blast. The charge is ignited, and the explosion deforms the ring against the anvil, resulting in a conventional type explosive weld. Because a ring is used, this processes could be used for welding CRA-lined pipe materials.

In shielded active gas forge welding, the pipe faces are dressed to form grooves on the internal and external surfaces of the pipe, and the bore is sealed and purged with hydrogen to remove oxides. The weld area is heated by induction, and the pipe ends pressed together to forge the weld. The function of the grooving is to accommodate the bulging that results during the forging of the weld.

Deep penetration welding uses either a laser or an electron beam. The very high energy beam creates a narrow keyhole, filled with vapor, which is moved around the pipe to form a single weld through the wall thickness. Carbon dioxide lasers can produce sufficient power to weld up to 25 mm wall. Electron beam welding can weld up pipe wall up to 40 mm when in vacuum, but this thickness reduces as the quality of vacuum reduces. The overall energy requirements are similar for electron beam and GTAW, whereas GTAW operates at ~12 V and 250 A, an electron beam operates at 30 kV at 0.2 A. The welds produced by deep penetration welding are very thin compared to conventional welding. The electron beam welding rate is high compared to GTAW, approximately 12 times the speed, up to 6 mm/second.

It is too early to predict which of the alternative welding procedures will become an accepted part of marine pipeline technology. Investment and prior track record do not appear to be pointers. For example, McDermott has been unable to persuade operators to use the flash butt welding process despite a large investment and the long history of successful use of this technique for onshore pipelines. Friction welding is attractive and is used for production of small diameter, non-oil and gas industry piping systems. Homopolar welding was the subject of a joint industry study and offers a suitable high current power source. Laser and electron beam welding are well established for specialized weldments, e.g., hip joints, and have good industrial track records. Explosive bonding is perhaps the most intriguing method.

5.19 Underwater Welding

5.19.1 Options

Connections sometimes have to be made underwater. In the context of construction, welds under water may be needed to join different sections of pipeline to connect a pipeline to a platform riser, or to add a tee or wye connection to an existing line. In the context of repair, discussed further in chapter 16, the need arises from design deficiencies and accidental damage by dropped objects, anchors, jet sled collision, corrosion, and fatigue.

The underwater options available to the engineer are mechanical connectors and welding. The need for deepwater connections beyond the depths accessible to divers has stimulated extensive development of remote-operated connection systems; and they are reliable and widely applied. Figure 5–11 depicts categories of welding options:

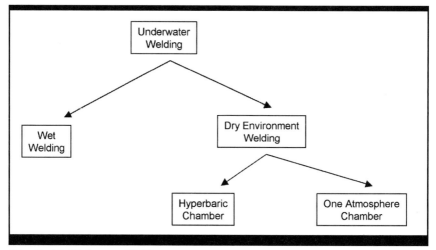

Fig. 5–11 *Underwater Welding*

In *wet* welding, the diver is suited and in the water, and the weld is made actually under water although the term is sometimes used for welding in a small air/gas filled box placed over the work area. The technique is generally only suitable for relatively shallow water. Wet welding is performed by divers trained as welders because access to the location and the decompression cycles are normal diving operations. A safety concern is that the diver is at risk from stray currents from electrical leakage. The quality of the weld is markedly reduced compared to surface welding. The wet environment increases the quantity of hydrogen that enters the weld, and the rapid cooling results in an adverse metallurgy. Wet welding is suitable for temporary or repair welds and for welds that do not require full strength and toughness.

Hyperbaric welding is undertaken in a dry atmosphere but at a pressure equal to the hydrostatic head at the depth of the weld. Compressed air diving is only feasible to a depth equivalent to eight atmospheres absolute (70 m depth) because of the increased toxicity of oxygen at higher partial pressures. Above eight atomospheres, it is necessary to use a helium-oxygen environment. In oil industry practice, compressed air systems are usually only used to 4 bara (30 m) for safety reasons. Hyperbaric welding is almost invariably carried by divers trained as welders, as most hyperbaric welding is done using saturation diving techniques. The welds are made in the dry chamber, and the main safety issue is the toxicity of the gases that the diver-welder breathes. Cleaning the habitat of the dangerous fumes is expensive. One option is for the welder to wear a breathing mask during the actual welding operation.

One-atmosphere welding is undertaken in a dry atmosphere at one atmosphere pressure. Welders can work in a shirtsleeve environment, and welders can be trained as divers using this option. The transfer to the chamber is by specialized one-atmosphere equipment. The welding procedure is the same as that for the surface welding, allowing high quality welds to be made identical in quality to those made on the lay barge.

The alternative, non-conventional welding techniques described in section 5.18 could all be developed to work under water. The only non-conventional technique ready for under water use is explosive welding, which is examined in section 5.19.5. Development of the other non-conventional techniques for underwater applications is unlikely to happen until they have achieved acceptance for applications above water, a much larger market.

5.19.2 Wet welding

Wet welding is reported to have been performed first in the 1930s. For pipeline installation and maintenance, wet welding is used for cosmetic and temporary repairs or for cases in which the quality of weld is not significant, e.g., retrofitting of sacrificial anodes where the prime requirement is to ensure electrical continuity. To reduce water pick up and hydrogen charging of the weldment, the electrode is coated with a special flux that is varnished or lacquered. The main drawbacks of wet welding are as follows:

- Reduced arc visibility which makes some welding procedures very difficult
- High levels of porosity in the weldment
- Lack of fusion to the work piece
- High quench rates resulting in high hydrogen contents in the weld and poor metal properties, particularly a reduced toughness

When the wet welding is done with the electrode and work piece immersed in water, the weld arc is struck; and the heat generated produces a vapor cloud that separates the weld from the water. Welding rates are modest at 3–4 mm/second. The vapor is essentially a mixture of the gases from the electrode covering and the decomposition of the water. Because of the high-energy input, the bubble shield produced is relatively stable. Gas metal arc welding processes appear less effective than the use of coated electrodes because the gas interferes with the bubbles formed by the vaporization of the water. The bubble formed by the volatilization of the water is stable until it grows to a critical size at which point it breaks away and a new bubble forms. Typical bubbles are 1–2 cm in diameter, and about 40–100 cc/second of gas are formed depending on the type of electrode used. The bubble is composed of 60–80% hydrogen, 10–25% carbon monoxide and ~5% carbon dioxide. The bubbles also contain some nitrogen and mineral vapors from the coating.

Compared to air welds, wet welds show higher hardness which increases with steel grade. The hardness also increases the crack sensitivity particularly as the carbon content increases. Consequently, the procedure is only applicable to modest strength steels and, preferably, to low carbon steels.

5.19.3 Hyperbaric welding

Hyperbaric welding is done in a dry habitat at a pressure equal to the local hydrostatic pressure. It is sometimes termed habitat welding. In modest depths and for repairs, the habitat may be a simple open bottom box that is pressurized to drive down the water below the area to be welded. Such approaches are used for repairs and pipe jointing. At present, the maximum depth for hyperbaric welding is about 50 bara (500 m). The welds are generally done by divers trained as welders, and they must be qualified to weld at the appropriate depth.

At high pressure the length of a stable arc is much reduced, typically from 10–15 mm at atmospheric pressure to 3–5 mm at 30 bara. Welding voltages and welding machine characteristics must be adjusted to compensate. The geometry of the weld bead is also altered with maximum effect at about 5 bara. Increasing the arc temperature alters the fluidity and surface tension of the weld metal so that large electrodes become practically useless. Consequently, the welding must be done with small electrodes, and many small weld beads applied compared to surface welding.

For pressures above 4 bara, the environment will be a helium-oxygen mixture. A common gas mixture contains 3% oxygen to give a partial pressure of about 0.5 bara (compared to the surface partial pressure of 0.2 bara). Helium, though an inert gas, has a thermal conductivity about six times that of air. The rate of cooling is approximately 20% greater than in atmospheric air over the range 800–500 °C. Consequently, the weld cools rapidly, and the weld area must be kept heated to prevent thermal stress problems and cracking.

GTA welding has proved problematic since argon is narcotic at high pressures. It is also a slow process and sensitive to magnetic flux. As pipeline repairs are likely to be done on pipelines that have been inspected by magnetic flux pigs, this factor can be important. Telluric currents may also occur on long pipelines—depending on sun spot activity—resulting in magnetization of the pipeline. However, given adequate demagnetization, GTA welding has been used for root and hot pass welds. Most projects have used SMAW with basic electrodes and flux cored wire.

Hydrogen absorption increases with pressure. Combined with more rapid cooling, there is increasing hydrogen retained in the weld, and the critical cracking stress is reduced. Carbon and oxygen contents of the weld increase with pressure, approximately doubling for a five-fold increase in pressure. The manganese, nickel, and molybdenum concentrations decrease with pressure while silicon concentration appears to be unchanged. The overall effect of pressure is to reduce the impact toughness of the weld and HAZ. Special welding consumables are used to compensate, but some loss of toughness appears inevitable. The increase in carbon results from the increased solubility of carbon arising from the carbonate coat on the electrode:

$$CaCO_3 = CaO + CO + \frac{1}{2} O_2 \tag{5.3}$$

$$CO = C_{Fe} + O_{Fe} \tag{5.4}$$

$$CO + \frac{1}{2} O_2 = CO_2 \tag{5.5}$$

$$H_2O = 2H^+{}_{Fe} + O_{Fe} \tag{5.6}$$

where

X_{Fe} = the element (e.g., carbon) dissolved in the iron weldment and HAZ

GMA welding has also been developed mainly to avoid the need for preheating and to reduce the concentration of adsorbed hydrogen. The toughness of the weldment is improved compared to SMAW because there is no carbon produced. It is also claimed that alignment and fit up are less exacting with GMAW. Not surprisingly GMA welding is used for more exacting jobs and at the higher pressures than SMAW. However, the welding equipment requires extensive modification to provide the necessary welding characteristics.

5.19.4 Atmospheric welding

Atmospheric welding at depth requires highly sophisticated equipment and the ability to transfer the divers from the surface to the habitat at atmospheric pressure. The technique permits welders to work in a shirtsleeve environment, and it avoids the need to retrain divers. One major advantage is that the welder can operate all year round compared to hyperbaric welding where the saturation divers are limited to five or six saturation dives each year.

The clear advantage of one-atmosphere welding is the quality of the weld, which should equal that of surface weld. It is the only feasible technique for high strength steels (above X65) and great depths. The present systems are available to weld at depths of 1,500 m. When considering costs, it should be recalled that the welding is only a small, albeit vital, part of the overall procedure. The pipe to be joined will require to be sealed using stoppling pigs (and possible secondary seals for safety), cutting to length, cleaning, bevel preparation, pipe alignment, and pre-heating prior to the welding. After welding, the weld must be inspected and then coated.

The most critical part of the system is the seal around the pipeline to ensure that the one atmosphere environment can be maintained safely. The seals may be temporary or permanent, and all have restrictions. Some pipelines have been pre-designed with fixed seal systems pre-installed on wyes and tees to reduce the preparation time to establish the habitat. In other cases, the habitat is treated as a disposable item and is left on the seabed after the weld is completed.

5.19.5 Explosive welding

Explosive welding is similar to the process used for bonding CRA material to carbon steel for the fabrication of clad vessels and pipe. The plate to be welded to the base plate is termed the *flyer plate* and is deformed by the high pressure of the explosion. As with bonding of CRA to steel plate, the flyer plate must be stepped off the base plate to allow it to develop sufficient velocity so that it impacts the base plate above a critical collision velocity. Excessive velocity must also be avoided as it can result in tearing and damage to the flyer plate. A schematic of the system is given in Fig. 5–12.

The quality of the weld depends on the collision velocity, V_o, related to the mass of metal that has to be accelerated and the stand-off distance. The collision pressure for steel has to be in the range 60,000–80,000 bar to create a good weld. The stand-off distance is required for the flyer plate to achieve V_o. Several reflections of the pressure wave are required to achieve this, and it is typically about half the thickness of the flyer plate. (The schematic in Figure 5–12 is not to scale.) The stand-off distance has to be free of water and other liquids that would reduce the critical velocity. The collision angle is critical for a particular combination of materials and thicknesses and is adjusted to ensure that the collision point velocity remains subsonic. The anvil is also carbon steel and is used to stiffen the base plate so that the shock propagation effects fall within the limiting tensile strength of the flyer plate material. The rate of the explosion (the detonation velocity) has to be low; and special explosives are used to ensure this. The low detonation velocity ensures that the welding velocity is in the subsonic range.

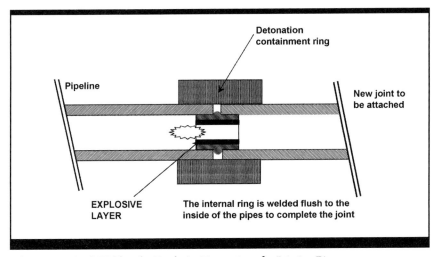

Fig. 5–12 *Girth Welding by Explosive Expansion of a Jointing Ring*

Ductile materials are the best to weld with this method. The higher the grade of steel used, the lower the ductility. However, in general, low carbon steel has adequate ductility at all grades. There is an increase in hardness of the steel adjacent to the weld area An increase of ~40 VHN is typical. Stainless steels increase more, typically 80 VHN. The long-term fatigue performance and corrosion resistance of explosion-welded joints remains unknown at present.

For welding of an offshore pipeline using the male-female connection system, the following steps must be completed:

- Machine of the outer diameter of the pipe to be welded to remove scale and ensure a consistent surface finish.
- Place the explosive charges and dewatering bags. These bags reduce the shock loading on the pipe and prevent unwanted deformation.
- Mechanically connect the sections of pipe. It is usual for the female section of the pipe to contain adequate material thickness to obviate the need for a separate anvil.
- Inflate the dewatering bags, and dewater the weld annulus.
- Detonate the charge.
- Remove dewatering bags.
- Inspect.

This method of joining requires good joint preparation but minimal operator skill and would be suitable for application by divers with minimal training.

5.20 Welding In-Service Pipelines

It is sometimes necessary to weld repair sleeves and hot tap branch connections onto in-service pipelines without interruption of flow. The economic benefits are obvious, but the weld procedure and safety aspects demand special attention during and after welding. High cooling rates occur when welding onto an in-service pipeline because the flowing contents quickly remove heat from the pipe wall. The high cooling rates promote formation of hard, heat-affected zone microstructures, and the welds are susceptible to hydrogen cracking during or soon after welding and to sulfide stress cracking if the pipeline is in subsequent sour service.

The Welding Procedure Specifications (WPSs) and Procedure Qualification Records (PQRs) need to be qualified to the requirements of the various welding codes that include API 1104, API 1107, ASME Section IX, BS 4515, BS 6990, and CSA Z662. Procedures are normally developed using laboratory mock-ups and full-scale pipes before validation of field welds onto live gas pipelines. Usually, the same WPSs and PQRs apply to both hot tap branch connections and sleeve repairs, i.e., both the fillet and groove welds. Procedures may also need to satisfy the 22 HRC requirement of NACE MR0175 for sour service.

The welding procedure for a particular situation is determined based upon four factors:

1. Hydrogen level of the welding process
2. Pipe wall thickness
3. Expected cooling rate of the weld
4. Chemical composition

These procedures rely on several factors, singularly or in combination, to produce satisfactory welds: heat input control (between 15 and 40 kJ/in.), preheat temperature control at 95 °C (for thick wall pipe only), a temper bead deposition sequence, or use of austenitic weld metal as a hydrogen *sponge*. The procedure under development should include consideration of carbon equivalent (usually 0.50 or less per international formula), pipe diameter, flow rate of product, the combination of pipe wall, branch wall, sleeve wall thickness, the welding procedure to be used (e.g., GMAW or SMAW) using low hydrogen or cellulosic electrodes, the weld orientation— up-hill and down-hill—the nature of the gas or liquid pipeline contents.

References

1   API 5L PSL 2: "Line Pipe." American Petroleum Institute

2   ISO 3183: "Petroleum and Natural Gas Industries: Steel Pipe for Pipelines Part 3." International Organization for Standardization.

3   API 1104: "Welding of Pipelines and Related Facilities." American Petroleum Institute.

4   BS 4515: "Welding of Steel Pipelines on Land and Offshore." British Standards Institution.

5   Boekholt, R. (1990) "Mechanised Welding Methods In Pipeline Construction." Proceeding of the Pipeline Technology Conference, Oostende, Belgium, October.

6   Parlane, A. J. A and J. R. Still, J. R (1988) "Pipelines for Subsea Oil and Gas Transmission." *Mat. Science and Tech.*, 314.

7   Jones, R. L. (1998) "Welding of Carbon Steel Pipelines." Proceedings of the TWI Conference.

8   Jones, R. L. and P. N. Hone. (1990) "Advances in Techniques Available for Girth Welding of Pipelines." Proceedings of the Pipeline Technology Conference, Oostende, Belgium, October.

9   A. C. Davies, Ed. (1998) *Welding Science & Technology, Parts 1 & 2.* Cambridge, UK: Cambridge University Press.

10 BS 499: "Glossary for Welding, Brazing and Thermal Cutting." British Standards Institution.

11 R. Pargeter, T. G. Gooch. (1994) "Welding for Sour Service" Proceedings of the Update on Sour Service: Materials, Maintenance and Inspection in the Oil & Gas Industry., IBC, London, October.

12 J. F. Lancaster. (1993) *Metallurgy of Welding.* London: Chapman & Hall.

13 M. F. Wheeler, N. J. Baker, T. D. Shipley. "An Engineering Critical Assessment—A Case History." (S.G., SPE 14024/1) Society of Petroleum Engineers, Aberdeen.

14 NACE MR-0175: "Sulfide Stress Cracking Resistant Metallic Materials for Oilfield Equipment." National Association of Corrosion Engineers.

15 Gunn, R. N. (1993) "Weldability and Properties of Duplex and Superduplex Stainless Steels." Proceedings of the Conference of International Offshore and Polar Engineering, Singapore, June.

16 Anderson, P. C. J. (1998) "Welding of Duplex and Superduplex Stainless Steels." Proceedings of the TWI Conference.

17 Laing, B. S. (1993) "Enhanced Techniques For Duplex Pipelines." Proceedings of the Technology Update—Duplex Stainless Steels, Middlesbrough, March.

18 Belloni, A. (1997) "Full GMAW Proved for CRA Pipeline Welding." Proceedings of the 5th World Conference on Duplex Stainless Steels, KCL Publishing.

19 BS 2600: "Radiographic Examination of Fusion Welded Butt Joints in Steel—Methods for Steel 2mm up to and Including 50 mm Thick." British Standards Institution.

20 BS 6072: "Method for Magnetic Particle Flaw Detection." British Standards Institution.

21 BS 4069: "Magnetic Flaw Detection Inks and Powders." British Standards Institution.

6 Flexible and Composite Pipelines

6.1 Introduction

The first flexible pipeline was installed across the English Channel during World War II to transport fuel from England to France and support the D-day landings. These pipelines were based on telegraph cable technology and were composed of a lead tube protected by tape, armoring wires, and an outer sheath. The modern type of flexible pipeline was developed in 1970.

Flexible pipelines are increasingly being used for small diameter, short distance flowlines; as jumpers from wellheads and wellhead manifolds to rigid flowlines; as expansion loops; and as static and dynamic risers. See Figure 6–1. The pipelines are used for transport of crude oil and gas, dead oil and water, and for test lines, kill lines, gas, lift, and chemical injection lines. Flexible pipelines have also been used in the North Sea for long distance transport of untreated well fluids, e.g., AGIP Toni and Shell Nelson.

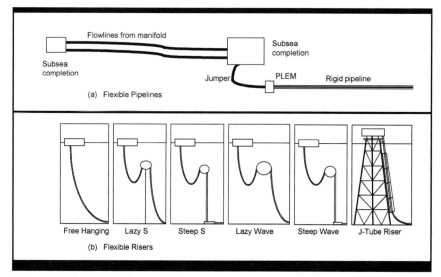

Fig. 6–1 *Flexible Pipelines and Risers*

Though the pipeline material cost is high, about five to six times the cost of an equivalent steel pipeline, the pipe is cheaper and faster to lay and can be installed from modified barges or drill ships. Using specialist vessels, the pipe can be installed at rates up to 500 m/hour. This has advantages for remote locations where the cost of mobilization of a conventional lay barge would represent a significant percentage of the total project. It is also feasible to recover the pipeline after use for inspection, refurbishment, and further service.

Standards relating to flexible pipes are not as comprehensive as for rigid pipelines: API[1], Veritec,[2] and IP[3] guidelines are available.

Sections 6.2 through 6.7 relate to flexible pipelines. An alternative that is much less used is to make a pipeline from a nonmetallic composite, and that option is examined in section 6.8.

6.2 Fabrication

6.2.1 General

The design and fabrication of flexibles are organized differently from those of steel pipes. The manufacturer invariably carries out the detailed design of a flexible and is often responsible for the installation as well. With steel pipes, on the other hand, the operator, or a design consultant working for the operator, carries out the design; and the manufacturer is only rarely engaged in the installation.

Flexible pipes are composites constructed from sequential concentric layers of metals and polymeric thermoplastic materials. Each layer has a specific function, and the layers used depend on whether the pipe is bonded or non-bonded. Submarine pipes and risers are bonded. Composite layers are illustrated in Figure 6–2. The layers are applied sequentially, and as each layer is added from inside outward, the pipe is reeled to allow the pipe to be run continuously through each fabrication stage. The essential layers from the inside to the outside are the carcass, the liner, layers that take the loads induced by the internal pressure and the longitudinal force, and an outer sheath.[4]

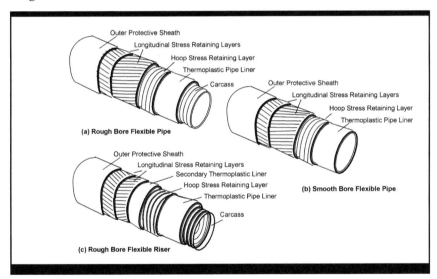

Fig. 6–2 *Fabrication of Flexible Pipelines and Risers*

Additional outer layers may be used to reduce gas permeability for use with sour fluids, to increase flexibility by allowing movement between steel windings for dynamic risers, or externally to improve thermal insulation. The following details describe the layers of material.

6.2.2 Carcass

The carcass is a helically wound interlocking metal strip that is permeable to the transported fluids. The carcass is not gas or fluid tight; however, when exposed to hydrocarbon fluids, the liner swells and seals the carcass to an extent. It also limits the thermal and chemical swelling of the

liner in service. The function of the carcass is to prevent collapse of the thermoplastic pipe liner as a result of gas expansion or hydrostatic pressure. At low operating pressures, it may contribute to the pressure retention properties of the pipe

Under operating pressure, some of the gas and low-density fractions of the hydrocarbon fluid will permeate through the plastic pipe liner into the voids in the metal reinforcements. Should the pressure in the pipeline reduce rapidly, these fluids will expand, possibly causing blistering and collapse of the plastic liner. One primary function of the carcass is to prevent this collapse. Stabilized crude oil or water lines do not necessarily require a carcass; the minimum gas-oil ratio (GOR) at which a carcass is considered necessary is around 300. The carcass also permits use of certain soft pigs through the flexible pipe for wax and sediment removal. Pipe fitted with a carcass is termed *rough bore,* and pipe without a carcass is termed *smooth bore.*

The carcass is manufactured from strip steel, cold-formed into an interlocking S-shape. The strip is then wound around a mandrel to form a continuous but flexible tube. Typical steels used are carbon steel to AISI 4130, austenitic stainless steels to AISI 304, 304L, 316, 316L, and duplex stainless steel to UNS S31803. The compositions of the steels used are given in Table 6–1. The carcass is electrically isolated from the end fittings to avoid galvanic corrosion at the end fittings.

Table 6–1 *Characteristics of Typical Carcass Materials*

Material	Composition (maximum %)					Mechanical Properties	
	C	Mn	Ni	Cr	Mo	UTS MPa	%
4130 CS	0.33	0.9	---	0.8–1.2	0.15–0.2	621	18
304 SS	0.03	2	8–10	17–19	---	540	40
304L SS	0.03	2	9–11	17–19	---	490	40
316 SS	0.07	2	10–12.5	16–18	2–2.5	560	40
316L SS	0.03	2	10.5–13	16–18	2–2.5	510	40
Duplex SS	0.03	N 0.2	4.5–6.5	21–23	2.5–3.5	790	25

Because the pipe is formed around a mandrel, the nominal diameter of a flexible pipe is the actual internal diameter. Hence, a 6-in. diameter, flexible pipe is actually 6-in. internal diameter; whereas, a 6-in. diameter rigid steel pipeline would be 6.625-in. outside diameter, and the internal diameter would be reduced by the wall thickness. Flexible pipes are available in diameters from 1–16 inch.

The carcass presents a rougher profile to the fluids than does rigid steel pipe, and this rough profile increases the resistance to flow. A rule of thumb factor of around 4 is often used as a rough guide for calculation of the pressure drop when making comparison with rigid steel pipe. The relationship developed by Coflexip is more precise and relates the internal roughness height k to the internal diameter D, using the notation given in chapter 9:

$$k / D = 0.004 \qquad (6.1)$$

A typical rigid steel pipe has a roughness height k of 0.015 to 0.02 mm, and therefore, if D is 250 mm, k/D is 0.00006 to 0.00008. This increased pressure drop can reduce the throughput from a primary drive reservoir and consequently, may have a significant effect on the production profile; but this reduction becomes of less significance as the pipe diameter is increased.

Rough bore pipelines (those with an internal carcass) can be pigged by spheres, polypigs, most bi-directional gauging plate pigs, and cleaning pigs, including those fitted with steel brushes. However, to minimize risk, polypigs are generally preferred for cleaning flexible pipelines. One risk with sold-bodied pigs is tipping of the pig and consequent damage to the carcass and liner. The use of long-bodied pigs will reduce this tendency, but in all cases, the pig supplier should have verified the serviceability of the pigs in the flexible pipe configuration. There is, of course, no purpose in using corrosion measurement pigs in these pipelines.

In smooth bore pipelines there is no carcass. Such pipelines are used for stabilized crude oil and for water injection lines. The type of pigs used must be restricted to avoid damage to the plastic liner. Special plastic bladed pigs and brushes are available. Though sensitive to mechanical damage, the high-density polyethylene (HDPE) and nylon-based smooth bore pipes are more resistant to erosion than steel pipe, showing typically about one-seventh the damage in water-sand slurries. Fluorocarbon materials are relatively soft and sensitive to erosion.

6.2.3 Liner

The liner contains the hydrocarbon fluids. Liners are fabricated from high-density polyethylene, nylon and fluorinated polymers. The factor that determines the service life of the flexible line is its degradation that occurs as a result of reaction with components in the hydrocarbon stream. The operating temperature is a major factor in this rate of degradation and consequently the choice of polymer depends on the service temperature. For low temperatures or low water content fluids, HDPE and polyamide (nylon) liners are used. These materials are suitable to about 65 °C and 95 °C, respectively, though the precise limit depends on manufacturing details and should always be confirmed by the manufacturer. At higher temperatures (to 130 °C) and high water cut fluids, a more thermally stable liner is required. Suitable materials in these conditions are extrudible fluorinated polymers, such polyvinyldifluoride (PVDF). The minimum temperatures for these materials are –50 °C for HDPE and –20 °C for nylon and the fluorinated polymers. Table 6–2 lists the mechanical properties of liner materials.

Table 6–2 *Properties of Thermoplastic Liner Materials*

Material	Density (kg/m³)	Thermal Tolerance (°C)		Thermal Conductivity (W/m °C)	Tensile Strength (MPa)	Bending Modulus (MPa)
Nylon 11	1050	Oil	100	0.33	350	300
		Water	65			
High Density Polyethylene	940	Water	65	0.41	800	700
Fluorocarbon	1600	Oil	130	0.19	700	900
PVDF		Water	130	0.19	700	900

Thermoplastics are permeable to low molecular weight hydrocarbons. In operation, there is a continued diffusion of gases through the liner into the void spaces in the steel pressure containment layers. The rate of diffusion can be calculated using the relationship:

$$q = \frac{KS\Delta p}{t} \tag{6.2}$$

where

q = the gas permeation rate (cm³/s)

K = the permeability coefficient (cm³ gas/(cm² plastic bar s))

t = the sheath thickness (cm)

S = the surface area of the thermoplastic sheath (cm²)

Δp = the differential pressure across the plastic (bar)

The permeability coefficient is characteristic of the plastic liner materials used and the particular gas; values for methane at 50 °C are given in Table 6–3. The permeability coefficient varies with temperature.

Table 6–3 *Permeability of Liner Materials to Methane at 50 °C*

Material	Permeability Coefficient K ($cm^3/(cm^2$ bar s))
High Density Polyethylene	1.5×10^{-8}
Low Density Polyethylene	5×10^{-8}
Nylon	0.45×10^{-8}
PVDF	0.1×10^{-8}

The gas pressure in the armoring builds up and escapes through one-way release valves fitted in the end connectors or, in an emergency, through bursting discs fitted in the outer protective sheath. The release valves are set to vent the gas at 3–5 bars above the local hydrostatic pressure. Permeation of gas through the outer sheath is very low because of the lower temperature and small differential pressure. Bursting disks are formed by incomplete drilling of a 16-mm hole through the outer sheath to leave a 2-mm thick residual membrane in place. This membrane will open when the differential pressure exceeds about 10 bar. The location of release valves and dimensions of the bursting disks are illustrated in Figure 6–3.

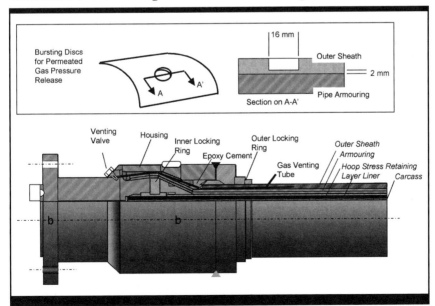

Fig. 6–3 *Sheath Pressure Relief Systems for Flexible Pipelines*

Because the permeability coefficient is gas-specific and increases with decreasing molecular weight and molecular dimension, the permeation of carbon dioxide and hydrogen sulphide is higher than the permeation of methane. Consequently, there is a slight enrichment of the corrosive gases in the armoring voids. This factor must be taken into account when evaluating sour service criteria.

With all plastics, there is a risk of blistering and explosive decompression when the internal pressure in the pipeline is suddenly reduced. The gas dissolved in the plastic expands faster than it can diffuse out of the plastic, resulting in internal blistering and mechanical breakdown of the plastic. Nylons and the fluorocarbons have low gas solubility; therefore, this form of damage is unlikely. However, it is generally good practice to restrict the rate of depressurization.

6.2.4 Pressure containment

The internal pressure is contained by a sequence of concentric steel wires wrapped around the thermoplastic pipe liner. One early version of flexible pipe used textile reinforcement. The first layer is wound on a short pitch and contains the majority of the radial (hoop) stress resulting from the internal pressure. This layer resists the external hydrostatic pressure during installation when the pipe is empty and the forces imposed by shrinkage of the pipe diameter that occurs during installation when the outer cross-wound wires tighten under the extension of the pipe as it is suspended from the lay barge. The hoop layer is formed from shaped steel that provides interlock between the adjacent coils.

To withstand external loads and impacts, and to restrict longitudinal extension, the inner pressure containment layer is overlaid with a minimum of two layers of armoring. These wires may be circular or flat steel strips that are wound crosswise and helically on a long pitch around the inner steel armor layer. The opposing helices are used to balance the extension forces, and the wrapping angles used depend on the service pressure of the pipeline and range from 15–35 °C. The schematic in Figure 6–2 illustrates these layers.

The thickness of steel used, the pipe diameter, and the design pressure determine the yield strength. Generally high strength, high carbon content steels are used to minimize the weight of steel required; but the strength of the steel may need to be restricted for sour service. The ultimate tensile strength of the steels will be in the range 800–1400 MPa. There are upper limits to the design pressures that typically range from >100 MPa for 2-in. pipe to 20 MPa for 16-in. diameter pipe. The sour service restriction reduces the design pressure by about 25–30%.

The armor wires are electrically continuous and are welded to the end fittings to ensure electrical continuity so that the cathodic protection system will be effective along the full length of the pipeline. To reduce fretting damage to the steel armor wiring, a plastic interlayer is applied between the steel layers. This is particularly important for risers that suffer cyclic fatigue loading. The thermoplastic used is usually the same plastic as used for the pipe liner.

6.2.5 Outer sheath

The steel armoring is protected from external corrosion by overlaying with a hot-melt extruded plastic sheath. HDPE is the most commonly used material because it has good adhesion, extensibility, abrasion resistance, electrical properties, and low water absorption.

Cathodic protection is provided to protect the pipe at areas of damage to the outer sheath. Typically a coating breakdown value of 1% is recommended for the design calculation of a CP system; but because the construction is with wire, the steel area exposed equates to an equivalent 3% bare steel area. The end fittings are often electroless nickel-plated and, in this case, the total area of the end fittings must be taken equivalent to bare steel area when calculating the CP current demand. It is usually advantageous to coat the end fittings over the nickel plate with an epoxy coating to reduce the drain on the CP system.

Sacrificial anodes cannot be placed along the length of the pipeline because of the risk of damage and tearing of the outer sheath by the anode during installation. Instead, the bracelet anodes are clustered at the end fittings and daisy chained together for electrical continuity. A typical arrangement is shown in Figure 6–4. The design basis for CP systems is given in chapter 8.

If additional thermal insulation is required, additional layers of insulating foam may be applied as pre-formed bracelets, secured by an outer wrap or as single or multiple layers of adhesive-backed tape helical wrapped around the pipe. The thermal insulation achieved depends on the density and thickness of the foam. The minimum bend radius for the pipeline occurs when the sheath diameter is extended by 7.5%. As the thickness of thermal insulation is increased, the bend radius must be increased considerably. Heat tracing can be incorporated into the outer armor cabling and is used in conjunction with thermal insulation.

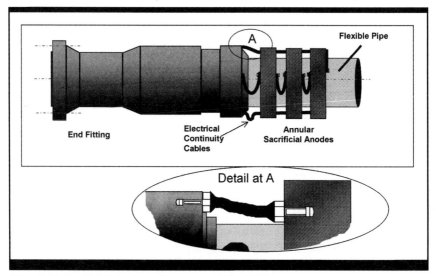

Fig. 6–4 *Sacrificial Cathodic Protection of Flexible Pipe*

6.2.6 Pipeline lengths

The lengths of pipe that can be fabricated in one piece depend on the diameter and weight of the pipe. The pipe is wound onto reels or carousels, and there are limits to the volume of the reels and carousels. Usually the crane loads within the fabrication shops also limit the size and weight of reels for shipment to site and laying. Generally, weight is the main factor determining the length, though for the larger diameter pipelines, the limitation may be the volume of the pipe. Flexible pipelines are generally heavier than rigid steel pipelines for the same diameter and pressure. Very high-pressure pipelines contain more steel windings and, hence, may need to be fabricated in shorter lengths. Typical lengths and weights available from one manufacturer are given in Table 6–4. For certain pipes limited by volume, the reel size can be increased slightly.

Short lengths of pipe may be connected to produce long flowlines. The maximum single length installed to date in the North Sea is about 30 km (Shell Nelson) though there are several developments with overall longer combined lengths of flowlines (e.g., Shell Draugen with 50 km and AGIP Toni and Saga Snore with 34 km each).

Table 6–4 *Continuous Lengths of Coflexip Flexible Pipe*

ID in.	Max Design Pressure (bar)	Max Collapse Pressure (bar)	Max Single Length (m)[1]	Empty Weight in Air (kg/m)	Empty Weight in Water (kg/m)
2	1,380	500	17,000[2]	11 to 37	5 to 27
4	1,170	400	12,000[2]	20 to 102	5.5 to 72
6	1,010	300	6,450[2]	29 to 208	1.5 to 144
8	815	300	15,000	43 to 263	-4 to 173
10	670	300	11,000	62 to 331	-10 to 207
12	580	300	7,000	102 to 401	-8 to 238
14	500	230	5,000	133 to 464	-15 to 255
16	440	190	4,600	172 to 528	-22 to 270
19	200	67	2,600	240 to 420	-38 to 134

[1] Packing and storage limitations come from weight or volume of the flexible pipe. Two methods can be used for packing the pipe: reel or carousel.
[2] Packing onto reel.
Weight in water can also be a limitation during installation. Product improvement and manufacturing facility developments may in time make this table out-of-date

6.2.7 End connections

The connections of the flexible pipe to the fixed ends, or of flexible lengths of pipe to each other, represent an area of potential weakness. Great care is taken in designing and fitting the connectors. The fixing method is primarily mechanical, using wedge-shaped fittings in the end fitting supplemented with epoxy cement to fill the voids.

The carcass is electrically isolated from the steel windings and armoring through a plastic insert at the connector. The steel armor layers are electrically connected to the end fittings to ensure cathodic protection is effective. In most cases, the connections between end fittings are made by bolting through rotating flanges, though fixed flanges are available. The end connectors are generally carbon-manganese steel (e.g., AISI 4130) and are protected from internal and external corrosion by a 150 μ coating of electroless nickel. The optimum corrosion resistance and mechanical properties are obtained when the electroless nickel contains phosphorous and is heat treated to increase diffusion of the nickel into the substrate steel and to relieve coating internal stresses.

For severe service, these couplings can be internally weld overlaid with Inconel 625 or a similar corrosion resistant alloy. Alternatively, solid corrosion-resistant alloy connectors are available, e.g., duplex stainless steel.

Quick release connections are fitted to the upper ends of risers. These connectors are hydraulically activated.

When the connections between pipes are made offshore, they must be tested before the pipe is laid on the seabed. Testing is done by a reverse pressure technique where an external hydraulic pressure is applied to the made-up joint. Once the complete pipeline is laid, it is hydrotested in the conventional way.

The bend that can be tolerated by a riser must be restricted to avoid damage to the pipe at the point of transition from the flexible pipe to the end connector at the riser hanger. Conical bend stiffeners are attached to the end fittings to ensure that the riser is not overstressed by an over tight bend. Similar stiffeners can be used for pipelines if there is a risk of overbending the pipe during installation.

6.3 Internal Corrosion

The carcass is in contact with the produced fluids and, hence, may suffer corrosion. Carcasses are constructed from carbon steel, austenitic stainless steels, and duplex steels. More exotic materials such as Inconel 625 could be used. Since the carcass is not a true structural member, it can tolerate some corrosion, but excessive corrosion might affect its function. Localized pitting can be tolerated, but longitudinal corrosion along the length of the pipeline or continued pitting would reduce the strength of the carcass to resist collapse.

Corrosion rates of the carcass would be expected to be similar to the rates of the same material as a rigid pipe. There would be some increase because of the enhanced roughness leading to increased turbulence. Chemical inhibition may also be less efficient as the crevices between the carcass interlocks would be less well inhibited than the exposed material. As a generality, therefore, an increase of around 20% would be used to evaluate the corrosion in comparison to a rigid pipe material. Alternatively, the corrosion inhibitor efficiency would be downgraded to 70% from 85%. Background to these aspects is covered in more detail in chapter 7.

Stainless steel is resistant to internal corrosion; and though these carcasses will suffer localized pitting, they should show little or no distributed corrosion. However, these materials suffer from severe pitting if the fluids are aerated and from chloride stress corrosion cracking if chloride-containing fluids are aerated and the temperature is above 50°C.

For these reasons, the introduction of aerated seawater and brines or other chloride-containing service fluids should be controlled. If there is any risk that aerated seawater could be introduced into a hot pipeline, then a duplex stainless steel carcass should be specified. The corrosion resistance of the materials for carcasses is in the order 1430 < type 304 < type 316 < duplex.

The degradation of the thermoplastic pipe liner depends on the composition of the transported fluids and the operating temperature. A breakdown of material performance is given in Table 6–5. The material performance with regard to temperature is of the order of HDPE < nylon < fluorocarbon. This ranking should be viewed with caution, however, as it varies with specific fluid composition.

Table 6–5 *Resistance of Thermoplastic Materials*

Material	HDPE	Nylon	Fluorocarbon
Crude Oil	poor	good	good
Sour Gas	poor	good	good
Stabilized Oil	modest	good	good
Seawater	good	good	good
Methanol	good	modest	good
Acids	good	poor to modest	good

6.4 Sour Service

For sour service, the steel reinforcement is generally specified to meet the requirements of NACE MR-0175[5] (the use of ISO 15156[6] could have major advantage as the non-cracking domain is more extensive than defined in the NACE practice). The critical level of hydrogen sulphide in the fluids is usually quoted as 100 ppm. Hydrogen sulphide readily migrates through the plastic liners and into the gas diffusion space. As described previously, the presence of sulphide on the metal surface results in enhanced hydrogen permeation into the steel as occurs with rigid pipeline steel. Compliance to NACE MR0175 restricts the tensile strength of the steel that can be used and consequently, for a given pressure rating, the cross-section of the armoring must be increased. Using an increased cross section increases pipe weight and this has cost implications because the length of individual pipes that can be fitted onto a reel is reduced, a situation that increases the cost of fabrication and the speed of installation.[7] Without an increase in cross-section of the armoring the use of lower strength steel would reduce the maximum design pressure by around 25–30%.

From a corrosion viewpoint, the risk of sulphide stress cracking (SSC) would be expected to be much less than for rigid pipelines for the same concentration of hydrogen sulphide. The hydrogen sulphide concentration in and around the reinforcement steel results from the migration of the sulphide through the liner. The system pressure is limited to the hydrostatic head plus about 5 bar, the maximum setting for the gas venting valves; hence, the partial pressure of hydrogen sulphide will be low compared to the partial pressure of hydrogen sulphide inside the pipe. This markedly reduces the risks of SSC unless the fluids being transported are very high in hydrogen sulphide and at high pressure with a permeable liner when the pipeline is in deep water. There may be occasion when an additional permeability barrier is more cost effective than the use of lower strength steel conforming to NACE MR0175.

6.5 External Corrosion

The steel in the pressure-retaining armoring is susceptible to corrosion once the outer sheath is perforated. In normal conditions, the armor wires are dry because the outer sheath is integral, so no corrosion occurs. If seawater enters, the acid gases that have permeated through the pipe liner will combine with the water and any residual oxygen to produce a potentially corrosive condition. The corrosion profile along the pipe away from the coating damage will alter fairly rapidly. At the defect, there will be an active area because oxygen is available, but at this area the cathodic protection system will be active. Further from the damage area, there will be a background corrosion rate related to the concentration of the acid gases in the seawater; the cathodic protection current will not penetrate to this region.

The extent and duration of the corrosion will depend on the degree of damage sustained by the outer sheath. Small areas of damage are expected to self-seal because the cathodic protection current will induce the formation of calcareous deposits that, in conjunction with corrosion products, will plug the small coating defects. Larger areas of damage are less likely to self-heal and likely will continue to corrode. The rate of corrosion would be specific to the particular operating conditions and service of the pipeline but is not expected to be severe, as the larger exposed area will allow a higher cathodic protection current density to enter. Large areas of coating damage would be identified by cathodic protection surveys.

CP is used to supplement the protection afforded by the outer sheath. At present, the available standards and recommended practices do not give information on the expected breakdown of flexible pipe outer sheaths. As noted earlier in this chapter, one supplier uses a coating breakdown value of 1%, which is calculated to equate to an exposure of 3% of steel. CP design is based on this value. Other CP design values are, typically 80–120 for rigid pipelines and 25–50 mA/m^2 for exposed and buried conditions respectively. Current density values are adjusted for flowline operating temperature. Sacrificial anodes are fitted to the end connectors at the ends of the pipe lengths. Details of CP design are given in chapter 8.

6.6 Failure Modes of Flexible Pipes

The general failure modes for these pipes are:

- Disbondment of the bonded components
- Fretting and wear of the internal parts
- Corrosion and fatigue failures

The likely mode of failure can be decided from the expected service of the pipe, though disbondment can occur in all service environments.[8] Unfortunately, disbondment cannot be easily evaluated by existing inspection techniques. Once the pipe is in service, it is difficult to inspect; therefore, great effort is made to ensure that the pipe is selected correctly to have proven performance under the expected conditions and enters service as free of defects as possible. Many regulatory authorities require flexible pipes to be hydrotested each six years to prove serviceability.

Fatigue can occur in flexible risers. Movement leads to fretting between the steel layers and to localized wear of the wires. The increase in stress at the wear points reduces the fatigue endurance at these thinned sections. A combination of lubrication and incorporation of anti-fretting layers of thermoplastic material between the steel armor layers reduces wear. The life expectancy of a riser is generally limited to 15 years, then the riser would be removed for inspection and, if necessary, replacement.

6.7 Inspection of Flexible Pipe

In-service inspection is limited with flexible pipes because conventional inspection techniques have been developed for rigid pipelines and are not easily transferred to composite materials. Instrumented pigs cannot be used, and ultrasonic testing does not provide accurate information because of the multilayered nature of the construction of the flexible pipes. One alternative approach has been to instrument the pipeline so that its condition can be inferred from data given by the instrumentation. Where possible, the instrumentation is installed at the critical areas identified during the engineering studies. Typical instrumentation gives indirect data and includes:

- Pressure drops or flow monitoring
- Load cells
- Pressure sensors
- Inclinometers
- Non-destructive examination of the end fittings.

Conventional inspection techniques that have been modified to allow application during pre-service testing and for in-service inspection include:

- Visual inspection of surfaces
- Hydrostatic pressure testing
- Soft pigging to confirm no obstructions in the bore
- Non-destructive examination of couplings and fittings
- Modeling of the effects of structural loading.

Several techniques are under evaluation for inspection of composites, and these may be applicable to flexible pipes. The techniques include:

- Thermography (conventional and flash)
- Real time radiography during construction
- C-scan ultrasonic inspection
- Acoustic emission[9]
- Radio-isotope leak detection

- Holography and laser imaging
- Impedance measurements
- Fiber optic sensor filaments

Corrosion of the carcass can be inferred from application of conventional corrosion monitoring applied to accessible sections of the pipeline. Usually these locations are at the wellhead or on the receiving platform. External damage can be assessed by use of conventional CP monitoring techniques. Monitoring techniques are discussed in chapter 15.

6.8 Composite Pipelines

Another possibility is to make the whole pipeline out of one of the composite materials that now have become available, such as epoxy reinforced with glass fiber, carbon fiber, or silicon nitride. This option makes it possible to exploit the high strength of the fibers and at the same time to eliminate corrosion.[10]

Materials of this kind are widely used in the automotive and aerospace industries and in specialized applications where a high strength-to-weight ratio is important, and are finding new applications in offshore platforms, ships and military vehicles. Much research has been done. With few exceptions (mainly sea water injection systems), composites have not been adopted by the petroleum pipeline industry, either offshore or onshore. A 28-in. fiber-reinforced epoxy pipeline for crude oil was constructed in Algeria to replace a steel pipeline that had failed due to corrosion and is understood to operate satisfactorily. Large-diameter glass-fiber reinforced epoxy pipes are used as outfalls and intakes, but they operate at modest pressures and at ambient temperature.

The reasons why composites have not found favor is related to manufacturing costs and to limited manufacturing capacity. For a given pressure and diameter, the cost to manufacture a composite pipe is higher than the cost for manufacturing an equivalent steel pipe; but that cost might, of course, change if a broader market for composites were to develop.

Joints between composite pipes remain an area of concern. The joints can be made in various configurations from glued and pinned sleeves, each method has drawbacks and inspection procedures are limited.

A more attractive possibility is to combine the corrosion resistance of composites with the strength and economy of steel. One option is to expand a plastic liner into a previously welded steel pipe. Another option is to wind high-strength strip around a polymer or composite pipe. These options appear more attractive than pure composites for high-pressure applications, at least in the medium term. However, this may be unduly pessimistic: few would have forecast that polyethylene would so rapidly displace cast iron for low- and medium-pressure gas distribution service.

References

1   API RP 17B: "Recommended Practice for Flexible Pipe." American Petroleum Institute.

2   JIP/GFP-02: "Guidelines for Flexible Pipe Design and Construction." Veritec.

3   "IP6 Pipeline Safety Code Supplement S25." Institute of Petroleum

4   *A Glossary of Flexible Pipe Terminology.* Coflexip Publication.

5   NACE Standard Materials Requirement MR0175: "Sulfide Stress Cracking Resistant Metallic Materials for Oilfield Equipment." National Association for Corrosion Engineers.

6   ISO 15156: "Petroleum and Natural Gas Industries—Materials for use in H2S containing environments in oil and gas production Parts 1, 2 & 3." International Organization for Standardization.

7   Marion, A. (1994) "The Design of Flexible Pipes for Sour Service Application." Proceedings of the Update on Sour Service: Materials, Maintenance & Inspection in the Oil & Gas industry, IBC, London, October.

8   Chaperon, G. R., H. P. Boccaccio, and M. J. Bouvard. (1991). "A New Generation of Flexible Pipe." (Paper 6584) Proceedings 23rd Annual Offshore Technology Conference, Houston, TX.

9   Sugier, A., J. Mallen, P. Marchand, and A. Marion. "Monitoring of Flexible Risers by Acoustic Emission." Coflexip Report, Paris.

10 Oswald, K. J. (1996) "Thirty Years of Fiberglass Pipe in Oilfield Applications: A Historical Perspective." *Materials Performance*, May, 65–68.

7 Internal Corrosion and Its Prevention

7.1 Introduction

Approximately 40% of the world inventory of 3×10^6 km of oil and gas pipelines is at or approaching nominal design life, and corrosion is appearing as a major limiting factor in continued pipeline operation. At present, corrosion represents between 20–40% of recorded incidents and failures of pipelines, though the information on corrosion associated with incidents is not always clear. The recorded failure rate of pipelines also varies widely among countries, and that variation presumably reflects technical, geographic, and cultural differences. The lack of uniformity in recording the reason for loss of pipeline integrity leads to the inconsistencies in the figures for integrity loss incidents given in Table 7–1.

Table 7–1 *Comparisons Between Onshore and Offshore Pipeline Failure Statistics*

(a) *Comparison of Integrity Loss Incidents*

Location	Reason for the Incident %			
	Construction	Material	Third Party	Corrosion
Onshore	4	9	40	20
Offshore	6	8	36	41

(b) *Breakdown of Onshore Pipeline Incidents*

Region	Average Incidents 10^3 km-year	Cause of Defect Associated with Incident %			
		Construction	Operational	Third Party	Corrosion
USA Gas	0.26	13	26	40	20
USA Oil	1.33	12	45	22	22
W. Europe Gas	1.85	10	47	28	16
W. Europe Oil	0.83	23	4	48	28
E. Europe All	4.03	13	13	57	18

(c) *Breakdown of North Sea Pipeline Incidents*

Average Incidents 10^3 km-year	Breakdown of Information %						
	Pipeline			Reason for Incident			
	Rigid	Flexible	Fittings	Impact	Anchor	Corrosion	Other
0.14	65	8	26	19	13	13	24

Weakening of the pipeline by corrosion will reduce the resistance of the pipeline to external forces and will accentuate materials and fabrication weaknesses, which effect perhaps explains the conflicts in the incident data. The prevention of corrosion does require attention throughout the life cycle of the pipeline, from design through fabrication and commissioning, to operation. Once corrosion processes are established, it becomes increasingly difficult to mitigate their effect on pipeline integrity. The approach often taken that corrosion mitigation is not required in the early phase of a pipeline life is only acceptable if the corrosion status is frequently reviewed.

The budget that should be assigned to corrosion prevention is difficult to evaluate because of uncertainty over the life expectancy of the pipeline and its production profile. Over the pipeline life, changes in flow and fluid composition modify the corrosion processes. In several cases, the composition of the produced fluids has been found to be sufficiently different from the initial drill stem test results to discount the original corrosion philosophy. The capital cost of corrosion prevention for a carbon steel pipeline is 10–20% of the materials cost (coating, cathodic protection, and inhibitor injection equipment) and later perhaps 0.3–0.5% of the operating revenue for the subsequent inhibitor costs and the inspection and monitoring costs. The extra steel provided as corrosion allowance is not included in these values.

Over-attention to corrosion is not necessarily cost effective. An 80:20 Pareto type rule seems to apply: 80% of the corrosion can be prevented for modest cost, but preventing the remaining 20% would cost many times more. Depending on the point in the life cycle of the pipeline, savings on corrosion prevention could be introduced by reducing the level of effort expended on corrosion prevention. However, several major transmission lines have been required to function for significantly longer than the original design life, and it is fortunate that they had been maintained in good condition.

With respect to the types of corrosion failure anticipated, there is a paucity of data. The published literature concentrates on coatings, cathodic protection, chemical inhibition, and studies of the various cracking mechanisms. There has recently been an upsurge of interest in microbiological corrosion. However, these areas reflect academic interests and subsequent funding, rather than engineering necessities. One oil company's breakdown of its actual corrosion failures averaged over several offshore fields (i.e., including topside processing plant and pipelines) is given in Table 7–2.[1]

Table 7–2 *Relative Occurrence of Internal Corrosion*

Corrosion Failure Mechanism	Percentage
Carbon dioxide corrosion	32
Combined velocity and carbon dioxide corrosion	5
Chemical attack	1
Combined corrosion and fabrication defects	3
Microbiological corrosion	13
Corrosion of threaded items	11
Corrosion in dead legs	16
Erosion	8
Mechanical associated corrosion failures	2
Corrosion fatigue	1
External corrosion	7

Internal corrosion processes depend on the service of the pipeline. Section 7.2 of this chapter describes corrosion mechanisms. Following are four separate cases of corrosion that illustrate these mechanisms.

1. Sweet corrosion caused by the presence of carbon dioxide dissolved in the fluids; also called carbonic acid corrosion. This form of corrosion is typically slow, initially localized, and includes a form of pitting attack that may, depending on flow conditions, convert to a form of grooving corrosion. It is sensitive to operating pressure and temperature, water cut, water composition, flow rate, and the presence of solids. (See section 7.3, followed by sections 7.4–7.6 on oil and 7.7 and 7.8 on gas.)

2. Sour corrosion is caused by hydrogen sulphide in the fluids. Corrosion may be rapid failure by cracking of the pipeline steel. The hydrogen sulphide can also exacerbate pitting corrosion. (See section 7.9.)

3. In water injection pipelines, the corrosion results either from the presence of oxygen in the water or microbiological activity of the sulphate-reducing bacteria (SRB). Some aquifer waters may be oxygen free but contain carbon dioxide and/or hydrogen sulphide. (See section 7.10.)

4. Microbiological corrosion resulting from the activity and growth of SRB in the pipeline. The SRB flourish in water and crude oil pipelines and produce localized overlapping pits mainly at the bottom of the pipeline. (See section 7.12.)

7.2 Corrosion Mechanisms

7.2.1 The electrochemical basis of corrosion

Corrosion is an electrochemical phenomenon and, thus, strictly concerns metals. Non-metallic materials do not corrode in the strict sense, but nowadays the term is used as shorthand for describing the degradation of engineering materials. High temperature oxidation, for example, is considered as a corrosion process. The breakdown of a metal involves the reversion of the metal to an ore, perhaps similar, to the ore from which it was derived. This is not surprising because the main engineering metals represent a considerable investment of energy in their finished form. Because the process liberates energy, corrosion is inevitable: metal degradation is merely a matter of time. The best that can be done is to reduce the rate of reversion of the material.

The basis of the corrosion process is illustrated schematically in Figure 7–1, for a generalized case and Figure 7–2 for sweet corrosion. Metal dissociates as positively charged ions leaving electrons behind. The removal of electrons is critical to the kinetics of the corrosion reaction. Because the corrosion process can proceed in a fluid that is only loosely describable as water (e.g., oil emulsions, blood, strong brines, damp rust deposits), the term electrolyte is used to describe the external conductive medium. No corrosion will occur if an electrolyte is absent. Sites where metal goes into solution are termed *anodes*, while the areas where electron discharges occur are termed *cathodes*. These processes are not necessarily fixed at particular areas and may alternate. When anodes and cathodes regularly alternate, the result is general corrosion. If they become fixed then the anode will suffer continued attack, resulting in pitting or selective dissolution. Fixed anodes and cathodes are also typical of *galvanic corrosion*, which is the term used to describe the corrosion resulting from the electrical connection of two dissimilar metals.

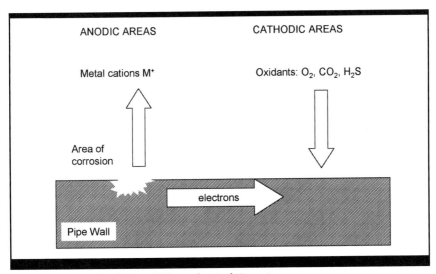

Fig. 7–1 *Generalized View of Electrochemical Corrosion*

Anodic reactions are the dissolution of the metal into the electrolyte as charged ionic species. Each metal molecule may yield one, two, three, or six electrons. The charged positive ions are solvated by water molecules and usually react with anionic species to form deposits, e.g., hydroxides, oxides, sulphates, and carbonates. The electrons left in the metal lattice are discharged at the cathodes by one of several possible reactions. For pipelines, the principal cathodic reactants are oxygen, carbonic acid, dissociated hydrogen sulphide, organic acids, and sometimes (inadvertently) oilfield chemicals.

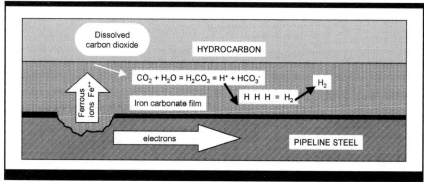

Fig. 7–2 *Sweet Corrosion Mechanism*

7.2.2 Corrosion morphology

Corrosion takes several forms, illustrated in Figure 7–3.

A chemical plant review indicated that about 30% of corrosion was general. However this form of corrosion is relatively rare in pipelines. Uniform and general corrosion are *safe* forms of corrosion in that they are readily measurable and control can be imposed. Sacrificial anodes used for cathodic protection of pipelines are designed to corrode in a uniform manner.

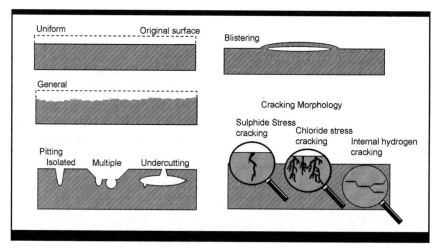

Fig. 7–3 *Corrosion Morphologies*

Localized corrosion is very common in pipelines. It is really a localized form of general attack and results from small variations in environmental/metallurgical conditions that become amplified by the corrosion process. Though prevention of this form of attack is relatively straightforward, it may be difficult to locate for measurement.

The distinction between localized corrosion and pitting is often confused.[2] True pitting is isolated attack where the main area of the metal is relatively unaffected. Pits in carbon steel tend to be hemispherical and often several pits overlap to produce a scalloped area of damage. Pits in corrosion resistant alloys (CRAs) are deeper than they are wide and represent a very small total surface area. There may be clusters of these. Crevice corrosion occurs in occluded areas, e.g., at partially detached coatings, gaskets and flange gaps, and occurs at similar rates to pitting. Many CRAs are susceptible.

Intergranular corrosion is rare in carbon steel pipelines unless there has been an inappropriate weld procedure. Sulphide and nitrate attack can cause this form of damage, though only sulphide damage is relevant in pipelines. Some CRAs show this form of damage as sensitization, e.g., type 316 stainless steel, resulting from chromium depletion in the heat affected zones (HAZ) of welds or as selective dissolution.

Stress cracking results from the conjoint action of stress and a specific set of environmental conditions. Pipeline carbon manganese steels can crack internally in sour service (hydrogen sulphide) or externally in carbonate soils. Type 316L, used as a liner, may crack in hot aerated chloride environments. Stress corrosion cracking is a high-risk form of corrosion attack and is prevented by correct materials selection, fabrication, and operation.

Blisters occur in steels in sour service resulting from the formation of hydrogen at unfavorable internal metallographic features in the steel. Corrosion reactions release hydrogen, and some of it can migrate as the atomic form into the steel where it can combine to form hydrogen gas at inclusions. The hydrogen cannot escape and collects to produce very high pressures resulting in the formation of blisters that distorts the steel microstructure.

Corrosion fatigue is rare on submarine pipelines but has been observed to occur on spanning pipe sections. Risers are at risk of low-cycle, high-stress fatigue from wave loading; therefore, design must obviate this risk. Cyclic stress has a significantly more deleterious effect on carbon steel in the presence of corrosive agents compared to air. There is no fatigue or endurance limit of steel in seawater, though this limit can be recovered by applying cathodic protection to the steel. Sulphide is a particularly aggressive species because it encourages hydrogen embrittlement of the steel, resulting in a reduction in fatigue resistance.

7.3 Sweet Corrosion

7.3.1 Corrosion mechanism

Sweet corrosion is described schematically in Figure 7–2. It occurs in systems containing carbon dioxide and the absence of or only a trace of hydrogen sulphide. The change from sweet to sour corrosion is not clearly defined. Sulphide cracking of steel occurs when the partial pressure of the hydrogen sulphide exceeds 0.34 kPa (about 50 ppm H_2S at a typical operating pressure of 7 MPa). Hydrogen blistering occurs at a partial pressure of 0.69 MPa (about 100 ppm of H_2S at 7 MPa). The generic types of corrosion inhibitor used for sweet systems are not suitable for sour service, and the change over concentration of hydrogen sulphide would be about 100 ppm.

Carbon dioxide is a highly soluble gas that produces acidity in solution: carbonated drinks have a pH of ~3. The acid formed can discharge electrons on the metal surface in several ways, and for this reason carbon dioxide is more corrosive than mineral acids of the same molarity. Corrosion increases as:

- Concentration of carbon dioxide increases
- System pressure increases
- Temperature increases

The corrosion process occurs in steps. Carbon dioxide dissolves in the water to form carbonic acid that dissociates to hydrogen ions and the bicarbonate anion. The hydrogen ions remove electrons from the metal surface, and the carbonic anion may also discharge an electron to form carbonate.

$$CO_2 + H_2O = H_2CO_3 \quad = \quad H_2CO_3 = H^+ + HCO_3^- \quad (7.1)$$

$$H^+ + electron = H \quad (7.2)$$

$$HCO_3^- + electron = H + CO_3^= \quad (7.3)$$

$$HCO_3^- + H^+ = H_2CO_3 \quad (7.4)$$

$$H + H = H_2 \quad (7.5)$$

On a bare metal surface, corrosion starts at a high rate; but the rate falls rapidly, within 24–48 hours, as films of corrosion product form on the metal surface. Surface films are formed by the reaction of the corroding steel with bicarbonate ions to form iron carbonate, termed *siderite*. The films are visible and appear as a pale brown tarnish on the metal. This is enough to reduce the corrosion rate by a factor of 5 to 10, depending on the local conditions. It is this stable corrosion rate that is used to evaluate the corrosion allowance for a pipeline. The reactions involved are shown schematically in Figure 7–2. Any process that aids formation and stability of the iron carbonate film will reduce corrosion, and any process or action that removes the films or prevents their formation will increase corrosion. In practice, the surface film is not a simple carbonate but includes hydroxides and oxides, and the steel surface may become enriched in the alloying elements in the steel.

A very simple rule of thumb used in low pressure, low temperature fields was:

- Partial pressure of carbon dioxide below 1 bara: low corrosion
- Partial pressure of carbon dioxide 1 to 2 bara: modest corrosion
- Partial pressure of carbon dioxide above 2 bara: high corrosion.

The partial pressure (psia/bara) is calculated from the molar, or volume, percentage multiplied by the total system pressure (bara).

$$\text{Partial Pressure } CO_2 = \text{System Pressure x mol \% } CO_2/100 \qquad (7.6)$$

This equation applies for systems below the bubble point where a free gas phase will be present. The European Federation of Corrosion documents suggest that at pressures above the bubble point the effective partial pressure can be determined from the bubble point pressure of the fluid at operating temperature and a determination of the mole fraction of carbon dioxide in the gas phase at the bubble point. The partial pressure is calculated as:

$$P_{bubble} \text{ x mole fraction } CO_2 \qquad (7.7)$$

For North Sea crudes and other similar light fluids the P_{bubble} can be estimated from the relationship:

$$P_{bubble} = P_s \exp[1/(t+273)] \qquad (7.8)$$

where

P_s is approximately 70 atm for these crudes

t is the temperature in °C.

Modest corrosion would be accommodated by increasing the pipe API 5L schedule by one increment and high corrosion by two increments. With high-grade steel in corrosive service, the percentage of the wall thickness provided for corrosion (the corrosion allowance, CA) can be a significant percentage of the wall thickness. To minimize the CA demands a more rigorous approach to evaluating the corrosion rate than these simple rules of thumb. The most common methods in use at present are the de Waard and Milliams (1993) algorithms that have been developed over time.[3, 4, 5, 6] The present basic corrosion rate equation follows:

$$\log_{10}(CR) = 5.8 - \frac{1710}{t + 273} + 0.67\log_{10}(fCO_2) \qquad (7.9)$$

where

CR = corrosion rate (mm/year)

t = temperature (°C)

fCO_2 = partial pressure of carbon dioxide adjusted for fugacity (bara)

The fugacity term, f_a, can be approximately calculated from the empirical equation:

$$f_a = \exp\left(-(606.92-99.719\ln(273+t))p\times10^{-4}\right) \qquad (7.10)$$

where:

t = temperature °C

P = absolute pressure bara.

A more precise value of the fugacity factor can be taken from fugacity tables of carbon dioxide.[7] An alternative approach is to factor the calculated corrosion rate based on the partial pressure, i.e., using partial pressure rather than fugacity in the de Waard and Milliams equation, then to multiply that calculated corrosion rate by F_{system}, calculated from:

$$log10 \ F \ System = 0.67\left(0.0031 - \frac{1.4}{t + 273} \right)P \qquad (7.11)$$

There are limitations to the reliability of these corrosion rate predictions, and several allowances must be made. First, the constants were derived from experimentation in weak brine (around 0.5% chloride) and in systems with low agitation. A BP-Amoco modification makes an allowance for the brine strength by reducing gas solubility in high concentration brines in proportion to the change in Henry's constant. The effect only becomes significant at brine concentrations $\geq$ 10% TDS.

The basic algorithm is considered to be valid up to about 60 °C. Above 60 °C, the corrosion rate of carbon steel reduces. This occurs because the nature of the siderite films alters, and the films become more dense and tenacious.[8,9] The temperature in °C at which the siderite film alters can be calculated using the following formula:

$$Tf = \frac{2400}{6.7 + 0.6log_{10}fCO_2} - 273 \qquad (7.12)$$

The reduction in corrosion with temperature was determined empirically and found to follow a law of the form:

$$log_{10} \ R_{temp} = min\left\{ 1, (\frac{1440}{273 + t} \ log_{10} \ fCO_2 - 6.7) \right\} \qquad (7.13)$$

where

R_{temp} =   the reduction factor

t =   temperature (°C).

The corrosion rate goes through a maximum value just before the formation of the more protective surface scales, and this temperature (°C) can be calculated from equation (7.12). An alternative reduction factor used by some operators is as follows:

$$\log_{10} R_{temp} = \min\left\{1, \left(\frac{2500}{273 + t} - 7.5\right)\right\} \qquad (7.14)$$

The corrosion rate calculated from the de Waard and Milliams algorithm is multiplied by R_{temp} to give the rate in the presence of the modified films. This effect, however, cannot be relied upon always unless the flow rates are modest, the temperature consistently high, the ferrous ion concentration is low, and the salinity of the water does not exceed 4%. If there is any doubt, it is conventional to consider the upper rate of corrosion as that at the filming temperature and to design the pipeline for this corrosion rate even if the temperature is higher, i.e., the beneficial effect on the corrosion product film is discounted.

If hydrogen sulphide is also present above modest concentrations, then the corrosion rate will be reduced. Hydrogen sulphide is more soluble than carbon dioxide, and the two gases compete. The presence of hydrogen sulphide modifies the corrosion product film; the films become sulphidic at about 100 ppm hydrogen sulphide in the water phase. The corrosion that occurs when sulphide films are formed is about 50% of that which would occur if the carbon dioxide alone were present. Below a hydrogen sulphide concentration of 5 ppm, any beneficial effect on corrosion rate is generally ignored. Several design contractors apply a blanket reduction factor (F_{H2S}) of 0.5, independent of the hydrogen sulphide concentration. An alternative formula derived from laboratory experimentation and which agrees in order of magnitude with Arabian Gulf field data is valid from ~20–20,000 ppm:

$$F_{H2S} = 0.0314 \, \ln(H_2S) + 0.0283 \qquad (7.15)$$

where H_2S is ppm (by volume).

The de Waard and Milliams corrosion rate is multiplied by F_{H2S}. At higher hydrogen sulphide concentrations, the corrosion rates can be much lower than the 30% predicted by this moderating equation, but there are no proven design procedures available at present. At hydrogen sulphide concentrations above 2% and where the ratio of hydrogen sulphide to carbon dioxide is greater than 1, the corrosion rate is reduced by a factor of about 5. Despite a reduction in general corrosion, the pitting rates remain about the same order of magnitude as if there were no hydrogen sulphide. Suppression of pitting requires that the pipeline is kept clean by frequent pigging and is continuously inhibited.

In oil pipelines, the partial pressure of the carbon dioxide is calculated from the conditions in the final separator. There, the oil is in equilibrium (more or less) with the gas at the particular separator conditions. After the separator, the partial pressure of the gas is not increased by an increase in the oil pressure for transmission through the pipeline. The worst-case corrosion in an oil line is somewhere downstream of the pumps when the temperature is at its maximum but where the water has settled out. In gas pipelines and multiphase flowlines, the partial pressure of carbon dioxide is calculated from the line operating pressure.

7.3.2 Effect of pH

The original algorithms described in section 7.3.1 were developed for gas pipelines where the free water was the condensed water vapor and, consequently, water without any dissolved buffering agents present. In multiphase systems, there may be formation water present that contains dissolved salts that act to buffer the pH. If the water pH is higher (less acidic) than would be the case with no buffering agents present, then the rate of corrosion will be reduced. The effect is quite marked as the pH is a log factor, and a shift of 1 pH unit represents a reduction in hydrogen ion concentration by a factor of 10. Alternative corrosion calculation procedures emphasize pH and water chemistry, rather than the partial pressure of carbon dioxide.[10, 11, 12, 13]

A pH factor can be calculated from the empirical formula:

$$\log F_{pH} = -0.13 \, (pH_{measured} - pH_{calculated})^{1.6} \qquad (7.16)$$

where the calculated pH is derived from

$$pH = 3.71 - 0.5 \log (PP_{CO2}) + 0.00471 \, t \qquad (7.17)$$

Whenever possible, the measured pH should be obtained from well fluids at the operational pressure and temperature, though obtaining this pH can be difficult. Where an actual pH cannot be obtained, the *Oil and Gas Journal* pH, or similar, can be used. See Equation 7.19.

The pH factor and the temperature scaling factor both should not be used to moderate the predicted corrosion rate, as they are not additive. It is usual to calculate both and to use the factor that has the largest moderating value.

In multiphase pipelines that operate in stratified flow, the corrosion rate in the water phase at the bottom of the pipeline is the rate calculated using the algorithms referenced earlier. In submarine pipelines, there will be water condensation on the upper surface of the pipeline; and this water will be unbuffered. The maximum corrosion rate would be that calculated using the above Equation 7.0 without pH adjustment; the moderation by hydrogen sulphide would be relevant. Usually, however, the rate of water condensation is low, the water becomes saturated with iron corrosion products. and the corrosion rate is moderated. To accommodate this, it is usual to take the corrosion rate at the top of the line (12 o'clock) to be about 10% of the rate at the bottom (6 o'clock). At high condensation rates, the corrosion rate would not be moderated.

7.3.3 Sweet corrosion morphology

Sweet corrosion results in pitting and localized attack in flowlines and pipelines. The damage is usually most severe at the bottom of the pipe (6 o'clock position) where water layers preferentially form. The surface of the pipe becomes covered with a siderite film, but local breakdown of this film can occur.

Certain metallographic features (e.g., surface-emergent manganese sulphides, weld spatter, and mill scale/oxidation films) or turbulence at weld defects or within pits prevent the adherence of the siderite on the metal. At these areas, corrosion may continue at a higher rate than where the siderite films are stable. The loss of metal results in increased turbulence and continued corrosion. The corrosion spreads as an even metal loss downstream of the initiation point. The metal appears as if it had been selectively milled in bands. This form of attack is characteristic of sweet corrosion and is called mesa corrosion (from the Spanish for table and the tableland topography of the American Southwest).[14] The process is shown schematically in Figure 7–4.

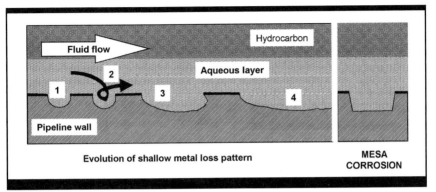

Fig. 7–4 *Sweet Corrosion Morphology*

A rough estimate of the corrosion morphology can be made from the calculated corrosion rate. The spread of corrosion downstream is approximately five times the rate of penetration, and the spread sideways is approximately twice the rate of penetration. The depth of water in the pipeline limits the lateral spread of corrosion. The longitudinal extent does not appear to be limited except, perhaps, by inclination of the pipeline. These values give some estimate of when the corrosion will be detected by the intelligent pigging and can give a guide to the applicability of defect dimensions in pressure retention calculations. Sweet corrosion may include isolated pitting.

For land pipelines the most extreme external corrosion frequently occurs at the 6 o'clock position, and the combined metal loss can result in rapid pipe perforation. There is no observed preferential position for external corrosion of submarine pipelines—so far.

7.4 Corrosion in Oil Pipelines

7.4.1 General

Empirical formulae are an alternative to the use of the de Waard and Milliams algorithm for high temperature oil flowlines. They are used because the corrosion rates obtained using the de Waard and Milliams algorithm are sometimes higher than the rates found in practice. Empirical formulae have been developed based on the assumption that the formation water will form protective scales on the metal surface. A typical empirical formula follows:

$$CR = 49.3 - \frac{8.06}{W_{pH}} + 0.03(\text{salinity}) \tag{7.18}$$

where

CR =   corrosion rate (mpy)

Salinity   =   salinity of the water (ppt)

W_{pH} =   the *Oil and Gas Journal* pH calculated from:[15]

$$WpH = \log_{10}\frac{[HCO_3]/61000}{ppCO_2} + 8.68 + \frac{4.05T}{10^3} + \frac{4.58T^2}{10^3} - \frac{3.07P}{10^5} - 0.477\sqrt{TDS} \tag{7.19}$$

where:

$[HCO_3]$ =   bicarbonate concentration in ppm

$ppCO_2$ =   partial pressure of carbon dioxide (psia)

T =   temperature in °F

P =   system pressure (psia)

TDS =   total dissolved solids as TDS/58,500 with the TDS in ppm.

Another factor that can reduce corrosion in oil pipelines is the wetting characteristic of the oil. Some oils form water repellent films on the surfaces of the steel, reducing corrosion. However, these films will only have a limited persistency, and the flow characteristics in the pipeline will need evaluation. Stratified flow will not allow the repeated formation of the oil films and corrosion would be expected.

Alternative models are available for making corrosion calculations.[16] The most widely known is the NORSOK model based on empirical data obtained by extensive flow rig testing while an alternative American commercial model is Predict. Both models are presently restricted to sweet corrosion. A later and more expensive model that takes hydrogen sulphide into account is the Electronic Corrosion Engineer. This model is restricted in throughputs and is more suitable for production flowlines than transmission pipelines. The corrosion rates calculated by these commercial models differ, sometimes by an order of magnitude. The models are not transparent and, therefore, should be used with caution and/or the corrosion rates validated.

7.4.2 Separation of water in pipelines

A long oil-service pipeline acts as a separator, and even small quantities of water will form a separate water phase at the bottom of the pipeline. Separators do not remove all the water from the oil. Some micro-droplets remain in suspension. There may be carry over of demulsifiers from the separation processes. Many corrosion inhibitors contain compounds that enhance emulsion breaking; therefore, the presence of inhibitor (including residuals from downhole treatment) may enhance water separation. During their transit along the pipeline, the droplets coalesce and eventually achieve a size that when they settle to the bottom of the pipeline, they form a water phase.

If the oil flow velocity is sufficiently high, the separated water is entrained with the oil resulting in the pipeline surface remaining oil wetted. Under such conditions, corrosion is unlikely to occur. Below a critical oil velocity, the oil-water shear forces are too low to sweep the water along the pipeline; and a semi-permanent water phase forms. If the water persists for long enough at the bottom of the pipeline, corrosion will occur. The velocity can be calculated if the physical properties of the water and oil are known.

The fluid properties vary with temperature; hence, for hot oil pipelines, it is usually necessary to perform a sequence of calculations over the range of temperatures determined to occur along the pipeline. A typical velocity value for water drop out in an oil pipeline at ambient temperature is 1 m/s. Some crude oils have viscosities that are not typical for their API values. These require caution in assessment.

Wicks and Fraser developed a method for calculating the critical entrainment velocity of oil in a pipeline.[17] Figure 7–5 describes the phenomenon schematically and indicates the velocities at which water will be entrained. The analysis indicates that the velocity increases with increasing pipeline diameter and that an oil flow rate above 1m/s may be safe for water cuts below 25%.

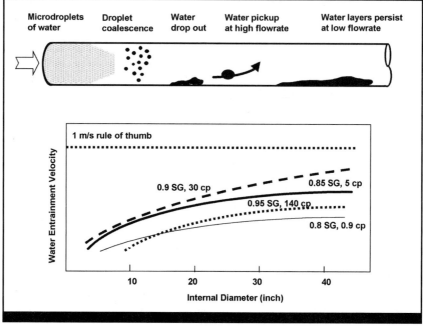

Fig. 7–5 *Water Entrainment in Oil Pipelines*

7.5 Effects of Flow on Corrosion in Oil Pipelines

7.5.1 Horizontal pipelines

Flow effects can be considered as very low flow, intermediate flow, and high flows.[18] At low flows, water may separate out to form a discrete phase; at intermediate flows, corrosion rates may accelerate slightly in proportion to flow rate; at high flows, erosion corrosion may occur. Pipelines are designed to operate in the intermediate flow regime.

In low flow rate pipelines, the corrosion rates can be high, and the corrosion tends to be pitting. Water layers and deposits may prevent inhibitor from reaching the pipeline wall. Most inhibitors are relatively ineffective in treating stagnant water layers compared to treatment efficiencies at intermediate flow rates.

In the intermediate flow range, the effect of the rate of flow on the corrosion rate is not well quantified. de Waard et al. have developed a model (using a Reynolds' analogy to account for flow effects), but to date this model has not been used to the exclusion of previous models. At lower rates of flow, the effect is small; corrosion does increase, but the siderite films form sufficiently fast to stifle excessive corrosion. At a critical value of flow, the siderite films are not formed rapidly enough; and the corrosion rate increases and becomes sensitive to flow, essentially marking the initiation of erosion-corrosion.[19] The shear stress imposed on the pipeline wall by the flow is calculated by the method given in chapter 9, Equation 9.12, determining the friction factor f from Reynolds' number Re and the chart in Figure 9–2.

The critical flow rate at the break point appears to depend on the corrosiveness of the environment including: temperature, carbonic acid concentration, ferrous iron concentration, and salinity. Norwegian research indicated that the uninhibited corrosion rate increases in proportion to $Re^{0.2}$ in the intermediate flow regime but in proportion to $Re^{0.8}$ above the critical rate.

Very high flow rates in topside piping are restricted by API RP-14E, which gives a formula for calculating the critical flow velocity at which erosion corrosion commences.[20] This formula is recognized as being overly conservative but is applied to pipelines in the absence of an alternative. The critical velocity is as follows:

$$U_{crit} = \frac{C}{\sqrt{\rho}} \qquad (7.20)$$

where

U_{crit} is measured in feet/s

ρ is the density in lb/ft^3

C is a constant that depends on the material.

For steel API RP-14E gives a value of 100, but field experience indicates this can be increased to between 100 and 125, between 300 to 350 for duplex, valued at 85 for cast steel, and at 195 for martensitic stainless steels. The relationship between density and API gravity is given in section C.6 in Appendix C.

For continuous flow, the lower value of the constant is used while for intermittent flow, the value may be increased. These values are for flow of solids-free fluids. If solids are present, erosion corrosion may occur at lower flow velocities. The constant value also appears to be sensitive to the corrosiveness of the fluids, being lower for highly corrosive fluids. The previously stated values give a reasonable first approximation.

For oil with a low water cut and a high velocity flow, water does not always separate from the oil as a separate phase; the water may remain in the main flow as distributed droplets. In these circumstances, the corrosion rates are very low. The critical flow velocity may be calculated using the algorithm of Wicks and Fraser.[17]

7.5.2 Effect of terrain/seabed profile

Multiphase flow is described further in section 9.4 of this chapter. Flow maps are generally derived for horizontal or vertical pipes representing ideal pipeline and riser cases. The maps are reliable for the riser, but in reality the seabed is rarely flat.[21] The flow regimes are markedly affected by pipeline inclination; the critical angle being 5°. At this inclination and above slug-flow becomes the dominant flow regime.[22] Plug flow is the alternative flow regime at lower gas velocities and high liquid velocities. At very high gas velocities, there is a transition to annular flow. The change in the flow regime has a marked effect on corrosion and its inhibition.

Slug flow is an intermittent flow regime that occurs at higher gas velocities. The slugs are large bubbles of gas with highly turbulent mixing zones where gas is entrained. The slugs impose very high shear forces on the bottom of the pipe that may be sufficient to remove the protective siderite films.

Test programs show that the corrosion rates may be orders of magnitude higher when slugs pass. The effect of slug frequency has been less well evaluated. The effect on corrosion is generally assumed to be directly related to the slug frequency, though this may not be correct at low slug frequencies as the time to reform a corrosion product film and/or a corrosion inhibitor film must be an important factor.

Work by Jepson indicated that at high slug frequencies of 50–90 per minute, there was a roughly linear relationship between slug frequency and corrosion rate.[21] Slug frequency increased with inclination and liquid velocity and decreased with increased gas velocity. Doubling the slug frequency resulted in about a 50% increase in corrosion rate. Increases in water cut increased the corrosion rate but did not alter the corrosion-slug frequency relationship. For a given inclination, there also appeared to be a critical water cut at which a stationary slug was formed.

An alternative approach is to evaluate the pipe wall shear stresses. If the stress is above 20 Pa, then the nature and application of inhibitor should be carefully evaluated.

7.5.3 Catenary risers

Catenary risers present an exceptional case of terrain effects because the inclination changes continuously from horizontal to vertical over a relatively long distance. Corrosion in these risers would be expected to follow a similar pattern to that noted in deviated production tubing, albeit production tubing is smaller diameter.[23] Analysis of corrosion indicates that the severity of corrosion is proportional to the sine of the deviation angle. In both vertical and deviated tubing, the worst corrosion occurs in the zone where the water hang up occurs. To move the water uphill requires a higher hydrocarbon flow velocity than is necessary in the horizontal case and, depending on the oil/gas flow velocity, there will be a particular zone where water hang up occurs. This zone moves slightly depending on operating conditions. Water accumulates until sufficient pressure builds up to cause periodic spasms of slugging.

In the catenary riser, water that is moved along the horizontal section of the pipeline by the shear forces imposed by movement of oil or gas necessarily cannot be moved up the riser because of the gravitational force resulting from the difference in oil/gas and water gravities. The corrosion in the riser would be expected, therefore, to be related to the local conditions (pressure and temperature) and where slugging occurs, multiplied by the slugging effect. Simple comparison with the deviated tubing case would suggest that the worst-case corrosion will be around the point of maximum curvature, i.e., 45°. In the absence of definite data, it would be prudent to assume that the de Waard and Milliams corrosion rate would be increased by 50%, accepting the effect of slugging as quantified by Jepson.[21] The same multiplier would apply to the availability method where a multiplier would be applied to the inhibited rate. Note, however, that the baseline corrosion rate in a riser will generally be low because the water chemistry will have been stabilized by corrosion upstream of the riser.

7.6 Solids in Oil Pipelines

The presence of solids, in particular sand, can alter the corrosion behavior by damaging the protective siderite films.[24] Corrosion can then be very severe, causing perforation in weeks. For a given flow, the damage is clearly more pronounced at the bends and other areas of high turbulence such as manifolds. It is wise to check the level of sand acceptable at the flow velocity expected at these areas. Usually small amounts of sand are tolerable, and typically 3–5 lb/1000 bbls oil would not be considered significant for a horizontal pipeline. Sand reduction methods need to be considered if the calculations suggest enhanced corrosion.

Erosion rates may be calculated from:

$$CR_{erosion} = \frac{K\,(0.65W)U^2\,\beta}{gP(\frac{\pi}{4}D^2)}.L \qquad (7.21)$$

where

$CR_{erosion}$	=	Erosion corrosion rate (mm/year)
K	=	Rabinowicz wear constant, and for steel pipelines is taken as 0.071.
W	=	Sand production rate (bbl/month); 1 bbl sand is 945 lbs
U	=	Average flow velocity (fps)
β =		Coefficient relating to the impingement angle: 1 for angles 10° to 60° and 0.5 otherwise, and is usually taken as 0.75 for pipelines.
g	=	Gravitational constant (32.2 f/s^2)
P	=	material penetration hardness, typically for steel 1.55 x 10^5 psi.
D	=	internal diameter of the pipe (in).
L	=	Correction factor to adjust for the units used (here L =1.36 x 10^8)

This equation can be modified to give the critical velocity (U_{crit}) at which the erosion corrosion is 0.25 mm/y (10 mpy):

$$U_{crit} = \frac{4D}{\sqrt{W}} \qquad (7.22)$$

Bends represent a special case. The present approach to predicting the erosion effect in a bend is to calculate a *stagnation length* for the particular geometry of bend. At a bend the flow must turn; but at the heel of the bend, the fluid forms a low flow zone locked at the surface by the force of the flow at right angles to the bend. An erosive particle must cross this stagnation zone to impact the pipe wall, and the distance it must cross is the stagnation length.[25]

If the stagnation length is long, the erosion corrosion rate will be reduced because the momentum of the particle will be reduced in its passage across the stagnation length. At low impact velocities, the protective scale remains intact; at high velocities, the scale is completely removed and general uniform corrosion occurs.

Experimental testing indicated that a sand production rate of 0.1 m³/day (200 bbl sand/month) in water would not cause excessive corrosion at 1.5 D bends if the flow velocity were < 5 m/s.[26] In methane the threshold velocities were very low, and no critical velocity could be reliably determined. The experimenters related this to the low density of the carrier fluid that would have a good correlation with pressurized gas. In crude oil the threshold values were very high, over 30 m/s. Because this would be above the API RP-14E critical velocity, the risk of sand erosion at elbows in crude oil pipelines can be discounted.

The main risk, therefore, would appear to be in gas and condensate lines. If sand production cannot be prevented, the design option would be to use the maximum diameter of bend or diameter of pipe in tees to augment the stagnation length. Even so, the metal loss may be too rapid to tolerate.

Though the metal loss calculation basis previously listed is loosely based on API RP-14E, it does highlight that the API RP-14E is not suitable for direct use in this context. The unmodified API formula predicts that the critical velocity decreases with fluid density, but experimental data show that, for a given rate of erosion, a higher sand content can be accommodated in a dense fluid compared to a light fluid.

7.7 Corrosion in Gas Pipelines

7.7.1 Corrosion rates in gas pipelines

Water occurs in gas pipelines because of condensation, determined by the water dew point of the gas, which is the temperature at which water condenses from the gas (not the same as the hydrocarbon dew point). If the gas is treated so that the water dew point is lower than the lowest gas temperature, water will not condense; and there will be no corrosion. Usually it is necessary to keep the humidity below 60–80% to prevent all corrosion.

Gas for transmission is generally dried to obviate corrosion problems and also to prevent the formation of hydrates. The gas temperature falls along the length of the pipeline because of the Joule-Thomson cooling described in chapter nine and may fall below the temperature of the surrounding water. A typical design dew point is 10 °C below the minimum design temperature.

Water contents need to be evaluated using the complete gas analysis as the higher molecular weight hydrocarbons reduce the dew point significantly. Appropriate allowance must also be made for high concentrations of carbon dioxide and hydrogen sulphide.

If the temperature falls below the water dew point, free condensed water will form on the top and sides of the pipe and run down the walls to the bottom. The temperature of the gas will fall rapidly downstream from the inlet, so corrosion rates tend to be high at the inlet and considerably lower for most of the pipeline length. Corrosion rates in gas pipelines are calculated using the de Waard and Milliams algorithm, equation (7.9), calculated at the local line operating pressure and temperature.

There is evidence that the water condensed on the top and walls of the pipeline becomes saturated with ferrous ion and less corrosive if the rate of condensation is low. The shift in pH because of iron saturation reduces the corrosion rate to about 10–20% of the calculated rate. This means that corrosion at the bottom of the pipeline (6 o'clock) position will be 5–10 times as severe as on the top and sides of the pipe. Caution is necessary in relying on this argument because high condensation rates may occur at low pipeline flowrates in cold waters.

7.7.2 Corrosion moderators in gas pipelines

The corrosion rate may be modified by operational conditions. Glycol drying is commonly used. The gas drying columns operate hot to improve efficiency. Gas entering the pipeline will be saturated with the equilibrium water content fixed by the design of the column and the exit temperature. If there is an upset in the glycol column operation, wet glycol may enter the pipeline. Water evaporated from the glycol/water mixture may condense at cooler areas in the pipeline. Some corrosion is possible where these condensed water films form. The rate of corrosion depends on the concentration of the water in the glycol: if it is below 10%, the corrosion rate becomes insignificant.

The water content in the glycol can be calculated as:

$$log_{10} W = 112.4 log_{10} P - \frac{2321}{T_{dpt} + 273} + A \, log_{10} P + B \qquad (7.23)$$

where

W =   water content in ppm vol

P =   total pressure (bar)

T_{dpt} =   dew point temperature (°C)

               for P < 27 bar A = -1.3   & B = 12.24

               for P > 27 bar A = -1.11 & B = 11.94

The equation is claimed to be valid up to 140 bar.

Even at high water contents, there is some reduction in corrosion if glycol or methanol is present. The residual corrosion is calculated by multiplying the de Waard and Milliams corrosion rate by the factor calculated from the following relationship for systems using ethylene glycols:[27]

$$log_{10} F_{glycol} = a \, log_{10} W - c \qquad (7.24)$$

where

W = % water in the water/glycol mixture

a = 0.7 and c is 1.4 for monoethylene and diethylene glycol (MEG & DEG)

a = 1.2 and c is 2.4 for triethylene glycol (TEG)

For a system using methanol, the flow pattern must be such that the fluids regularly and frequently wet the pipe walls. Methanol confers far less protection than the glycols.

7.8 Effects of Flow in Gas Pipelines

7.8.1 Fluid phase flow

The critical velocity defined in API RP-14E also applies in gas pipelines. The density of the fluid is the density of the gas at the operating pressure and temperature. Whereas in oil pipelines, the critical velocity is regarded as overly conservative. This is not the case in gas pipelines. Many corrosion inhibitors will not function efficiently at flow velocities exceeding 17 m/s, for at this velocity the shear forces on the pipe wall remove the inhibitor though leaving the surface water-wet as fresh water condenses constantly onto the wall. An alternative method of estimating the velocity limit for inhibitors is to calculate the pipe wall shear stress. The limiting shear stress is about 20 Pa. In gas pipelines, inhibitor addition to the gas stream can be ineffective, whatever the flow velocity, unless a high-pressure atomization technique is used.

Inhibitors are available for flow velocities up to 25 m/s, but high concentrations are required. Many operators periodically batch dose high velocity water-wet gas pipelines to ensure effective inhibition. This is done by driving a liquid slug, doused with inhibitor at high concentration, through the pipeline between spheres or bi-di pigs.

Gas carrying liquids may cause impingement attack at bends. This form of attack does not commence immediately. The repeated liquid impacts work-harden the surface of the metal and after a given time the metal surface begins to break down. One approach to calculating the corrosion rate under these circumstances is to use a similar equation to that used for sand entrainment in oil:

$$CR_{erosion} = \frac{Kv\,\rho\,V^2}{2Pg}\left[\frac{2\,\rho\,V^2}{27gPE_c^{\,2}}\right]^2\frac{1}{A} \qquad (7.25)$$

$CR_{erosion}$ = wear rate (mpy)
K = high–speed erosion coeficient, 0.01
v = impacting fluid volume rate (ft^3/s)
ρ = fluid density (lb/ft^3)
V = impact velocity of the fluid (fps)
P = hardness of the metal (1.55 x 10^5 psi)
G = gravitational constant (32.3 f/s^2)
E_c = critical strain to failure (0.1 for steel)
A = area of the pipeline affected (ft^2)

Given an erosion wear of 0.25 mm/y (10 mpy), the calculation can be simplified to give a critical velocity that should not be exceeded, which is equation (7.20) with C taken as 300.

7.8.2 Erosion corrosion in gas pipelines

Sand or other solids in gas pipelines result in high rates of corrosion. The rate is calculated for solids in oil pipelines, subject to the points in section 7.8.1. The low density of the gas means that the momentum of impinging particles remains high. The very erosive effect of particles in a gas flow can be used to in-situ grit blast long lengths of line prior to internally coating the line for refurbishment. The damage prevention options are as follows:

- Prevent entrainment of sand and solids in the gas phase
- Keep the flowing velocities low such that the protective films are not damaged
- Keep the flow velocity high so that uniform corrosion occurs, rather than pitting
- Design out problem areas by using more resistant materials of construction

Of these options, the prevention of solids entrainment is clearly the most attractive. This may mean internal grit blasting of the pipe to remove mill scale as well as paying attention to the completion techniques in the reservoir. Maintaining low flow velocities is not usually an option and may result in adverse performance of corrosion inhibitors. Maintaining a high flow velocity has the merit of allowing calculation of the life expectancy of the bend and is used in certain mineral processing industries that have a policy of regular replacement of bends, elbows and tees in their production plant, not an approach suitable for submarine pipelines. Materials selection is the fall back position if the entrainment option is not open or considered unreliable. Reliance on corrosion inhibitors to reduce the overall corrosion rate would be optimistic unless the problem is a spasmodic, discontinuous production of sand.

7.9 Sour Corrosion

7.9.1 Hydrogen sulphide

Hydrogen sulphide is a highly toxic and corrosive gas, having effects at ppm concentrations. It is soluble in hydrocarbons and water and will partition between these depending on local pressure, temperature, and pH. Other factors affecting solubility are the ratio of aromatic to aliphatic hydrocarbons in the crude oil and the salinity of the water. Solubility of hydrogen sulphide in highly aromatic crudes is increased by a factor of up to two.

Sour corrosion occurs in fluids containing hydrogen sulphide. The level of sulphide at which a fluid would be regarded as sour is normally defined by NACE MR-0175 as a partial pressure of hydrogen sulphide of 0.05 psia (0.34 kPa). The partial pressure is calculated from the ppm (volume basis) of hydrogen sulphide and the total system pressure:

$$\text{Partial Pressure } H_2S = (\text{ppmv } H_2S/10^6) \times P_{system} \qquad (7.26)$$

For pipelines operating above the bubble point, the partial pressure is calculated from the bubble point pressure at the operating temperature and the concentration of hydrogen sulphide determined from samples at this pressure and temperature. For North Sea crudes and other similar light crudes, the bubble point pressure can be calculated from the expression:

$$P_{bubble} = P_s \exp[1/(t + 273)] \qquad (7.27)$$

where P_s is typically of the order of 70 atm and t is in °C.

7.9.2 Sour corrosion morphology

Sulphide corrosion takes several forms:[28]

- Pitting attack from the deposition of solid sulphides
- Pitting attack at areas of breakdown of sulphide films formed on the pipeline surface
- Sulphide stress cracking (SSC)
- Hydrogen-induced cracking (HIC) and blistering
- Stress-oriented hydrogen induced cracking (SOHIC)

The interactions of sulphides and steels are complex. Figure 7–6 is a schematic of the different types of sulphide-related cracking corrosion.[29]

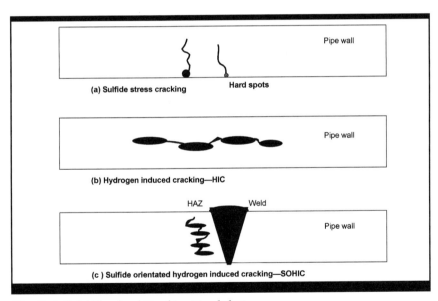

Fig. 7–6 *Sulphide-related Cracking Morphologies*

7.9.3 Sulphide pitting corrosion

Solid sulphides are formed by the reaction of the fluids with ferrous iron produced during corrosion or by reaction with heavy metals in the production fluids, though the latter would only occur when fluids are co-mingled. They are mostly iron sulphides, but manganese sulphide (pink) and zinc sulphide (white) may occur in pipelines and drilling mud. Each of the solid iron sulphides has a characteristic corrosiveness in that a given weight of a particular sulphide causes a given amount of steel corrosion.

After the iron sulphide has completed its quota of corrosion, it remains relatively inactive. It is supposed that inactivity results partly from hydrogen absorption into the sulphide lattice and partly from the formation of hydroxides that cocoon the sulphide. The SRB appear to be able to remove the hydrogen and reactivate the sulphides.

The relative corrosiveness of the different sulphides is given in Table 7–3. Iron sulphide exposed to acidic micro-aerobic environments may transform into the FeS_2 forms marcasite and pyrite; these materials are particularly corrosive.

Table 7–3 *Corrosiveness of Iron Sulphides*

Sulphide Species	Corrosion (gm) Per Mole of Iron Sulphide	Corrosion (gm) Per Sulphur Atom	Formula	%S
Pyrite	123.06	61.53	FeS_2	52.5
Greigite	50.12	12.53	Fe_3S_4	42.4
Smythite	78.04	19.51	Fe_3S_4	42.4
Mackinawite	10.08	10.08	$FeS_{(1-X)}$	35
Pyrrhotite	6.39	6.39	$Fe_{(1-X)}S$	36

The area of production of the sulphides may be upstream of where they settle and cause corrosion. For example, sour corrosion of production tubing would produce sulphides that could settle in the flowlines. This corrosion could be prevented by increasing the flow rate to reduce settlement and provision of corrosion inhibition downhole. At the pipeline terminal, there would need to be a vessel to collect the sulphide able to resist the corrosion, i.e., lined, coated, or provided with a corrosion allowance.[30] The use of corrosion inhibitors is often unsatisfactory as the sulphides present a massive chemically active surface area. One kg of freshly precipitated colloidal sulphide has a surface area of about 22 m².

Large quantities of sulphides may be recovered when wet, sour lines are pigged. The pig traps should be hosed to remove the sulphide to avoid pitting damage that is particularly severe if the wet sulphides are exposed to air because the sulphides oxidize to form sulphur and iron oxides.

Solid sulphides need careful storage because they are pyrophoric, i.e., they spontaneously ignite when dry. The best disposal policy appears to be storage under water to allow slow oxidation to sulphur and iron oxides. Dumping at sea around platforms should be avoided. The sulphides have a high chemical oxygen demand, which is environmentally destructive; and settlement of the sulphides onto the structural members of the offshore structures and sacrificial anodes would result in enhanced corrosion.

7.9.4 Sulphide films

In sour fluids with a low concentration of heavy metals, the hydrogen sulphide reacts with the metal surface to form adherent sulphide films. They are protective in the short term but can break down to reveal bare steel. Over time, breakdown occurs because the initial mackinawite converts to greigite and the change in density results in cracking of the sulphide film. Cracking tends to initiate at the grain boundaries where the initial sulphide films were weakest.

The iron sulphides form a galvanic couple when in contact with bare steel and cause rapid pitting attack. The bare steel area under the sulphides is a small anodic area compared to the large cathodic area of the sulphide. The corrosion rate may be too high for a protective sulphide film to be reformed.

In mildly soured systems, the films formed may be a mixture of siderite and iron sulphides. The percentage of sulphide in the films increases with increasing hydrogen sulphide concentration, and the films become completely sulphidic at about 100 ppm. Not surprisingly, this is the concentration at which blistering and hydrogen-induced cracking becomes problematical. At lower concentrations (<100 ppm H_2S), inhibitors selected for sweet corrosion mitigation are effective; at around 100 ppm it is necessary to consider re-selection of the inhibitor.

Low concentrations of hydrogen sulphide reduce the sweet corrosion rates, presumably by improving the tenacity of the siderite films. This effect is particularly noticeable at elevated temperatures. The protectiveness of the black siderite-sulphide films formed on downhole tubular goods is well recorded. The stability of the films is affected by salinity, temperature cycling, and metallographic details. However, this effect is unreliable in pipelines and may lead to pitting attack.

7.9.5 Sulphide Stress Cracking

Sulphide stress cracking is one form of Sulphide Stress Cracking (SCC) and was termed sulphide stress corrosion cracking (SSCC). SCC results from the joint action of stress and environment on a material. The stresses may be residual or externally applied or both together. Cracking is very specific to the environment. Table 7–4 gives examples of common SCC contexts. For example, brasses crack in ammoniacal solutions, and austenitic stainless steels crack in chloride environments at temperatures above 50–60 °C. The proposed mechanisms of SCC range from active-passive corrosion, coalescence of voids, and hydrogen embrittlement.

Table 7–4 *Stress Corrosion Cracking Environments*

Material	Environment	Other Factors
Carbon Steel	Carbonates in soils	Temperature + cathodic protection
Carbon Steel	Strong alkali	Temperature
Carbon Steel	Hydrogen sulphide	Hardness RC > 22
Austenitic Stainless Steels	Chloride	Temperature
Duplex Stainless Steels	Hydrogen sulphide	Temperature sensitive

Figure 7–7 illustrates schematically how SSC results from the combined action of stress and hydrogen generation in a sour environment acting on a high strength material. The minimum concentration of hydrogen sulphide for SSC is empirically established as a partial pressure of 0.05 psia, and free water must be present. SSC occurs when the material strength is above 725 MPa (105 ksi).

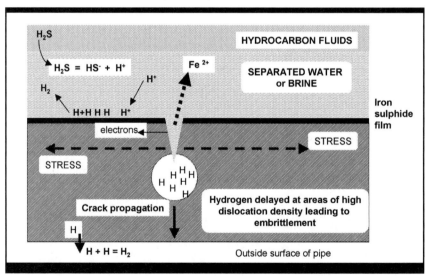

Fig. 7–7 *Mechanism of Sulphide Stress Cracking*

To permit non-destructive testing of materials, the limiting strength to avoid SSC is defined using hardness. In the context of pipeline steel in sour service, hard is defined as equal to or above Rockwell C 22; the equivalent Vickers Hardness value (10 kg weight) is 248; and the equivalent Brinell Hardness is 237.

The acid corrosion reactions (carbonic acid or hydrosulphide acid discharge) produce hydrogen on the metal surface. The hydrogen is formed in the following sequence:

1. Diffusion of the charged hydrogen ion to the metal surface
2. Discharge of the hydrogen ion by an electron to form atomic hydrogen
3. Surface migration of atomic hydrogen
4. Combination of atomic hydrogen to form molecular hydrogen
5. Solution or diffusion of the molecular hydrogen from the metal

Under normal conditions, about 98% of the atomic hydrogen generated combines to form gaseous hydrogen on the steel surface, 2% diffuses through the steel to the outer surface where it combines and dissipates as molecular hydrogen. On clean steel surfaces, the surface migration and combination of the hydrogen is rapid.

If there are iron sulphide films on the metal surface, the atomic hydrogen is held on the metal surface, resulting in up to 10% of the atomic hydrogen diffusing into the steel. The rate of hydrogen uptake is mainly determined by the concentration of hydrogen sulphide, the pH, and the temperature. Other parameters include carbon dioxide content (which affects the pH) water composition, fluid flow rate, and surface condition (presence and nature of rust, mill scale, inhibitor films).[31]

Atomic hydrogen migrates into the steel and concentrates at inclusions and *voids* in the steel. Voids are faults in the crystal lattice that can be filled by small interstitial atoms. Most voids occur at points of high stress where slippage of metal atoms has concentrated the dislocations in the crystal lattice as a result of yielding of the steel at high stress points.[32] The steel becomes embrittled because the hydrogen acts in a way similar to an alloying element and prevents further stress relaxation by movement of dislocations. Local micro-yielding does not occur, and the metal fails in a brittle manner once the stress exceeds a critical value.

The cracking process occurs in two stages: initiation followed by propagation. Unfortunately, neither process can be accurately quantified. For pipelines, materials of construction and fabrication methods are selected to avoid high strength (hard) materials or hardened areas susceptible to cracking. The controlling codes are NACE MR-0175 and ISO 15156, which are updated regularly to include details of the susceptibility of other materials to SSC.[33] For carbon manganese steels, the relevant hardness is Rockwell C 22 or Vickers Hardness value 248. This value is also quoted in other fabrication codes, e.g., BS 4515, though this BS code does allow a slightly higher hardness value of RC 26, which corresponds to VHN 275.

The critical level of hydrogen sulphide for SSC is defined as a partial pressure of 0.05 psia, so the concentration is sensitive to the system total pressure. At low pressures, the risk of SSC is small. The chart given in Figure 7–8 indicates the regions of risk of SSC for gas and multiphase systems.

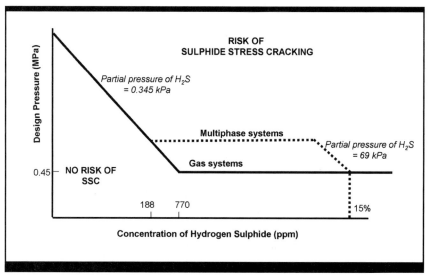

Fig. 7–8 *NACE MR 0175 Domains of SSC*

Recent research at TWI indicated that for pipe with a wall thickness above 0.4 in. (9.5 mm), the cap weld does not require to conform to the NACE value of RC 22 but can be raised to RC 30, equating to VHN 300. This relaxation would allow increased mechanization and speed of welding and may reduce overall welding costs, but to date this latitude has not been incorporated into NACE MR 0175.

ISO 15156 based on European Federation of Corrosion Publication 16 (1995) gives a less conservative evaluation.[34] Whereas NACE MR-0175 defines the domain of risk of SSC in terms of concentration of hydrogen sulphide and system pressure ISO 15156 defines the domains in terms of partial pressure of hydrogen sulphide and the pH. See Figure 7–9.

Embrittlement of steel by hydrogen is reduced at high operational temperatures because the diffusivity of hydrogen and the ductility of the steel are both increased. The risk of SSC is reduced, but this reduction cannot be relied on for pipelines because they may need to be shut down, causing the temperature to fall.

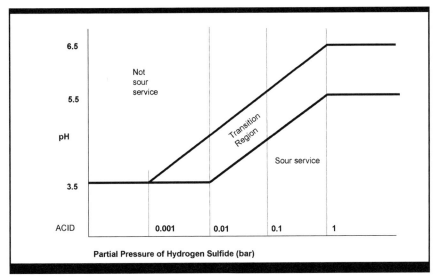

Fig. 7–9 *ISO 15156 Domains of SSC*

Line pipe steels are not of such strength that SSC is likely to happen in the parent plate except where the materials properties are altered by creation of hard zones adjacent to welds (both longitudinal and circumferential), subsequent to hot induction formation of bends and as a result of cold working.[35] It is against these local effects that precautions in fabrication are required. Other financial implications may occur when specifying valves and other pipeline components to conform to NACE MR-0175.

The welding procedure is the most important factor to control. It is rare that the requirement for a pipeline to conform to NACE MR 0175 has a significant impact on the cost of fabrication of a pipeline. EFC Publication 16 allowed welding consumables containing up to 2.2% nickel to improve toughness and strength provided hardness is limited to HV_{10} 250 in the root pass and 275 in the cap pass. The guidelines permit a relaxation of the hardness values of materials to be used for submarine pipelines protected by sacrificial anode CP, but caution is necessary if the pipelines are to be protected by an impressed current system because overprotection of the pipeline could charge the outer surface of the pipe with hydrogen. The acceptable hardness values for pipelines protected by impressed current CP are HV_{10} 260 in the root pass and 300 in the cap pass.

When measuring weld hardness values, it is usual to use micro-hardness techniques. For single QC tests, the hardness values should not exceed 10% of the specified value. If it does, additional measurements should be taken; and if any of these exceed the specified value, then the weld is not acceptable.

Cold working can resulting in work hardening and risk of SSC if the steel outer fiber deformation exceeds 5%. This is unlikely to occur with conventional fabrication techniques but may have an implication for novel reeling methods. To overcome the work hardening, the material needs to be heat treated by quench and tempering, normalizing or thermal stress relieving at above 620 °C. Thermal stress relief at 590 °C is inadequate. Micro-alloyed steels may require special treatments.

The pipe material itself is unlikely to require testing. Testing will be confined to weldments and post-formed sections of pipe (hot bends). Appropriate samples are cut from the pipe, and a range of methods used to apply a stress of 90 % yield (SMYS or actual) to the test pieces. Test methods include uniaxial tensiles, using smooth specimens, four-point bend, and C-ring testing.[36] For screening purposes, slow strain rate testing can be used with an extension rate of 2.5×10^{-5} mm/sec for a 25-mm gauge length.

The pipe specimen under test is submerged in a saline acid solution through which hydrogen sulphide gas is bubbled. The standard test solutions are detailed in NACE TM-0177 and are sodium acetate solutions with the pH adjusted to 3.5 for simulation of gas systems and 4.5 for oil systems (or, alternatively, actual or calculated pH values).[37] The pH values should be maintained within 0.1, and the temperature maintained at 23 ± 2 °C. This is the worst-case temperature for SCC: the risk of embrittlement reduces slowly at lower temperatures and more rapidly at higher tempera-tures. After the exposure period, the samples are examined metallo-graphically for cracking using 50x magnification. Slow strain rate samples are tested to failure, and the failure surface examined for evidence of brittle fracture and the percentage noted.

7.9.6 Hydrogen induced cracking

HIC is a form of blistering and is also called hydrogen pressure induced cracking, stepwise cracking, staircase cracking, and Cotton cracking (named after Henry Cotton, deceased). When corrosion occurs hydrogen ions are discharged on the metal surface to form atomic hydrogen that combines to form hydrogen gas. In the presence of iron sulphide films, the rate of combination of atomic hydrogen to form hydrogen gas is reduced, and up to 10 times more atomic hydrogen diffuses into the steel.

Manganese is added to steel to remove oxygen and sulphur and to improve toughness and strength. Despite best efforts, some sulphur remains in the steel combined with manganese as manganese sulphide inclusions. These are cast into the slabs of steel. Rolling stretches the inclusions into plates. Atomic hydrogen diffusing through the steel is adsorbed on the inclusions and combines to form hydrogen gas, at which point it cannot escape and accumulates to produce high pressures large enough to crack the steel internally. The hydrogen blisters grow horizontally and become interconnected by vertical cracking of the steel. Under low power microscopy, the blister-crack combinations appear as interconnected steps; and under higher magnification distortion of the metal structure can be seen at the steps. The cracking may take several forms, depending on where the inclusions are. Figure 7–10 shows typical morphologies.

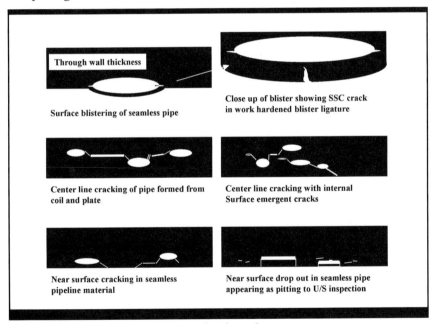

Fig. 7–10 *Morphology of Hydrogen Induced Cracking*

In contrast to SSC, HIC is a function of the metallurgy of the steel and is not affected by the level of stress. However, it can interact with SSC if the formation of a surface emergent blister work-hardens the surface ligament beyond Rockwell RC 22. Once the HIC cracking or secondary SSC cracks reach the inner surface of the blister, the sour environment can permeate into the blister leading to further cracking in the outer ligature of the steel. HIC damage occurs very rapidly if the steel is susceptible:

typically wet sour gas will cause extensive cracking within 48 hrs. After the initial flush of blisters and cracking, the process continues but at a decreasing rate.[38]

Seamless pipe is much less susceptible to HIC than pipe formed from plate, but it is not immune. The inclusions tend to be on the inner surface of the pipe, which reduces the effect on pipeline integrity but may interfere with ultrasonic monitoring and MFL intelligent pig inspection. Most companies impose restrictions on the sulphur content of the steel to prevent severe HIC of seamless pipe, typically 0.01%. For forgings, the sulphur content should be limited to 0.025% maximum.

A hydrogen sulphide partial pressure of 0.1 psia is generally taken as the value above which steels resistant to HIC will be required. Figure 7–11 is a schematic indication of the domains of severity of HIC for oil pipelines with fluid pH >4.5 and gas pipelines with fluid pH <4.5. The risk and severity of HIC increases with a lower pH. Condensed water in gas pipelines is usually more acidic than the formation water in oil pipelines for the same partial pressures of carbon dioxide and hydrogen sulphide because the formation water contains salts that buffer the pH. Pipeline material for sour gas service, therefore, normally needs a higher resistance to HIC than material for oil pipelines.

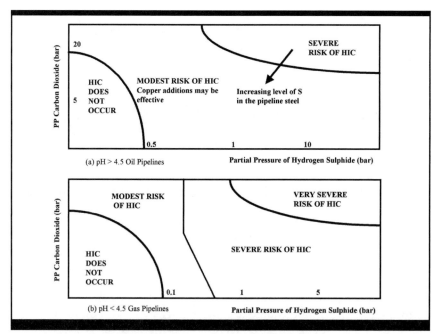

Fig. 7–11 *Conditions in which HIC is Important (Schematic)*

The steel metallurgy must be modified to increase resistance to HIC. The approach taken is to produce steel with a very low residual sulphur content by careful selection of primary ores and melt treatment. The shape of the manganese sulphide inclusions is also modified by secondary addition of calcium, which combines with the MnS and causes the MnS to form spheroids, rather than plates. As the sulphur content is reduced to increase resistance to HIC, so does the cost of the final plate; and it is, therefore, necessary to select a level of resistance to HIC appropriate to the service required.[39] Caution is needed when using the present NACE test procedures for steel selection, as these are very conservative.

The standard test method of testing steels for HIC is detailed in NACE TM-0284 and also in ISO 15156.[40] Testing may be of small samples cut from the pipe or of complete ring sections of pipe. Some testing has been done on complete joints pressurized up to operating pressure.[41] Small samples are taken from the start and end of each continuous casting sequence. Samples from slabs are taken from the highest and lowest pouring temperatures.

The faces of the samples are thoroughly cleaned and placed in the NACE TM-0177 test solution. The pH must be maintained within 0.1 units and the temperature at 23 ± 2 °C. The pH used is 4.5 for oil pipeline simulations or 3.5 for gas pipeline simulations. The concentration of H_2S should be at least 2,300 ppm. After exposure for 96 hrs, the samples are removed and inspected by sectioning and metallographic examination. Ultrasonic examination can be used to aid in determining where to section the samples, but doing so is not a good method for estimation of low levels of cracking. The crack lengths and locations are measured, and values of crack length ratio (CLR), crack sensitivity ratio (CSR), and crack thickness ratio (CTR) are calculated. Typical criteria, as ISO 15156, are as follows:

CLR ≤ 15%

CTR ≤ 3%

CSR ≤ 1.5%

Cracks close to the surface are ignored. The CTR is relative to sample thickness and can only be used as a comparative value.

In a full ring test, the inner surface of the pipe section is cleaned; and a mechanical jack is installed across the pipe diameter to impose a stress on the pipe. The top and bottom of the pipe are sealed, and NACE TM-0177 test fluid is introduced. Hydrogen sulphide is bubbled through the test solution. During and after the exposure period, the pipe is examined using ultrasonic inspection to evaluate the extent of cracking.

Pipelines suffering HIC attack may often be kept in service until a replacement line can be constructed, depending on the extent of cracking and the criticality of the line. Since the extent of the cracking throughout the length of the pipeline is unknown, it is usual to reduce the operating pressure because this reduces the corrosion rate by reducing the partial pressure of the corroding gases. However, for very severe corrosive conditions, this may have little more than psychological value. Each case must be considered on merit and attention paid to environmental and safety regulations.

Inhibition will reduce the corrosion rate and, therefore, the amount of hydrogen generated. Introducing the inhibitor in a liquid slug has been successful in markedly reducing the hydrogen migration into the steel. Recent developments in testing measure the volume of hydrogen diffusing through a steel membrane placed between a corrosion simulation cell and a measurement cell. As a result of the insights gained by use of these techniques, a new range of inhibitors has been developed. These inhibitors are particularly good at suppressing hydrogen entry into the steel by shielding the surface from the atomic hydrogen.

7.10 Water Injection Pipelines

Water is injected into oil reservoirs to maintain the pressure and to drive the oil towards the production wells. The principal corrosive agent in seawater is dissolved oxygen.[42] If aquifer or produced water is used, oxygen may be absent; but the water may contain carbonic acid and/or hydrogen sulphide. Their effects are discussed in sections 7.3 and 7.9 of this chapter.

Oxygen is removed from the seawater to be injected to reduce corrosion of pipelines and injection tubulars. The corrosion products of steel are very voluminous and could plug the injection wells. Oxygen is removed by gas stripping or, if the field has inadequate gas, by mechanical deaeration. Gas stripping, with the water running counter-current to the gas is an efficient process for removing oxygen but can result in acidification of the water if carbon dioxide is absorbed. Mechanical deaeration is achieved by extraction of the air from the water by vacuum on water percolating through a distribution column or, in some cases, by vigorous agitation of the water. Mechanical deaeration is less efficient than gas stripping and may need to be supplemented by chemical treatment with oxygen scavengers (sodium or ammonium bisulphite or sulphur dioxide). Both de-oxygenation processes make the water susceptible to growth of SRB.

Seawater is treated with chlorine to prevent the growth of marine organisms in the intake pipework and to oxidize dissolved organic material.[43] The concentration of residual chlorine entering the pipelines must be carefully controlled because it is an effective oxidant and can induce corrosion: four chlorine molecules are equivalent to one oxygen molecule. Residual chlorine will add to residual oxygen to create oxidant loading, which leads to corrosion. Using the equivalent oxygen concentration, the rate of corrosion of steel may be calculated using the Oldfield equation:[44]

$$CR_{ox} = \frac{0.0565C_0U_0}{Re^{0.125}\,Pr^{0.75}} \tag{7.28}$$

where

CR_{ox}	=	corrosion rate of the carbon steel (mm/year)
C_o	=	oxygen concentration (ppb)
U_o	=	flow velocity (cm/s)
Re	=	Reynolds number $\rho UD/\mu$ (see Chapter 9)
Pr	=	Prandtl number, listed for seawater in Table 7–5

Table 7–5 *Prandtl Numbers for Calculation of Corrosion Rates in Seawater Injection Pipelines*

Temperature	Prandtl Number	$Pr^{0.75}$
0	1313	218
10	875	161
20	600	121
30	410	91
40	296	71
50	213	58
60	160	45
70	117	36
80	91	29
90	69	24
100	55	20
111	41	16
120	33	14

Corrosion rates in water injection pipelines are quite high and, not surprisingly, most are thick-walled. The major corrosion occurs over the first few kilometers of the pipeline until the oxygen is depleted. Thereafter, the predominant form of corrosion in the injection pipelines is microbiological attack. Chlorine is not very effective against sulphate-reducing bacteria, and organic biocides are generally required to kill them.

7.11 Corrosion Inhibition

Corrosion can be markedly reduced by the addition of corrosion inhibitors.[45] The reduction of corrosion rate obtained by inhibition is typically 75–85% for gas pipelines and 85–95% for oil pipelines. The corrosion rate calculated using the de Waard and Milliams equation is multiplied by 0.25–0.15 or 0.15–0.05 to obtain the inhibited corrosion rate. These values are practical values that assume that the inhibitors will show a minimum of 95% efficiency in laboratory tests. The corrosion allowance is then calculated by multiplying the inhibited corrosion rate by the design life. Inhibitor efficiency is calculated as:

$$\text{corrosion inhibitor efficiency} = \frac{\text{corrosion rate without inhibitor} - \text{corrosion rate with inhibitor}}{\text{corrosion rate without inhibitor}} \times 100 \quad (7.29)$$

The alternative approach to evaluating the effect of inhibition is to assume that high quality inhibition can suppress corrosion into the range 0.1–0.2 mm/year and that failure to provide quality inhibition results in the occurrence of corrosion at the uninhibited rate. The annual corrosion rate is then based on the following equation that assumes an inhibited rate of 0.1 mm/year and the availability denoted by A (the uptime in which the inhibition injection system is operating, expressed as a fraction of the total time):

$$Annual\ corrosion = 0.1A + (1 - A)(uninhibited\ corrosion\ rate) \quad (7.30)$$

The corrosion allowance is calculated by multiplying the annual corrosion rate by the design life. Caution is necessary when taking this approach. Availability is generally restricted to 95%, which corresponds to

18 days per year downtime of the inhibition injection system. It is possible to design inhibition systems with 100% redundancy such that availability above 95% can be achieved. Other cautionary factors are operation at high temperatures, operation in a remote location where regular maintenance of inhibitor injection equipment is not possible, and high flow velocities where shear forces are close to inhibitor tolerance. A combination of the efficiency and availability methods usually provides a reasonable estimate of a corrosion allowance.

The efficiency of an inhibitor is markedly affected by the cleanliness of the pipe. Pipes containing a high level of debris (rust, mill scale, and solids from production) are more difficult to protect with inhibitor because the chemical is adsorbed onto the surfaces of the debris. This is a particular problem in sour systems where very large quantities of finely divided iron sulphides are formed. At about 100 ppm, the nature of the corrosion films alters from being predominantly iron carbonate to iron sulphide. Inhibitors for sour systems generally contain imidazolenes.

Inhibitors are also sensitive to flow rate and flow regime. Inhibitor efficiency is reduced in low and stagnant flow conditions.[46] In very high flow environments, the inhibitor efficiency is also affected: the inhibitor films are stripped from the metal surface by the shear forces exerted by the flow.[47] There is no typical flow rate—values of 17–20 m/s are often regarded as limits for many inhibitors, and candidate corrosion inhibitor must be tested. The critical pipe wall shear stress is reported to be 20 Pa; it can be calculated from the pressure drop. See chapter 9, Equations (9.4) and (9.12). For cases where stratified water layers will be persistent, the inhibitors recommended are generally oil-dispersible and water soluble so that the inhibitor partitions preferentially into the water phase. When water layers are not persistent, the more efficient oil soluble water dispersible or oil soluble inhibitors can be used, provided there is confidence that the oil phase will repeatedly wet the pipeline walls to ensure fresh inhibition of the metal surface.

Slug flow (described in chapter 9) reduces the effectiveness of corrosion inhibitors because the high shear forces arising from passage of the slug remove the inhibitor film from the pipe wall. One operator factors inhibitor efficiency in proportion to the slug frequency and assumes 90% inhibitor efficiency during periods when there are no slugs and 0% inhibitor efficiency during the passage of the slugs.

Based on total fluids, typical corrosion inhibition concentrations are 5–50 ppm for continuous addition and up to 250 ppm for batch dosing. Gas inhibitor dosage rates are in the range 0.25–0.75 litre/MMSCF of gas.

The effectiveness of the inhibitor is markedly affected by the concentration; therefore, the injection dosing rate must track the flowrate of fluids in the pipeline. Often the injection pump is set at an average value and not altered despite changes in operation.

Inhibitor is injected into an oil pipeline fluid stream using a quill. In gas pipelines, the method of addition of the inhibitor into the fluid stream needs careful attention. Most inhibitors must be diluted and injected by atomizer. Particular attention to inhibitor formulation is needed for hot gas pipelines as the diluent fluid may be flash evaporated, leaving the inhibitor as a tar-like gunk on the pipe wall. Inhibitor injection quills and atomizers can be inserted in and retrieved from the pipe through conventional 2-in. access fittings with a side entry pipe. Newer types of fittings include a non-return valve inside the fitting that can be inserted into normal 2-in. fittings.

There has to be a corrosion monitoring program to ensure that the inhibition program functions adequately. A program should include: regular analysis of the produced fluids (e.g., composition, iron counts, residual inhibitor), corrosion monitoring (e.g., weight loss coupons, electrical resistance), and periodic inspection using ultrasonic surveys or intelligent pigging. The frequency of the analyses and inspections is decided by the perceived risk resulting from corrosion and the measured corrosion rates, data that feeds back into the corrosion risk assessment.

7.12 Microbiological Corrosion

7.12.1 Bacterial growth and activity

Microbiological corrosion results from the activity and growth of SRB in pipelines.[48] In crude oil pipelines, the SRB flourish in stratified water layers and produce localized, overlapping pits predominantly at the 6 o'clock position. In water injection pipelines, the SRB colonies are distributed more widely around the pipeline, though the most severe corrosion is in the lower 30° section. In severe cases, pits may link up to form roughly continuous channels of corrosion along the bottom of the pipeline. Loss of integrity, therefore, may occur by either perforation or bursting. The sulphides produced by SRB activity and their rate of pitting attack are similar to those formed in sour fluids. (See section 7.9.) The risk of activity and growth of SRB in product pipelines can be evaluated from the parameters given in Table 7–6.

Table 7–6 *Growth Limitations for Sulphate-Reducing Bacteria*

Parameter	Critical Value
Water salinity	>15%
Carbon source	Not Present
Sulphate	<50 ppm
pH	<5 and >9.5
Temperature	>45 °C
Hydrogen sulphide	>300 ppm

The bacteria may grow in association with a wide range of other bacteria. SRB usually predominate in mature colonies because the dissolved sulphide they produce is toxic to other bacteria. Fatty acids are the food source for the SRB and are commonly present in formation waters in concentrations ranging from a few hundred ppm to 10,000 ppm. The SRB oxidize these acids to carbon dioxide, water, and reduced acids using the oxygen in the sulphate radical:

$$x.C_nH_mCOOH + y.SO_4^= =$$

$$a.CO2 + b.H_2O + x.C_{n-a}H_{m-2b}COOH + y.S^= \qquad (7.31)$$

Oxidation of organic material using sulphate is not a high-yield energy process compared to aerobic respiration. The bacteria must process large quantities of organic material and sulphate to obtain adequate energy for growth and reproduction. Because of the adverse energy gradient of the sulphate-reduction process, the bacteria have evolved control systems to prevent sulphate reduction in the presence of oxygen. Consequently, the SRB only grow in the absence of oxygen; however, oxygen does not kill them but merely inactivates them.

During active growth, the bacteria produce copious quantities of sulphide. Growth and activity in pipelines, however, does tend to be spasmodic as the bacteria take advantage of favorable local conditions. Sulphides are toxic to the SRB themselves, and there is an upper limit to the concentration of sulphide that the SRB can tolerate in the water. At neutral pH the upper tolerance is 300 ppm.

The bacteria can also utilize molecular hydrogen as an energy source, using the hydrogen to reduce sulphate to water and sulphide. The main corrosive agent is the sulphide; but, by utilizing hydrogen adsorbed on iron sulphides, the bacteria reactivate the sulphides and enhance the normal sulphide corrosion processes. Figure 7–12 is a schematic of the microbiological corrosion process.[49]

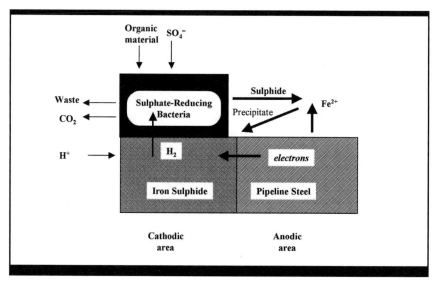

Fig. 7–12 *Corrosion Mechanism of Sulphate-Reducing Bacteria*

The solid sulphides generated by this mechanism act as an extended cathode. The microbiological corrosion rate tends to accelerate over time as the solid sulphides accumulate. There is normally an upper limit to the corrosion rate, which is fixed by diffusion through the sulphide mass; and the corrosion rate tends to settle to a linear rate. Corrosion pit depth is proportional to $(time)^n$, where the exponent n is between 0.9 and 1 (in contrast to 0.5 for oxygen corrosion).

7.12.2 Reservoir souring

Reservoir souring can occur in sweet systems where water injection is used for enhanced oil recovery. The growth of the bacteria in the reservoir causes a souring of the associated gas, and this may be sufficient to turn a sweet production system into a sour system. Cases have been recorded of initial hydrogen sulphide concentrations increasing from 1–2 ppm to 10,000 ppm. The very large increase occurs because the bulk of the hydrogen sulphide in the produced water is stripped into the produced gas in the separators. Though the sulphide concentration in the water may be ~30 ppm, the increasing volume of water and reducing volume of gas result in a large multiplier in concentration.

7.12.3 Counting the bacteria

Rapid enumeration techniques for counting bacteria have been developed but require consolidation, i.e., the bacteria are concentrated by filtration of large quantities of water. The conventional techniques are still the most widely used.[50] Bacteria in pipeline fluids are counted by serial dilution in a growth medium or by deep agar culture. The agar technique is useful for low population counts, though it is possible to concentrate the bacteria using a filtration technique (see NACE TM-0173). The serial dilution method and medium used are described in NACE E1/1 and API RP-38, though the latter document is not completely relevant to the offshore environment.[51] A 1-ml aliquot of the water sample is injected into a small vial containing 9 ml of growth medium. After shaking the bottle to distribute the bacteria, a 1-ml aliquot is taken from the vial using a new sterile syringe, and the aliquot injected into a second vial containing 9 ml of growth medium. The process is repeated up to six times. After all the dilution serial bottles are inoculated, they are incubated in a warm dark place and their condition reviewed each day. The growth medium contains dissolved ferrous iron, and the vials turn black when significant SRB growth occurs. In the deep agar technique, the 1-ml sample is dispersed in the liquid agar-based medium that is then allowed to cool. Each bacterium produces a colony that turns the local medium black.

After a fixed period (usually 10 days), the test is completed. The original number of bacteria is estimated by back counting from the last vial to turn black. Each dilution represents a 10^{-1} dilution of the original sample; if the first four vials turn black but the fifth and sixth do not, there must have been at least one bacterium in the fourth vial, so the original water sample contained at least 10^4 bacteria. If the tests are done in triplicate, statistical methods can be used to estimate the most probable number (MPN) of bacteria in the original water sample. With the agar technique, the number of black spots is recorded and represents a direct evaluation of the number of bacteria per ml of sample.

7.12.4 Treatment

If the number of the bacteria is high or is found to increase over time, the pipe may be treated with a biocide slug. Bacteria are not killed instantly and must be exposed for a minimum time to the biocide for the biocide to be effective. The time required depends on the concentration of the biocide. It is important to ensure the length of the slug and the concentration of the biocide are sufficient.

Many biocides are strongly acidic and affect the corrosion inhibitor films on the metal surface. After biocide dosing, it may be necessary to re-establish the inhibitor films by slug dosing with inhibitor or by increasing the inhibitor dosage rate threefold for a 24-hour period after the biocide treatment.

References

1   A. J. McMahon and Paisley, D. M. E. (1996) "Corrosion Prediction Modelling." BP Guidelines, Report Number ESR 96.ER.066, November.

2   ASTM G46: "Examination and Evaluation of Pitting Corrosion." American Society for Testing and Materials.

3   de Waard, C. & Milliams, D. E. (1975) "Carbonic Acid Corrosion of Steel," *Corrosion 31*, 131.

4   de Waard, C., D. E. Milliams, & U. Lotz. (1991) "Predictive Model for CO_2 Corrosion Engineering in Wet Natural Gas Pipelines." (Paper 577) Corrosion 91, National Association of Corrosion Engineers, Cincinnati, OH.

5   de Waard, C. & U. Lotz. (1993) "Prediction of CO_2 Corrosion of Carbon Steel." (Paper 14) Corrosion 93, National Association of Corrosion Engineers, Houston, TX.

6   de Waard, C., U. Lotz, & A. Dugstad. (1995) "Influence of Liquid Flow Velocity on CO_2 Corrosion: A Semi-Empirical Model." (Paper 128) Corrosion 95, National Association of Corrosion Engineers, Orlando, FL.

7   Newton, R. H. (1935) "Activity Coefficients of Gases." *Industrial and Engineering Chemistry*, March, 302–306.

8   Videm, K. & A. Dugstad. (1987) "Effect of Flowrate, pH, Fe_2+ Concentration and Steel Quality on the CO_2 Corrosion of Carbon Steels." (Paper 42) Corrosion 87, Houston, TX.

9   Dugstad, A. (1992) "The Importance of $FeCO_3$ Supersaturation on the CO_2 Corrosion of Carbon Steel." (Paper 14) Corrosion 92, National Association of Corrosion Engineers, Houston, TX.

10 Bonis, M. R. & J. L. Crolet. (1989) "Basics of the Prediction of the Risks of CO_2 Corrosion in Oil and Gas Wells." (Paper 466) Corrosion 88, National Association of Corrosion Engineers, New Orleans, LA.

11 Gunaltun, Y. (1991) "Carbon Dioxide Corrosion in Oil Wells." (Paper 21330) Joint SPE-BSE Conference, Society of Petroleum Engineers, Bahrain.

12 Jones, L. W. (1988) *Corrosion and Water Technology for Petroleum Producers*. 14–15. Tulsa: OGCI Publications.

13 Crolet, J. L. (1982) "Acid Corrosion in Wells (CO_2, H_2S): Metallurgical Aspects." (SPE 10045) Society of Petroleum Engineers, Beijing, March.

[14] Hausler, R.H. & H. P. Goddard, eds. (1984) *Corrosion in the Oil and Gas Industry.* Houston: National Association of Corrosion Engineers.

[15] Oddo, J. E. & M. B. Tomson (1982). "Simplified Calculation of pH and $CaCO_3$ Saturation at High Temperatures and Pressures in Brine Solutions." *Journal of Petroleum Technology,* 34, 1583–1590.

[16] Schmitt, G. (1983) "Fundamental Aspects of CO_2 Corrosion." (Paper 43) Corrosion 83, National Association of Corrosion Engineers, Houston.

[17] Wicks, M. & J. P. Fraser. (1975) "Entrainment of Water by Flowing Oil." *Materials Performance,* May, 9–12.

[18] Burke, Pat A., A. I. Asphahani, & B. S. Wright, eds. (1985) *Advances in CO_2 Corrosion: Vol. 2.* Houston: National Association of Corrosion Engineers.

[19] Eriksrud, E. & T. Sentvedt. (1983) "Effect of Flow on CO_2 Corrosion Rates in Real and Synthetic Formation Waters." Proceedings of the Advances in CO_2 Corrosion, 1, Corrosion 83, Symposium on CO_2, National Association of Corrosion Engineers.

[20] API RP14E: "Recommended Practice for Design and Installation of Offshore Production Platform Pipeline Systems." American Petroleum Institute.

[21] Jepson, W. P. (1996) "Study Looks at Corrosion in Hilly Terrain Pipe Lines." *Pipeline and Gas Industry,* 8, 27–32.

[22] Vedapuri. D. (1997) "Studies on Oil-Water Flow in Inclined Systems." Report Section 9, University of Ohio Multiphase Flow and Corrosion Project.

[23] Thomas. M. J. J. S & P. B. Herbert. (1995) "CO_2 Corrosion in Gas Production Wells: Correlation of Prediction and Field Experience." (Paper 121), Corrosion 95, National Association of Corrosion Engineers, Orlando, FL.

[24] Salama, M. M. & E. S. Venkatesh. (1983) "Evaluation of API RP14E Erosion Velocity Limitations for Offshore Gas Wells." (OTC 4485) Proceedings of the Offshore Technology Conference, Houston, TX.

[25] Shadley, J. R., S, A. Shirazi, E. Dayalan, M. Ismail, & E. F. Rybicki. (1996) "Erosion Corrosion of a Carbon Steel Elbow in a Carbon Dioxide Environment." *Corrosion,* 714–723, September.

[26] Shirazi, S. A., B. S. McLaury, J. R. Shadley, & E. F. Rybicki. (1994) "Generalisation of the API RP14E Guideline for Erosive Services." (SPE 28518), Society of Petroleum Engineers, New Orleans, September.

27 Van Gelder, K. (1989) "Inhibition of CO_2 Corrosion in Wet-Gas Lines by Continuous Injection of a Glycol-soluble Inhibitor." *Materials Performance,* 50–55, July.

28 Kane, R. D. (1985). "Roles of H2S in Behavior of Engineering Alloys." *International Metals Review,* 30, 291–301.

29 King, R. A. (1992) "Sour Service: A Review of the Chemical and Electrochemical Behavior of Sulphides." Proceedings of Environmental Management and Maintenance of Hydrocarbon Storage Tanks, BSI Conference, London.

30 Horner, R. A. (1996) "The Technical Integrity Management of Sour Service Ageing Facilities." Proceedings of the UK Corrosion Conference, London.

31 Kane, R. D. & D. J. Schofield. (1992) "Development and Application of NACE Standard MR0175 for Selection of Materials for Sour Oil and Gas Service." Proceedings of the Conference on Redefining International Standards and Practices in the Oil and Gas Industry, IIR, London.

32 Warren, D. (1987) "Hydrogen Effects on Steel." *Materials Performance,* 26 , 38–47.

33 NACE Materials Requirement 0175: "Sulfide Stress Cracking Resistant Metallic Materials for Oil Field Equipment." National Association of Corrosion Engineers.

34 ISO 15156: "Guidelines on Materials Requirements for Carbon and Low Alloy Steels for Hydrogen Sulphide Containing Environments in Oil and Gas Production." International Organization for Standardization.

35 NACE (1993) "SSC Resistance of Pipeline Welds." Report of NACE Task Group T-1F-23, Chair Bruno, T.V., Materials Performance, 58–64, January.

36 ASTM G30: "Making and Using U-Bend Stress Corrosion Cracking Specimens." American Society for Testing and Materials.

37 NACE TM0177: "Testing of Metals for Resistance to Sulfide Stress Cracking at Ambient Temperatures." National Association of Corrosion Engineers.

38 Kushida, T. T. Kudo, I. Tamato, T. Kobayashi, & I. Sakaguchi. (1995) "Evaluation of Line Pipe for Sour Service by Full Ring Test." Proceedings of the7th NACE-BSE Conference, Bahrain.

39 Cornelius, O. E. & F. Borouky. (2000) "Optimisation of Carbon Steels for Sour Service." ADIPEC 0961, Abu Dhabi, October.

40 NACE TM0284: "Evaluation of Pipeline Steels for Resistance to Stepwise Cracking." National Association of Corrosion Engineers.

41 Matsumoto, K., Y. Kobayashi, K. Ume, K. Murakami, K. Taira, & K. Arikata. (1986) "Hydrogen Induced Cracking Susceptibility of High-Strength Line Pipe Steels." Corrosion, 42, (3), 337–348.

42 NACE RP0475: "Selection of Metallic Materials to be Used in All Phases of Water Handling for Injection into Oil Bearing Formations." National Association of Corrosion Engineers.

43 NACE TM0173: "Methods for Determining Water Quality for Subsurface Injection Using Membrane Filters." National Association of Corrosion Engineers.

44 Oldfield, J. (1982) "Corrosion of Steel in Water for Injection." Proceedings of the 1st NACE-BSE Corrosion Conference, Bahrain.

45 NACE RP0175: "Control of Internal Corrosion in Steel Pipelines and Piping Systems." National Association of Corrosion Engineers.

46 Prodger, E. M. (1992) "An Overview of the Selection Procedures Employed for Oilfield Chemicals." Proceedings of the Conference on Redefining International Standards and Practices in the Oil and Gas Industry, IIR, London.

47 Harrop, D. (1992) "Inhibitor Test Methodologies." Proceedings of the Conference on Redefining International Standards and Practices in the Oil and Gas Industry, IIR, London.

48 NACE (1976) "The Role of Bacteria in the Corrosion of Oilfield Equipment." TPC 3. National Association of Corrosion Engineers.

49 King, R. A. & J. D. A. Miller. (1971) "Corrosion by the Sulphate-Reducing Bacteria." Nature, 233, 491–492, number 5320.

50 Tatnall. R. E., K. M. Stanton, & R. C. Ebersole. (1988) "Testing for the Presence of Sulfate-Reducing Bacteria." Materials Performance, 71–80, August.

51 NACE CCEJV E1/1: "Review of Current Practices for Monitoring Bacterial Growth in Oilfield Systems." National Association of Corrosion Engineers.

8 External Corrosion, Coatings, Cathodic Protection, and Concrete

8.1 External Corrosion

8.1.1 Corrosion mechanisms

Corrosion is the result of two separate reaction processes on a metal surface: the loss of metal and production of electrons at anodic areas and the consumption of these electrons at cathodic areas. Hence, the overall rate of corrosion on the pipeline external surface is dictated by the ratio of anode area to cathode area, the concentration of cathodic reactant and, to a lesser degree, the resistivity of the local environment, which determines the rate of transport of ions between the anodic and cathodic areas. The corrosion process is the dissolution of the iron of the pipeline at the anodic areas as charged positive ions into the seawater or seabed sediment. These ferrous ions react to form oxides and hydroxides and may form ferric salts if the water is well oxygenated. For the corrosion to continue, the electrons remaining on the metal surface must be removed by a cathodic reaction. Typical cathodic reactions are hydrogen evolution and oxygen reduction. Because seawater has an alkaline pH of 8.2 or above, the principal cathodic reaction in seawater is oxygen reduction; and the corrosion reactions are the following:

$$Fe \rightarrow Fe^{++} + 2 \text{ electrons} \qquad \text{anodic reaction} \qquad (8.1)$$

$$O_2 + 4H_2O + 4 \text{ electrons} \rightarrow 4 (OH)^- \quad \text{cathodic reaction} \qquad (8.2)$$

The overall reaction is:

$$2Fe + O_2 + 4H_2O \rightarrow 2Fe(OH)_2 \qquad (8.3)$$

This oversimplifies the corrosion reaction because it is uncertain which process occurs first: iron dissolution or removal of electrons, which must then be provided by iron dissolution. Iron hydroxide does not always form on the pipeline surface and/or can also react with oxygen in the seawater to form a range of iron oxides and hydroxides. The oxides found on pipelines are black magnetite (Fe_2O_3), white to green lepidiotite $(FeOH)$, and mixtures of brown to red ferrous and ferric compounds.

The higher the availability of the oxygen to the metal surface, the higher the potential rate of corrosion. Oxygen access to a bare metal surface increases as the temperature of the water decreases or as the flow rate over the surface increases. Pipelines are, therefore, at highest risk of corrosion in cold water moving at a high velocity. Table 8–1 gives theoretical corrosion rates for steel in seawater over a range of temperatures, and these rates are comparable with actual rates taken from corrosion of sheet steel piling.[1]

Table 8–1 *Potential Corrosion Rates of Steel in Seawater*

Seawater Velocity (m/s)	Centralise Potential Corrosion Rate (mm/yy)				
	Oxygen Concentration (ppm)				
	6	7	8	9	10
0	0.08	0.90	0.11	0.12	0.13
0.3	0.09	0.11	0.12	0.14	0.15
0.6	0.10	0.12	0.14	0.16	0.17
1	0.12	0.14	0.16	0.18	0.20
2	0.16	0.19	0.21	0.24	0.27

8.1.2 Anaerobic corrosion

Section 8.1.1 suggests that pipelines buried under seabed sediment ought to be at little risk of corrosion because the oxygen content of the water would be low. Usually this is true, but the exceptions are the presence of highly organic sediments and sediments containing sulphate-reducing bacteria (SRB). The most common organic sediments are found in coastal marshes, mangrove swamps, and some river delta mud, though areas of mud also occur in deep offshore water. Organic sediments contain high concentrations of organic acids that can dissociate near the pipe to form hydrogen ions that act as cathodic reactants to remove electrons and form hydrogen gas. The following shows this reaction:

$$C_nH_mCOOH \rightarrow C_nH_mCOO^- + H^+ \tag{8.4}$$

$$2H^+ + 2\text{ electrons} \rightarrow H_2 \tag{8.5}$$

These sediments may also encourage and foster the activity of the SRB. Corrosion by SRB is discussed in detail in section 7.12.

SRB utilize organic acids as their food source and oxidize them using the oxygen in the sulphate radical. The energy gain in this process is small, and so the SRB have to process large quantities of organic material and sulphate to obtain sufficient energy for activity and growth. Not surprisingly they produce large quantities of sulphide. The overall process is as follows:

$$\text{organic acid} + SO_4^= \rightarrow CO_2 + \text{acetates} + S^= \tag{8.6}$$

The sulphide may be of the form $S^=$ or HS^-, the latter more likely at seawater pH levels. The sulphides react with the pipeline steel to form metastable iron sulphide films. At first these films reduce corrosion, but in time they become less protective; and severe local attack occurs where the film breaks down. Corrosion rates in SRB active sediments are less severe than potential corrosion rates in aerated seawater but are still sufficiently high to perforate a pipeline within five or six years.[2]

8.1.3 Influence of environmental factors

Both the coating breakdown assessment and cathodic protection (CP) design require some knowledge of the seabed sediment conditions. Estimation of sediment corrosiveness is difficult and rather rough rules of thumb are used to characterize sediments. Several studies on soil parameters indicate that the factors most likely to predict sediment corrosiveness are resistivity, redox potential, and salinity.

The resistivity of seabed sediments is measured on cores by inserting four in-line steel electrodes into the sediment, imposing an alternating current (AC) voltage between the outer electrodes and measuring the voltage drop across the center two electrodes. Sediment resistivities range widely from 0.25 to 25 Ωm. Even nominally homogeneous sediments show a normal distribution of resistivity that is sufficient to result in uneven loads on the individual sacrificial anodes in a CP system. Additionally, the pipeline coating may be damaged more in some places and less in others. So that the load on the anodes is evened out, the spacing between anodes is 12–15 joints (150 to 200 m).

The redox potential reflects the balance between reducing conditions and oxidizing conditions. A high positive potential indicates that an environment is oxidative, and a negative or low positive potential indicates a reducing environment. A typical oxidative environment is +300 Mv in aerated water or sediment. The redox potential can be used to determine whether sediment is likely to support the growth of sulphate-reducing bacteria. A high risk of microbial activity is shown by a low positive potential, below +100 Mv. The redox potential is measured by driving a platinum-tipped stake into the sediment to the depth at which the pipeline will be placed or into a freshly recovered core taken from the depth of burial of the pipeline. The potential of the platinum is measured against a reference electrode. A sample of sediment is taken and the pH measured. The measured potential is then converted to a base potential of pH 7 giving the redox potential.

The concentration of salts in the environment and the local temperature affect the resistivity and pH, and, hence, both the potential corrosiveness of the environment and the coating degradation behavior. Environments with particularly high chloride or low sulphate levels compared to seawater are at higher than normal risk of corrosion as the corrosion products of iron will be more soluble. Activity of the SRB also alters with salinity and temperature. Salinity changes occur in shallow inshore waters in hot climates (e.g., the south eastern sector of the Arabian Gulf) and in estuaries where the seawater is diluted by river water.

8.2 External Coatings

8.2.1 Introduction

Oxygen corrosion, organic acid attack and microbiological corrosion can all be prevented by a combination of external coating and CP.

Though it is possible to protect a bare pipeline by CP alone, the cost of the protection current would be extremely high. It is much less costly to protect the pipeline with a coating and to apply CP to make up for coating defects, damage, and degradation over time.[3] This strategy was realized in the very early days of pipelining, and protective coatings of modified tar and pitch-based paints were used. Those coatings were found not to protect satisfactorily in the long term and gave way to tape wraps, which were developed for land pipelines and could be applied *over the ditch* just before the pipeline was lowered in.

It was soon realized that many tape-wrap coating systems had significant drawbacks. Almost all submarine pipelines use factory-applied continuous coatings, though tape wraps are sometimes used for river crossings. There has been continuous improvement in coating systems, and modern coatings are surprisingly sophisticated.

The purpose of the coating is to isolate the pipeline steel from the soil and seawater and to present a high resistance path between anodic and cathodic areas. To perform these functions, the coatings must have a complex blend of properties, among them:

- Low permeability to water and salts
- Low permeability to oxygen
- Good adhesion to the pipeline steel
- Adequate temperature stability
- Ease of application
- An acceptable unit price (since much coating is needed even for a modest pipeline)
- Flexibility to accommodate strains imposed during laying, reeling, or towing,
- Resistance to biodegradation

- Ease of patch repair at areas of damage
- Non-toxicity, environmental neutrality, safety in application and handling
- Ultraviolet (UV) stability for the period during storage
- Resistance to cathodic disbondment

Most of these requirements are common sense. They are discussed in more detail in the following.

Fewer coating systems are more suitable for marine pipelines than for onshore pipelines as tape wraps and solvent-based coatings are not used.[4,5] The principal coatings, in rough order of cost, are as follows:

- Asphalt
- Coal tar enamel
- Fusion bonded epoxy (FBE)
- Cigarette wrap polyethylene (PE)
- Extruded thermoplastic PE and polypropylene (PP)
- Elastomeric coatings: polychloroprene and ethylene propylene diamine (EPDM)

8.2.2 Asphalt and coal tar enamel

Asphalt, bitumen, and coal tar enamel (CTE) coatings are *flood* coatings applied as a molten material to a rotating length of pipe.[6] They are typically 5–6 mm thick, have relatively poor adhesion to steel, and are not strongly coherent. To ensure adhesion, the pipe surface must be adequately roughened during the cleaning process. These coatings are reinforced to secure greater cohesion, with a single or double layer of fiberglass matting introduced into the centre of the coating during the flooding of the coating onto the pipe. Figure 8–1 is a schematic of the process. An alternative process applies the molten coating to the primed pipeline surface and then wraps an outer reinforcement into it.

Asphalt and coal tar enamel coatings are invariably overcoated with a concrete weight coating that affords mechanical protection. Asphalt and CTE are the cheapest continuous coatings available. Asphalt enamels will tolerate 65–75 °C and CTE 70–80 °C, though it is important to verify the specification of the material. Some special order enamels tolerate higher temperatures.

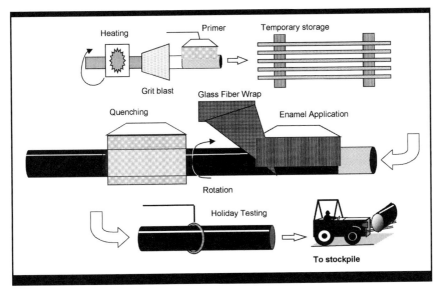

Fig. 8–1 *Flood Coatings*

8.2.3 Polyethylene wraps

Polyethylene wrap was developed in the late 1970s and has been adopted by German and Japanese producers. It is a moderately thick coating, typically 3–4 mm. The usual coating procedure is to apply an epoxy primer to the pipe, followed by an amorphous polyethylene based adhesive over which one or two layers of polyethylene sheet are applied as a cigarette wrap. The coating is termed a triple coating, 3PE or 3LPE. The epoxy primer is a very thin coating, typically 75 microns thick; hence, it does not completely cover the surface. Polyethylene has a high electrical resistance, very low moisture uptake, and a long life expectancy.

The upper temperature tolerance of polyethylene wrap is about 65 °C, a factor that reflects the tolerance of the adhesive. The main concerns from a materials viewpoint are undercoating corrosion, resulting from blisters, and inadequate CP current access because of the high electrical resistance. Polyethylene coatings may be covered with a concrete weight coating and are also suitable for use on reeled pipelines that cannot be coated with conventional concrete. The polyethylene coating is slippery, and when a concrete weight coating is applied on top, it is usual to provide anti-slip bands to reduce the risk of slippage between the concrete and the coating. The bands are 1 m lengths of sprayed material containing sand applied to both ends of each pipe length.

8.2.4 Thermoplastic extrusion coatings

Thermoplastic extrusion coatings are polyethylene or polypropylene and are moderately thick, typically 3–4 mm.[7] The pipe is coated with a primer coating of epoxy, applied by spray or fusion-bonded onto a heated pipe. The adhesive is then extruded over the primer, followed by the molten thermoplastic sheath. The epoxy primer is a thin coating approximately 75 microns thick. Figure 8–2 is a schematic of the application process.

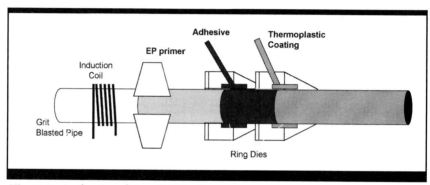

Fig. 8–2 *Application of Extruded Coatings*

A polyethylene coating can tolerate 65 °C, limited by the tolerance of the adhesive. Polypropylene coatings tolerate up to 105 °C. These coatings may be overcoated with a concrete weight coating and are also suitable as uncovered coatings for use on reel-laid pipelines. Polyethylene and polypropylene coatings are slippery and, when a concrete weight coating is applied, it is necessary to provide anti-slip bands to reduce risk of slippage of the concrete cover over the anti-corrosion coating. The three-layer coatings are more expensive than enamels, having a cost ratio of 1.8 to 2. Cigarette wrap coatings are cheaper than three-layer extruded thermoplastic coatings. The price has drifted upward over time, and they are now about 25% more expensive than fusion bonded coatings.

8.2.5 FBE coatings

FBE coatings are thin film coatings, 0.5–0.6 mm thick. They were developed separately in the United States and UK. The epoxy has a very strong chemical bond to the steel that provides good adhesion. FBE is also flexible and is the most commonly used coating for reeled pipelines. In the United States, the clean pipe is heated to 250–260 °C, and the epoxy powder is applied directly to the hot rotating pipe as a fine powder. The epoxy particles melt and flow over the hot pipe surface. Immediately after coating, the pipe is cooled by water quenching. In the UK and Europe, the clean pipe is coated first with a chromate etch primer before being heated for coating. The primer micro-etches the steel surface and increases coating adhesion. Improved adhesion permits a broader range of epoxy formulations to be used. Figure 8–3 illustrates the coating process.

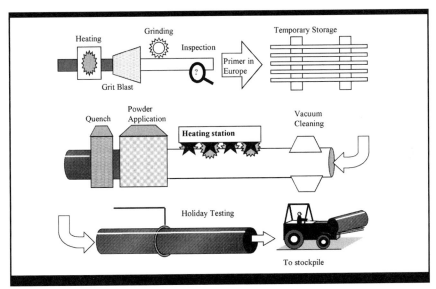

Fig. 8–3 FBE Application

FBE coatings in seawater can withstand temperatures between 85–95 °C. At these high temperatures, it is usual to apply double coatings of FBE applied in two passes through the coating machine while the pipe is hot. FBE coatings may be used with or without a concrete weight coating. Because FBE coatings are slippery, when a concrete weight coating is over them applied, it is common to provide anti-slip bands to reduce risk of concrete slippage.

8.2.6 Elastomeric coatings

Elastomeric coatings are used for high-temperature pipelines. One type of elastomer, polychloroprene is also widely used for covering risers because of its resilience to impact. On pipelines the coatings are 3–5 mm thick, and on risers they are typically 6–12 mm thick. Sheet polychloroprene (commonly known as neoprene) or EPDM is glued onto a cleaned (and for EPDM primer coated) pipe, and the assembly then autoclaved to vulcanize the elastomeric coating. Neoprene is suitable for temperatures up to 105 °C while EPDM can be formulated for service up to 120 °C. These coatings are suitable for use on reeled pipelines. Elastomeric coatings are expensive because of the autoclave treatment required.

EPDM coatings may be applied directly to the steel, but in deep water this method may lead to untoward wastage of the CP system. EPDM coatings have a high carbon black content to provide UV resistance while the material is in the stockyard. When installed at depth, the hydrostatic pressure compresses the elastomeric coating; and the coating resistance is reduced because the carbon black particles contact each other. The coating may act as if it were a continuous carbon surface. The sacrificial anodes, pipe, and carbon coating are electrically connected and may act as a mixed metal couple. Carbon is a highly electrochemically active cathode and will cause enhanced corrosion of the sacrificial anodes. When these are consumed, the coating may enhance corrosion of the steel pipeline at areas of coating breakdown. Recently developed EPDM coatings have a lower carbon black content and contain inert fillers that reduce the risk of the coating becoming conductive at depth.

8.2.7 Thin film coatings

Conventional solvent-free, thin film coatings are used for special conditions. Thin film coatings are not widely used, not because the coating application is slow, but because the pipe must be stored in a dry, dust free area while the coating cures. During the cure period, the coating is very easily damaged. Thin film coatings are used for special conditions. Applied in two passes, the coatings are sprayed on and are 0.5–1 mm thick. At present modified epoxies are the most commonly used as these have good adhesion to the steel and set rapidly. Polyurethane is being investigated as an alternative and can be formulated to have higher flexibility, cohesive strength, low-temperature strength, and a faster cure time.

Thin-film coatings are used for short lengths of pipe for tie-in lines, for riser tube turns and spools, and also for the protection of bundle outer pipe which is often fabricated into triple pipe lengths (up to 40 m) before coating. These triple joints are too long to be passed through a conventional coating plant. For all these applications, the coatings are not covered subsequently with concrete weight coatings. Thin film coatings are inexpensive to apply, but the overall coating cost can be high because of the hold up and storage space required for curing.

8.2.8 Coating application

8.2.8.1 Surface cleanliness. For a coating to be effective, it needs to be applied to a clean surface with the correct surface roughness profile. The surface must be dry, grease-free, and dust-free to ensure that the coating sticks. Some coatings have been developed for application onto wet surfaces (even applicable underwater) and onto rusty surfaces, but these speciality products have a limited and uncertain life expectancy.

Surface cleanliness is measured by the degree of removal of rust and the absence of soluble salts. Removal of rust is achieved by grit blasting the coating with chilled shot, copper slag materials, garnet, or dry sand.[8, 9] No cleaning material is ideal. Chilled shot is expensive but less efficient because it can smear the metal surface over dirt. Copper slag can leave a *shadow* of very fine slag on the metal surface and also has a negative environmental impact. Garnet is intermediate between shot and slag and is more expensive though environmentally neutral. Sand is light in weight and, consequently, relatively slow and creates a silicosis hazard. If the sand is not completely clean, some salts can be introduced to the metal surface, and minor corrosion (blooming) may occur.

The quality of the blasting is measured as the degree of whiteness of the metal, and this is estimated visually. There are instruments that have been developed to measure surface cleanliness, and undoubtedly these will eventually be used on pipelines to automate the cleaning process. The most commonly encountered standards for cleanliness are the following:

- Swedish Standard SIS 05 5900 now ISO Standard 8502
- British Standard BS 4232
- German Standard DIN 55 928(4)
- American Standard ASTM D2200

The most commonly used is the ISO Standard.[10] It classifies the cleanliness of a metal after blasting into three main grades, from SA 1 to SA 3. It is usual to specify either SA 2½ minimum and to impose a high level of inspection or to specify SA 3 (a white metal surface) and to reduce the inspection level for surface preparation for shop-applied coatings and to accept Sa 2½ for field preparations of surfaces for the field joints. The clean surface is highly active, and it is important that the pipeline surfaces are prevented from flash rusting before the coating is applied. The local environment must be low in humidity and/or a primer applied to the steel surface. In many coating yards, the pipe enters the coating line immediately after cleaning and inspection before rusting or blooming can occur.

The requirement for a salt-free surface is less obvious. Pipeline coating yards are often located on the coast to expedite shipment of the pipe in and the coated pipe out. It is possible for the pipe surface to be heavily contaminated with sea salts. Pipe produced at inland plants may be surface contaminated with sulphates from atmospheric pollution in contrast to chloride contamination. If the salts are not completely removed, the coating will be applied over the salts on the metal surface. When the pipe is installed on the seabed, seawater eventually will penetrate to the metal surface and dissolve the soluble salts. The water at the steel surface then may be more saline than seawater, and the coating will act as a semi-permeable membrane. The osmosis will draw more water through the coating, resulting in the formation of water blisters. Coatings are permeable to oxygen, and some corrosion can occur inside the blisters. If SRB are present, the hydrogen sulphide they produce can permeate into the blisters and cause more serious damage. Unless the coating over the blister fails, the resistance path may prevent sufficient CP to suppress this corrosion.

8.2.8.2 Surface profile. Many coatings rely on a rough surface to improve the key for adhesion, and so the surface profile is important. For the thick flood coatings, e.g., asphalt coatings and coal tar enamels, there is a need for a minimum profile for keying, but the upper roughness level is of less importance. However, for thin coatings (for example fusion bonded epoxy coatings that are about 0.5 mm thick), a very rough surface profile must be avoided because rough spikes of metal may reach through the coating and create an unacceptable level of holidays (holes).

Profile is not easily measured. The most common technique for shop and field use is the production of plastic impressions made by extruding quick setting plastic material onto the pipeline surface.[11] Once it has set, it is removed. Then the profile obtained is compared to standard roughness profiles that are available as a set of plastic discs. Electric devices have been developed that can give a quantitative valuation of a surface. One type of profile meter incorporates a probe, similar to a phonograph needle, which is pulled across the surface. The oscillations of the needle are amplified and recorded on a chart recorder or digitized for computer analysis. Light reflectance techniques have also been evaluated. Many coating inspectors rely largely on feel and a visual assessment of light reflectance as quick guides and will only call for a quantitative procedure if their experience indicates that the profile may be incorrect.

8.2.8.3 Coating adhesion. Cleanliness, profile, and freedom from surface salts help to ensure that the coating sticks to the pipe surface. However, the failure of a coating to adhere can also result from incorrect formulation of the coating itself. To prevent this, it is usual to have each batch of coating material tested. See section 8.2.8.6.

Adhesion of the coating must be ensured when the pipe is coated. It cannot be retro-fixed. It is usual to require regular inspection of the adhesion of the coating to the pipe to ensure a good adhesion. The most popular technique is to cut the coating through to the base steel with a crosshatch and to try to pry off the coating. A version of this technique is to cross-hatch the coating with a multiple blade knife, cutting a sequence of squares, then to cover the area with adhesive tape and to pull the tape off. The number of squares removed is a measure of the adhesion. An alternative quantitative technique is to glue a small stud to the coating with epoxy, and then to pull the stud off the surface with a calibrated spring-loaded device. A technique used for multilayer coatings is to cut V-shaped grooves in the coating and to inspect the cross section with a microscope to determine how well the layers adhere to each other.

8.2.8.4 Coating thickness. The thickness of the coating must be within prescribed limits. Coatings protect because they are thicker than the maximum expected flaw size. Thin coatings make it more likely that a coating flaw will be surface emergent (e.g., gas bubble, spike of metal). On the other hand, an over-thick coating is wasteful.

Wet coating thickness is measured during the application using a *comb*, a thin metal plate with notches of different depths along its edges. The comb is pressed into the wet coating, and the thickness read from the deepest notch wetted by the coating.

The dry film thickness is a more important parameter. It is measured with electrical eddy current or magnetic test equipment after the coating has cured. The eddy current meter uses a coil probe that imposes an oscillating voltage onto the steel and measures the induced current flow. The induced current is reduced in proportion to the square of the coating thickness, and the instrument is calibrated for the range of expected coatings. The magnetic technique also provides a quantitative estimate of coating thickness but is generally regarded as less accurate. These devices include a magnet, which is pressed onto the coating surface, and a counter spring, which is tensioned until the magnet springs off the surface. The degree of counter tension is used to estimate coating thickness.[12]

8.2.8.5 Holidays. Large flaws in the coating are detected by eye, but the coating must also be inspected to detect the more numerous small coating defects that are termed *holidays*, technical jargon for the small holes that expose the pipeline steel and usually arise because of poor profile control.[13, 14]

Thick coatings are tested with high voltage spark discharge instruments. The coating is tested by passing a loop around and along the pipe with a high voltage imposed between the loop and the pipeline steel. This test voltage across the coating is usually set at 5 V per micron and will be typically 30–33 kV for a 6 mm coating. An alternative to a loop is a fine wire brush. Holes are detected by sparking which triggers an alarm on a test instrument.[15]

For thin coatings, low voltage techniques may be used. In this technique, a 30–35 V source is used with one terminal connected to the pipeline and the other connected to a sensor, which is a sponge wetted with water containing a mild detergent. As the wet sponge is passed over a defect, the water flows into the defect; and a flow of current triggers an alarm. The techniques are illustrated in Figure 8–4.

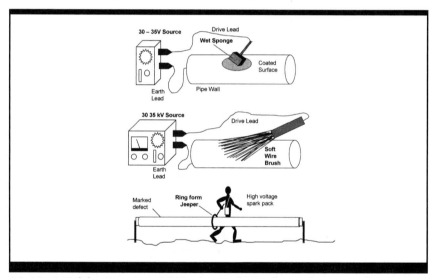

Fig. 8–4 *Holiday Detection: (a) Wet Sponge (b) Sparking with Wire Brush (c) Sparking with Ring Jeeper*

The inspector marks the location of the holes and holidays with a crayon. Small holidays are repaired using hot sticks or some other patch technique. If the total area of the holidays is excessive, the pipe will be stripped and recoated.

8.2.8.6 Coating formulation. The composition of the coating must also be checked to ensure that the consecutive batches of coating are to an adequate standard. Several chemical and thermal tests are used. Thermal tests are the most important quality control (QC) tests. A typical thermal test involves a slow increase of temperature while material properties such as extensibility and heat capacity are measured. Each coating formulation has a characteristic fingerprint. It is also usual to conserve small samples of each batch of coating for later review should problems arise. It is important in the coating specification to ensure that there are adequate lines of communication so that problems arising between the coating supplier and the coating applicator do not prejudice the coating quality and schedule.

8.2.8.7 Cathodic disbondment. Coatings may be sensitive to the alkalinity generated by the CP system. The sensitivity increases with operating temperature, and it is usual to check that the selected coating has adequate tolerance.[16, 17, 18, 19]

8.2.9 Coating defects

Coatings can have various kinds of defects. Blistering occurs when flaws in the coating or salts on the metal surface can result in water migration to form sub-coating blisters. With impressed CP, the pH in the blisters can rise to 12 or 13. This high pH may result in additional coating detachment. Blisters begin small and are detectable at about 20-mm2 area. They grow over time and sometimes link together, and clusters of blisters may represent half the pipeline surface over lengths of 2–3 m. FBE field joint coatings often show micro-blistering around the girth weld because the heating of the pipe for the application of the FBE field joint coating increases hydrogen outgassing. These micro-blisters are inconsequential in service.

Some coating flaws may not be detected during inspection. Metal flakes or grit nearly protruding through thin film coatings. Cracks or bubbles or intercoat disbondment may cause these flaws. Flaws may be single or in localized patches. The holidays are often small, about 1–2-mm^2 area, but they may grow in the same way as blisters.

Bagging is the result of the stretching and gathering of the coating at the 3–5 o'clock and the 7–9 o'clock positions on the pipeline. Bagging may occur for 1–4 m along the pipe length and 50–100 mm circumferentially. Bagging does not occur if the anti-corrosion coating is protected by concrete.

Crown cracking is caused by soil stresses and rock dump settlement. The coating stretches and splits across the crown of the pipe (12 o'clock position), often over the longitudinal weld that is an area of coating weakness. Crown cracks may run for 5–10 m in length and be 40–50 mm wide. Crown cracking does not occur if the anti-corrosion coating is protected by concrete.

Weld exposure occurs when failure to dress or clean welds adequately leads to coating disbondment and metal exposure along the welds. If the blast cleaning guns are not correctly positioned, protuberant welds may cast a shadow and leave an area unblasted. Weld exposure is a particular problem for thin film coatings.

If the pipeline is laid onto sharp material on the seabed, the coating at the bottom of the pipe (6 o'clock position) may be damaged.[20, 21] The damage occurs as individual holes in the coating or as a localized run of holes. Each hole will represent a bare metal area of around 10–200 mm^2. Damage to the coating may be caused by pipeline handling equipment, slings and chains, tensioners, and dropped objects. Concrete coating protects the anti-corrosion coating against these kinds of damage unless the impact to the pipeline is very severe.

8.3 Cathodic Protection

8.3.1 Introduction

The coating is the primary barrier to corrosion of a pipeline. CP prevents corrosion at areas of permeable, missing, or damaged coating and can be thought of as an electrical way of making the metal more stable thermodynamically. This increase in stability is achieved by providing electrons for the cathodic reactions to replace the electrons generated by the normal corrosion processes. Electrons are transferred into the pipe by draining an electrical current from it, and the electrical circuit is completed using an anode through which the electrical current flows into the environment. The electric current may be supplied by a generator (impressed current system) or by coupling the metal to a baser metal to form a primary battery (sacrificial anode system). Either system can be used for pipelines on land, close to shore, or at river crossings; but in practice long submarine pipelines are protected only by sacrificial anodes.[22, 23]

For CP, the number of electrons must equal the electrons that would normally be removed by the cathodic reactions, e.g., oxygen reduction or hydrogen ion discharge. Seawater is alkaline, pH 8.2–8.4, and the principle reaction is oxygen reduction. Sulphate-reducing bacteria may flourish in seabed sediments, and the biogenic sulphides they produce can remove electrons.

CP can only work if there is a continuous and sufficiently conductive medium (called the electrolyte) around the object to be protected. Seawater and seawater saturated sediments provide an excellent electrolyte. All electronic reactions occur on the pipeline surface exposed to the electrolyte and, consequently, CP applied through the seawater only protects the outside of a pipe. Internal corrosion is not affected. To protect a pipe internally by CP, it would be necessary to install anodes inside the pipeline. Standard textbooks[24, 25] and reviews[26, 27] should be referred to for general information on CP.

The CP system is conservatively sized to accommodate at most 25% coating breakdown over the design life of the pipeline, though a more normal value is 5%. Coating breakdown is assumed to be evenly spread over the length and area of the pipeline, though in practice breakdown is often found to be localized. Periodic inspection of the CP of the pipeline is performed to identify such areas and to check that the CP system is providing protection.

8.3.2 Electrochemical basis of CP

The corrosion process involves metal dissolution as positively charged ions at the anodes. The electrons left in the metal lattice are discharged at the cathodes by reaction with oxygen, hydrogen ions, or other reactants.

Anodic reaction:

$$M \rightarrow M^{n+} + n \text{ electrons} \tag{8.7}$$

Cathodic reactions

$$\text{(i)} \quad 2H^+ + 2 \text{ electrons} \rightarrow H_{ads} + H_{ads} \rightarrow H_2 \tag{8.8}$$

$$\text{(ii)} \quad O_2 + 2H_2O + 4 \text{ electrons} \rightarrow 4 \text{ OH}^- \tag{8.9}$$

$$\text{(iii)} \quad H_2S + 2 \text{ electrons} \rightarrow H_2 + S^= \tag{8.10}$$

If electrons are provided by an external power source, the metal dissolution process (Equation 8.7) is not required to provide the electrons for the cathodic reactions. This is the basis for cathodic protection: the metal is flooded with electrons sufficient to oxidize the oxygen and hydrogen ions migrating to the metal surface. Electric current is the reverse flow of electrons, and the provision of electrons to the metal is regarded as a current drain from the metal. Though simple in theory, in practice CP can be complex because sufficient electrons (but not too many) must be provided at all areas of the metal to be protected.

The electrochemical potential of the metal gives information on the degree of saturation of the metal with electrons. As the concentration of electrons in the metal is increased, the potential becomes more negative because electrons carry a negative charge. The measurement of potentials is performed using a reference electrode, which has a stable potential, against which the metal potential is measured using a voltmeter. For each engineering material, there is a critical potential at which complete cathodic protection is achieved. For example, for steel the protection potential is -800 mV (referred to as silver-silver chloride reference electrode (SSCE)). This is termed the CP potential. For some metals, the *CP protection* potential can vary with the pH. The potentials for protection of steel in seawater for several reference electrodes are given in Table 8–2.

Table 8–2 *Reference Electrode Potentials*

Reference Electrode Type	Potential vs Hydrogen (mV)	Protection Potentials for Steel (mV)*
Hydrogen	0	-550
Lead-lead chloride	+325	-850
Copper-copper sulphate	+318	-850
Silver-silver chloride	+250	-800
Calomel (mercuric sulphate)	+242	-800
Pure zinc (>99.96% Zn)	-780	+250

* Values are rounded to the nearest 50 mV

The electric current required to achieve the protection potential varies with the local cathodic reactions, and hence in seawater will alter with factors that affect these reactions, e.g., local salinity and pH, aeration, flow rate, and temperature. A buried pipeline will require much less current to achieve the protection potential because the oxygen access is limited, but the anodes can supply less because of the higher resistivity of the soil cover compared to water. The extent to which a pipeline will be buried is often uncertain; therefore, it is prudent consider both the exposed and buried options. Rock dump cover is generally not considered as burial because there is free circulation of seawater within the rock matrix.

It is fortunate that the thermodynamic behavior of steel is such that the protection potential of steel is a constant value independent of the environmental pH over a wide pH range. The numerical value of the protection potential depends on the reference electrode used for the measurement. The absolute potential is the same whatever the reference electrode used; it is merely the numerical expression of potential that changes. This is similar to the measurement of the temperature of a room, which is a different number in Centigrade and Fahrenheit.

In *Codes of Practice and Standards,* the protection potentials are rounded to the nearest negative 50 mV for convenience.

8.3.3 Sulphate-reducing bacteria and CP

Sulphate-reducing bacteria (SRB) are described in section 7.12. The reactions between hydrogen sulphide and steel occur over a wider range of potential than the oxygen reduction reaction. In seabed sediments where active SRB are present, the protection potential needs to be more negative; and there needs to be more current than would be required for a buried pipeline is non-active sediment. The adjustment found to prevent SRB attack is an additional -100 mV potential depression, i.e. -900 mV SSCE.

In most cases, the effect of SRB activity is not significant because the sacrificial CP system design is conservative. A new pipeline has (a) a quality coating, and (b) the anodes are designed to provide a high current density in the final year when the coating breakdown is a maximum. Through most of the pipeline life, the anodes can provide more current than the pipeline requires; consequently, the potential of the pipeline is close to the potential of the anodes: -1050 to –1100 mV SSCE. Late in the design life, the pipeline requires more current because the coating has deteriorated, and the anodes will produce less current because the anode surface area has decreased. Sometime in the last quarter of the pipeline life, the potential of the pipeline-anode system will start to rise (become less negative). At this time, the organic material in the sediments will be depleted, and the SRB at low activity. The pipeline will have calcareous deposits at the exposed areas, which will isolate the pipeline from the SRB action.

Simulation studies showed that if steel is cathodically protected to -800 mV SSCE before exposure to SRB, the steel remained protected despite subsequent vigorous bacterial growth. If exposure to SRB occurred before cathodic protection was applied, the steel corroded, and the potential then had to be depressed to -900 mV (or more negative) to stop the corrosion.[28]

Consequently, a CP design to -900 mV SSCE is required only if it is known that the pipeline will be exposed to bacterially active sediments. The bacterial activity of the sediments can be evaluated by measuring the bacterial population in cores or grab samples of sediments. Where obtaining core or grab sample is not feasible, an estimate can be made using a factor evaluation, as given in Table 8–3.[29] The scores are added up, and a total of 10 or more indicates a sediment likely to harbor highly active SRB. Organic contents can generally be evaluated from the level of fish life, plankton, and other organisms in the water column. Nitrogen (N) and phosphorous (P) occur from agricultural run off and from upwelling cold waters.

Table 8–3 *Evaluation of Microbial Activity of Sediments*

Factor	Score
Sediment	
mud	4
sandy-mud	2
sand or rock	0
Organic content	
If mud: high	3
medium	2
low	0
If sand: high	1
medium/low	0
Water depth	
Shallow (below 200 ft)	2
Deep (over 200 ft)	0
Availability of nitrogen and phosphorus	
High organic content + N&P	2
Low organic content + N&P	1
Low N&P	0
Seabed temperature	
above 10 °C	2
below 10 °C	0

Studies of zinc, aluminum alloy, and magnesium anodes in cultures of SRB showed that the anode materials were corroded by the biogenic sulphides.[30] Extrapolating the wastage to a design life of 25 years indicated that zinc anode material was corroded by about 10–14%, aluminum alloy anode material by 40%, and magnesium anode material by 14–25%. It is prudent to allow for this corrosion wastage when designing anodes for use in bacterially active sediments.

Table 8–4 lists protection potential ranges for different environments against common reference electrodes.

Table 8–4 *Protection Potentials Ranges for Steel*

Environment	Reference Electrode		
	Cu/CuSO$_4$ (mV)	Ag/AgCl (mV)	Zinc (mV)
Free seawater environment	–850 to –1000	–800 to –1050	+250 to +0
Anaerobic environments	–950 to –1100	–900 to –1050	+150 to +0
High strength steels UTS > 700 N/mm^2	–850 to –1000	–800 to –900	+250 to +100

8.3.4 CP and coatings

CP can be (and often is) applied to bare metal, but the cost of power to perform this CP current is high. For example, steel offshore jackets are often not coated and are protected from corrosion only by CP, but this approach has both a cost and weight penalty. Pipelines are always coated, and the CP is only required for protection at areas of permeability and faults in the coating. As the coating deteriorates over time, more CP current is required.

The electric current required depends on the corrosiveness of the environment but is around 12-15 mA/m^2 of *bare* pipe in sediments to around 100–120 mA/m^2 exposed in seawater.[31,32] The total current demand is the product of the current density and the bare area of pipeline—pipeline area times coating breakdown percentage. On a perfectly coated pipe, the total current demand would be close to zero. Typically on a new submarine pipe with a good coating, the total current demand would be 1–2 A per km, but over time this can rise to 20 A. The total current demand is a measure of the quality (or lack of quality) of the protective coating.

8.3.5 Impressed current cathodic protection (ICCP)

Impressed current CP is the term used to describe a case in which the electric current for protection is provided by an external power supply. For a buried onshore pipeline, a generator, either a local utility or a national grid, provides the electricity. Alternating current (AC) current is converted to low voltage direct current (DC) current by a combination transformer and rectifier, usually referred to as the *T/R*. The pipe is connected to the negative terminal of the T/R, and the positive terminal is connected to buried noncorroding metal anodes. Electrons are introduced into the pipe, and *leak* out at the bare areas where the cathodic reactions occur. Because CP systems are usually installed and operated by electrical engineers, it is usual to refer to conventional electrical current flow, rather than electron flow.

In conventional terms, the electrical current is passed into the anodes, runs through the soil to the pipe, and returns through the pipe to T/R.

ICCP is rarely used for submarine pipelines. Estuarine crossings, short pipelines close to shore, and outfalls are occasionally protected in this way, but often they are isolated from the onshore pipeline and protected by sacrificial anodes. There has been considerable interest in solar- and wind-powered CP systems in conjunction with battery storage systems for these small ICCP systems.

The potential of the pipe will vary along the pipeline length away from the drain point and will depend on the geometry of the pipe, the soil and sediment resistivity, and the extent and location of coating damage. The potential is always most negative at the drain point and becomes less negative further along the pipeline, an effect termed *attenuation*. Over time as the coating degrades, the attenuation of the potential increases and the total current to protect the pipeline increases.

In extreme conditions the areas near the drain-point may need to be over protected to ensure that the remote sections of the pipeline are adequately protected. When overprotection occurs, the coating may be damaged. Pipeline operators often place a limit on the acceptable negative potential, which depends on the protective coating. See section 8.2.8.7. Asphalt-based coatings tend to have lower limits of about –1250 mV while FBE coatings may tolerate potentials of -2000 mV SSCE. The impact of potential on pipelines is shown in Table 8–5.

Table 8–5 *Under and Overprotection Potentials for Steel Pipelines*

Steel Potential mV (Ag-AgCl reference)	Cathodic Protection Status
-550	Intense corrosion
-551 to -650	Corrosion
-651 to -750	Some protection
-751 to -950	Zone of cathodic protection
-951 to -1050	Slight over protection
-1051 to -1200	Increased overprotection
< -1200	Blistering of paint and embrittlement of high strength steels

8.3.6 Sacrificial anode CP

Submarine pipelines are almost always protected by sacrificial anodes. The most commonly used design procedures follow either NACE or DNV guidelines.[31, 32, 33] The anodes are formed by casting a sacrificial material around a steel former, which provides both strength and electrical continuity. The anodes are attached to the pipe at regular spacing and are electrically connected with welded or brazed cables between the anode former and the pipeline. Most anodes are cast as half shell bracelets and these form a tight fit around the pipeline over the corrosion protection coating. For very large pipelines, the anodes may be cast as segmented blocks, which are assembled to form a complete bracelet assembly.

The sacrificial material is a base material, which creates a battery between itself and the steel pipe. The anode corrodes and provides the electrons. Anode materials are listed in Table 8–6. Magnesium is used for onshore pipelines but is not efficient for subsea pipelines because it corrodes in seawater and only about half the electric current is available to provide CP. Zinc used to be the most widely used material, but over the last decade aluminum alloys have gained popularity because aluminum alloy provides more electricity per unit weight than zinc and using aluminum alloy anodes is less expensive. Anodes are not pure materials but are relatively complex alloys; the alloying elements are required to ensure even corrosion and to prevent passivation.

Table 8–6 *Sacrificial Anode Materials*

Anode Alloy	Environment	Anode Potential (NEG V)	Current Capacity Ah/kg	Consumption Rate kg/Ayr
Al-Zn-Hg	Seawater	1.0 – 1.05	2,600 – 2,850	3.1 – 3.4
Al-Zn-In	Seawater	1.0 – 1.1	2,300 – 2,650	3.3 – 3.8
Al-Zn-In	Sediments	0.95 – 1.05	1,300 – 2,300	3.8 – 6.7
Al-Zn-Sn	Seawater	1.0 – 1.05	925 – 2,600	3.4 – 9.5
Zn	Seawater	0.95 – 1.03	760 – 780	11.2 – 11.5
Zn	Sediments	0.95 – 1.03	750 – 780	11.2 – 11.7

The DNV RP B401 guidelines are conservative about the anode efficiency in comparison to manufacturer's data (aluminum and zinc alloy anodes efficiencies of 2000 and 700 A-hr/kg respectively compared to 2400 and 760 A-hr/kg). A designer may include a minimum value for anode efficiency into the anode material specification.

The electrical current required to protect a submarine pipeline is greater than that for a buried pipeline. This is not surprising, as seawater is very corrosive, highly conductive, and contains about 8 ppm of oxygen. The corrosion products formed by natural corrosion in cold seawater are chlorides, which are not protective compared to corrosion products formed on buried onshore pipelines, which are hydroxides, oxides, and sulphates. Typical CP protection currents are given in Table 8–7.

Table 8–7 *CP Current Requirements*

Location	Current Density (mA/m^2)		
	Initial	Mean	Final
N. North Sea	150 – 250	90 – 120	120 – 170
S. North Sea	150 – 200	90 – 100	100 – 130
Arabian Gulf	130 – 150	70	80
India	130 – 150	70	90
Australia	130 – 150	70	90
Brazil	130 – 150	70 – 80	90 – 110
Gulf Mexico	110 – 150	60 – 70	80 – 90
West Africa	130 – 170	70	90 – 110
Indonesia	110 – 150	60	80
Buried	50	40	40
Risers in open shafts	180	140	120
Risers in sealed shafts	120	90	100
Saline mud	25	20	15

The temperature of the water is generally the most important factor as it dictates the concentration of dissolved oxygen. The water flow rate may also be a factor if the pipeline is not buried or concrete weight coated. NACE guidelines take seawater flow rates into account whereas DNV guidelines do not.

Concern is often expressed about current interference resulting from mixing anodes fabricated from different sacrificial materials. Mixed systems are commonplace. Aluminum alloy anodes are used on structures and zinc anodes on the pipelines. CP surveys indicate that the mix of anodes does not affect the protection of either item. This is not unexpected because the difference in voltage between the different anode materials is small.

8.3.7 Coating breakdown

The protective coating on the pipeline restricts the access of oxygen to the pipeline and, consequently, reduces the current demand. For simplicity in CP design, it is assumed that the protective coating is 100% effective except at areas of coating breakdown, i.e., where the pipeline is effectively bare. In reality, though some bare areas do occur, the bulk of the protection current passes through the coating because all organic coatings are permeable to oxygen to some extent. When the oxygen arrives at the steel surface, it will remove electrons. This appears as a current flux through the coating. As the coating ages, the resistance to permeation decreases and a higher oxygen flux occurs resulting in a higher current flux through the coating. Some CP design procedures do use total surface area as the design criterion, but this approach is not widely used for submarine pipelines.

Typical effective bare areas for the different types of coatings are given in Table 8–8. The DnV RP B-401 takes a different approach and assumes that for weight coated pipelines, the mean and final coating breakdown values are 5% and 7% for the first 30 and 20 years, respectively, rising thereafter by 2% and 4% per year, respectively. For pipelines that are not weight coated, the mean and final coating breakdown value is 2% with increases of 0.75% and 1.5% per year, respectively.

Table 8–8 *Coating Breakdown Values*

Coating Type	Percentage Breakdown		
	Initial	Mean	Final
Thick coatings	≤ 1	5	10
Vinyl coatings	≤ 2	20	50
Epoxy coal tar	≤ 2	5 - 10	10 - 20
Fusion bonded epoxy	1 - 2	5 - 10	5 - 20
Polypropylene (25 yr)	0.5	2	5
Polyethylene sheathing on flexible pipes (20 yr)	0.5	1	3
Polychloroprene (30 yr)	0.5	2	5

For pipelines in deep water the increase in coating breakdown is reduced slightly to 0.6% and 1.2% per year respectively. However, not all operators accept these values, and they base the expected coating break-downs on historical data obtained from CP surveys of similar operational pipelines. It is of particular concern at present that there are no commonly accepted coating breakdown values for the more recently applied types of coatings: polyethylene, EPDM, and polypropylene coatings. Coating breakdown values, which are reported to have been used for CP designs, are also given in Table 8–8. Note, however that these values have no provenance, and many designers will use the DnV values.

8.3.8 CP current demand

The current density required to reduce the steel potential to the protection potential depends primarily on the oxygen content of the seawater, which is strongly related to temperature. Typical empirical values are given in Table 8–7. Water flow rates are of secondary importance if the pipeline has a weight coating. The weight of anodes required is based on the mean current density that is calculated from the pipeline geometry and the mean coating breakdown percentage.

The final current will be higher than the mean current because the final current density and coating breakdown are higher. This final current demand is used to design the bracelet anode dimensions. In some circumstances the weight of anode material necessary to provide suitable anode dimensions will be higher than that calculated using the mean current demand.

8.3.9 Calcareous deposits

In seawater the pH is around 8.2–8.6, and hydrogen ion reduction is not a significant cathodic reaction. The oxygen reduction reaction is the dominant cathodic reaction and produces alkali at the metal surface. The hydroxide reacts with the calcium and magnesium in the seawater to form deposits on the metal surface. The sequence is:

$$O_2 + H_2O + electrons \rightarrow 4OH^- \qquad (8.11)$$

Followed by the reaction of the alkali with the cations in the seawater:

$$Ca^{2}+ + CO_3^{2-} \rightarrow CaCO_3 \qquad (8.12)$$

$$Mg^{2+} + 2OH^- \rightarrow Mg(OH)_2 \qquad (8.13)$$

The products formed are white-brown scales, and they are beneficial in that they reduce the access of oxygen to the metal surface and, therefore, reduce the protection current required. High current densities favor the formation of magnesium hydroxide. Calcium carbonate scales are much more protective than magnesium hydroxide scales, but they form slowly and at low current densities. On pipelines, magnesium hydroxide scales will probably be formed initially, with some later conversion to a mixed scale.

In cold seawater the scales form very slowly, if at all, and only at high current densities. In warm waters, the scale forms readily and at lower current densities. Scale should form readily on hot pipelines and, if the rate of coating breakdown were modest, the high current densities at bare areas would ensure the formation of the magnesium hydroxide scales.

Because of the formation of scales, the current densities for protection change over time. Table 8–7 illustrates the changes in current density. The initial current density is high. It declines over a period of weeks to months to the moderate mean current densities, reflecting the formation of the deposits. Toward the end of the CP system design life, the current density will fall further, perhaps sufficiently to prevent formation of continuous calcareous deposits. Consequently, the CP final current demand will increase compared to the mean current density. For pipelines, the mean and final current densities are the relevant parameters. The initial values are given because it may be that a pipeline coating is severely damaged during installation, and the capability of the CP system must be calculated to check whether it is adequate to protect the pipeline.

8.3.10 Temperature effects

Hot pipelines in cold water need a higher current density than do cold pipelines. Anode performance is also reduced at high temperatures. Several hot risers were severely corroded in the early 1970s because the current provision was inadequate, and several pipelines have had to be retrofitted because of premature failure of sacrificial anodes that operated hot. Research on this topic has been fairly extensive, but resultant data are somewhat conflicting. Most designers use the guidelines given in DNV RP B-401. In this document, to simplify the CP design, the fluid bulk temperature is used rather than the pipeline steel skin temperature.

The CP protection potential changes slightly as the pipeline operating temperature increases above 25 °C. For each degree above this temperature, the protection potential becomes more negative by 1 mV, i.e., a pipeline operating at 50 °C would require a protection potential of -825 mV compared to -800 mV SSCE. The required CP protection current density is also considered to increase by 1 mA/m2 for each degree rise in operating temperature above 25 °C.

The effect of operating temperature on the anode materials is more drastic. The zinc alloy used for anodes contains small quantities of iron, which forms a zinc-iron intermetallic compound. This intermetallic precipitates preferentially at the grain boundaries. Somewhere in the temperature range 50–60 °C, there is a reversal in potential between the zinc-iron intermetallic compound and zinc, so that the intermetallic becomes anodic to the zinc. Preferential corrosion of the intermetallic occurs, the cast alloy material fragments into a loose cluster of zinc grains, and the anode efficiency falls to 10% or less.

Consequently zinc alloy anodes are not used for pipelines that operate at temperatures higher than 50 °C. The worst case occurs with buried pipelines because the anode temperature rises to the pipeline temperature. In free seawater, the anode surface is cooler than the pipeline, and a higher operating temperature may be tolerated. For a pipeline operating at 50 °C, it would be reasonable to consider the use of zinc anodes if the pipeline were exposed on the seabed or if a thermal insulation sleeve were placed between the anodes and the pipeline. It is unlikely that zinc anodes would be used on a hot buried pipeline.

Aluminum alloy anodes are also affected by temperature but to a lesser degree. Testing indicated a reduction in anode efficiency of 25 A-hr/kg for each degree centigrade rise above 25 °C. For a pipeline operating at 50 °C this would imply the current capacity of the aluminum alloy would reduce to 1775 A-hr/kg compared to the nominal capacity of 2400 A-hr/kg. When designing a CP system, the weight of each anode would need to be increased by 25%. Thermal insulation between the pipeline and the anode would reduce the anode temperature and improves the anode efficiency.

8.3.11 CP system design

The design of a CP system has several steps:

- Decide the CP potential
- Estimate the final current density
- Supply the mean and final currents

Estimation of current demand involves determining the required protection potential, CP current density and coating breakdown percentage. The first step is to estimate the protection potential required. A potential of -800 mV SSCE is normally used, or -900 mV if highly active SRB sediments are expected. Given discussion in section 8.3.3, the lower potential for SRB may not be considered at the design stage but is relevant to CP inspection.

The second step is to estimate the mean and final current densities needed to lower the pipeline potential to the protection potential. Typical values are given in Table 8–7. If the pipeline is not weight coated and exposed on the seabed, then the effect of seawater flow may need to be considered. The third step is to estimate the expected percentage of bare area on the pipe, and how it will alter over the life of the pipe. This requires experience in how a particular coating behaves over time in the particular environment. Typical values are given in Table 8–8. Concrete-coated pipelines show little coating degradation over the normal design life; the principal degradation appears to occur at the field joints. From these values, the total current demand can be calculated for each year of the pipeline life and for the overall life.

The supply side involves a decision as to how to best supply this total current: impressed current cathodic protection (ICCP) or sacrificial anode cathodic protection (SACP). If ICCP is used, then the T/R and groundbed must be sized to cope with the maximum current required over the pipeline life and the groundbed location selected.

If sacrificial anodes are used then the total tonnage of anode material is calculated, and the size and spacing of the individual anodes decided. For submarine pipelines, the bracelet anodes are always bought as specials made to fit the pipeline. Bracelet anodes are made by casting the anode material around a steel former. To provide mechanical strength, the former must be cast within the anode body; therefore, whatever the former design, it is not possible to obtain 100% utilization of the anode material. When the anode has corroded down to the inner ligature of the casting around the former, the anode material loses electrical connectivity with the former. About 20% of the anode material will be unusable. The initial mass of anode material must be increased to allow for this loss. Determining the initial mass of anode is calculated by dividing the mass of anode material by the utilization factor U (typically 0.8). A theoretical anode requirement of 10 kg would result in a mass of 12.5 kg of actual anode.

The number of anodes is determined by the spacing of the anodes along the pipeline. Usually anodes are spaced in the range of one anode for each 10–16 lengths of pipe. This equates to an anode spacing of 120–200 m. The weight of each anode is then determined by dividing the total anode weight between the numbers of anodes required.

Generally, the anode is designed to have a thickness equal to the concrete weight coating. The anodes are installed at the center of the linepipe sections, and their thickness is the same as the weight coating. The uniform outside diameter reduces the risk of damage to the weight coating and anodes when the pipeline traverses the tensioners (traction rollers or belts) during installation. Given that the bracelet anode thickness is equal to the weight coating thickness, the required length of the anode can be calculated from the weight and density of the anode material. Practical anodes must be at least 250 mm in length and at least 35 mm thick. If the calculated anode length is too short, the spacing between anodes must be increased. This is the first step in anode design.

Towards the end of the CP system design life—usually coincident with the pipeline design life, the coating breakdown percentage will have increased. The protection current density will also have increased because the protective quality of the calcareous deposits will have reduced as the CP potential becomes more positive. Therefore, the final current demand will be greater than the mean current demand. The anode must be able to supply this final current demand.

The larger the exposed surface area of the anode, the lower the resistance to the environment and the greater the current the anode can supply. Typical formulae indicate that the anode resistance is inversely

proportional to the square root of the anode area. For a bracelet anode, the formula used is one developed by McCoy:

$$\text{Anode resistance} = 0.315 \, \rho e \sqrt{A} \qquad (8.14)$$

where

ρe is the local environment resistivity (ohm m, Ω m)

A   is the area of the anode (m^2).

The resistivity depends on the local environmental conditions. For seawater, it varies with salinity and temperature. For example, in the North Sea, seawater resistivity is about 0.3 Ωm while in the Arabian Gulf, the resistivity is about 0.2 Ωm. Table 8–9 gives typical resistivity values for a range of seawaters and seabed sediments. For critical pipelines, the actual resistivity can be measured from sediment cores if they are available.

Table 8–9 *Resistivity Design Values for Anodes*

Environment	Resistivity (Ω m)
Seawater (t < 10 °C)	0.3
Seawater (10 - 20 °C)	0.25
Seawater (t > 20 °C)	0.2
Sand	1.1 – 1.6
Soft clay and mud	0.6 – 0.75
Stiff clay	0.75 – 1.1

It is assumed that the anode will be corroded to 20% of its initial weight. From this final weight, the final area of anode is calculated. From this final area, the anode output is calculated. This current output must be equal to or greater than the final current required. If it is not, the anode size must be increased.

The weight of the anode must be sufficient for the anode to last the required design life while the final surface area of the anode must be large enough to allow the anode to supply the final current. Often these two requirements conflict, and more anode material must be provided to satisfy the area requirement for final current output.

The anodes supplied by the manufacturer will be slightly different in length and mass to the theoretical design because the actual installed anodes will contain slots for attachment and electrical connection, and these vary slightly between manufacturers as they depend on the construction details of the formers.

8.3.12 Anode manufacture and geometry

Pipeline anodes are cast as two half shells that fit snugly together around the pipe. The two sections are bolted or welded together. If the pipe is covered with a concrete weight coating, the thickness of the anodes is selected to be close to the thickness of the concrete so that the pipe profile is constant. If there is no weight coating, the anodes are profiled with rounded or tapered ends. The anodes are kept in place by the weight coating. For non-weight coated pipe, the anodes are held in place by overlapping the former so that the anode grips the pipe.

The zinc or aluminum is cast onto a steel lattice, called the former or insert. The steel skeleton ensures mechanical strength and aids electrical flow through the anode. The anodes are electrically connected to the pipeline by copper braided wire (pigtails), one end connected to the steel insert and the other brazed or welded to the pipeline.

The design of the former is important, and the pipeline engineer should review the proposed designs. It is prudent to avoid T-bars and other angular shapes that can create areas of high stress in the casting that might lead to cracking of the anode as it shrinks during casting. Rounded bar appears to give a better performance. The insert shape and location determine the utilization factor of the anode. When the anode has corroded down to the former, any further corrosion will electrically disconnect the rest of the anode. The residual material is wasted. The volume of the ligature of anode material between the steel former and the pipe should be estimated and should be less than 20% if the 0.8 utilization factor is to be achieved.

The copper braid pigtail is connected to the pipe either by thermite welding or by pin brazing. Thermite welding, termed CAD welding, uses a mixture of aluminum powder and iron oxide. This mixture is ignited and creates a very high temperature. (Thermite was used to make incendiary bombs.) When the reaction is at its peak, a gate is opened in the crucible holding the thermite charge and the molten material drops onto the pigtail and pipeline material and forms an electrical connection between the pigtail and the pipeline steel. After the weld is made, the pot is removed, and the weld tested by striking it with a hammer. It is usual to test the size of thermite charge needed for a particular pipeline material to ensure that it is adequate to provide a solid connection and does not produce a hard spot in the outer surface of the pipe, which could increase risk of fatigue or cracking in a sour service pipeline. The size of charge varies with pipeline wall thickness and the steel carbon content.

Pin brazing is an electrical process in which a very high, short duration electrical current is discharged through a stud into the pipeline and returns to the pin brazing through a contact shoe. The current causes brazing material on the stud face to melt and electrically connect the pigtail to the pipe. After brazing, the connection is tested by striking it with a hammer. As with a thermite weld, it is important to check that the pin brazing procedure does not cause a hard spot in the pipeline steel that could increase the risk of fatigue crack initiation or for sour service pipelines, the risk of initiating sulphide stress cracking.

On some pipelines, the steel insert has been directly welded to a doubler plate on the pipe. Usually the former is not welded directly to the pipe because the steel in the insert is high carbon, and the weldment could result in a hard spot in the pipeline surface. The inspection facilities available during anode installation in the coating yard are usually limited.

The anodes are fitted in the coating yard to the centers of selected pipes over the corrosion protection coating. If pipelines are fabricated as double-jointed lengths (24 m total length), the anodes may be installed across the field joint. The anodes are covered for protection, and the weight coating is applied. Alternatively, the anodes can be installed after the pipe has been weight coated. The weight coating can be cut while it is still *green*, and the anode installed in the slot. A small area of the corrosion protection coating on the pipe is removed, the pigtails are brazed to the pipeline, and the coating is repaired.

The slots in the anode and between anode and concrete weight coating are filled with mastic. It is good practice to place the pigtails in the slots between the half shells, rather than install the pigtails to the side of the anodes. A wider slot between the bracelets is required to allow this. Aluminum alloys should not abut directly to the concrete because the alkali in the concrete will react with the aluminum. The abutting faces need to be coated or a gap left between the anode and concrete to allow for a mastic infill.

For a pipe installed by reel barge, the anodes are installed on the barge as the pipeline is unreeled into the sea. There is a size and weight restriction on the anodes because they must be manually handled. The size limitations are 600 mm length and 100 kg weight, but these limitations should be checked with the particular reel barge operator because they can vary between lay barges. The installation of the pipe must stop when the anodes are fitted, and frequent stops delay the pipeline installation, causing increased expense.

To reduce the number of anodes and, hence, the number of stops, aluminum alloy anodes are favored.

The anodes are also usually installed in clusters, up to five anodes installed at each halt of the reel barge. The installation cost is reduced because the installation proceeds with fewer stops. However, this approach increases the number of anodes required because the clusters have a slightly higher output resistance than the sum of the individual anodes, and consequently the cluster provides lower current. The anode resistance of a cluster is calculated using the total area of the anode cluster in the McCoy formula.

8.3.13 Quality control of sacrificial anodes

To ensure anodes are fit for purpose, it is accepted practice to specify comprehensive inspection supported by laboratory testing.[34, 35, 36] Inspection includes visual inspection of the moulds for cleanliness prior to casting, inspection of anodes for cracks and surface blemishes, dimensional and weight checks, and inspection of the pigtails for continuity and sheathing integrity. The steel formers should also be inspected prior to insertion into the moulds. Some destructive inspection is also required; randomly selected anodes are sectioned, and the casting examined for porosity and adherence of the casting to the former.

Supporting laboratory testing is required to ensure conformance to metallurgical composition and microstructure. The laboratory testing should include electrochemical tests in which small samples of anode material from each heat are tested for current capacity.[34] The electrochemical tests should be conducted at the expected operating temperature and maximum current output. The most commonly used test procedures are specified in the NACE Standards and DNV RP B401. Anodes to NACE specifications are the most commonly available.

8.3.14 Retrofit of sacrificial anode CP systems

Retrofit is the term used to describe the installation of additional anodes when a CP system is deemed inadequate. Retrofitting a pipeline may be required when a CP system design life is exceeded, when the anodes on a new pipeline have been damaged during installation, and when there is a change of use of a pipeline and the new service is more severe than anticipated. For example the new use may include an increase in operating temperature.

A new pipeline will have small anodes installed at regular intervals along its length. This approach is not required when retrofitting. If the CP survey indicates inadequate potentials close to an offshore structure, anodes may be attached to the structure and/or installed on the seabed on large anode rafts connected to the pipeline by mechanical methods.[38] Anode rafts may be fabricated from several platform standoff anodes attached in parallel to a steel support frame or from a length of pipe fitted with six or seven bracelet anodes. A batch of anodes, either on a structure or raft, can provide cathodic protection for a distance up to 5 km. The distance the CP can be *thrown* depends on the condition of the protective coating and the conductivity of the pipeline. The distance is calculated from the attenuation of potential:

$$E_x = E_A \, exp(-\alpha x) \tag{8.15}$$

where

E_x is the potential at distance x from the anode package

E_A is the anode potential

α is the attenuation constant.

The attenuation constant is derived from the combination of current permeation through the protective coating and the pipeline resistance.

$$\alpha = \sqrt{rg} \tag{8.16}$$

where

g   is the permeability of the coating and can be approximated from knowledge of the voltage drop across the coating (taken to be 400 mV), the CP mean protection current density per unit length of pipe, the thickness of the coating and the coating breakdown percentage.

r   is the conductivity of the pipeline and is calculated from the pipe material resistivity (15×10^{-8} Ω m for steel), the annular area of the pipe wall and the distance x.

Thick wall pipe has a lower resistance and, hence, the attenuation is less, allowing the CP current to be thrown further. A poor quality coating increases permeability and increases attenuation thus reducing the distance that the CP current can be thrown. Generally the attenuation distance is in the range 1–5 km.

The life expectancy of an anode raft can be calculated from the mean output of the cluster of anodes. However, anodes that are close together, as on a raft, have a reduced output compared to the same number of widely separated anodes. The anode cluster output can be calculated approximately by using the McCoy formula for anode resistance where the anode area (A) is the plan area of the cluster of anodes.

8.3.15 CP surveys

8.3.15.1 Survey interval. Periodic inspection of the pipeline CP system is necessary to ensure that the system is functioning correctly. Only superficial external corrosion can be tolerated because there is no corrosion allowance provided for external corrosion. CP surveys may also be used to estimate the residual life of sacrificial anodes. This information may be required if the life of the pipeline must be extended or its operating conditions change, e.g., an increased operating temperature.

There is no accepted optimum timing for CP surveys. A relatively common approach is to inspect the pipeline shortly after installation, usually within the first year of service to ensure that the anodes are functioning and to resurvey about half way through the design life of the CP sacrificial system.[39] Subsequent surveys would depend on the findings from the resurvey. The long delay from initial to second survey is acceptable because, over that period, the coating on the pipeline should remain good and because the anodes are designed for protection of a significantly deteriorated coating, the anodes are capable of providing adequate protection current. Adjacent functional anodes can compensate for an occasional defective anode.

An alternate approach is to survey within the first year of service if installation of the pipeline were judged to have adversely affected anode performance. Usually anodes would be damaged as a result of bad weather and severe pipeline movement during installation. Otherwise the first CP survey would be four to five years after installation. Subsequent CP surveys would be every five years thereafter or as determined by the findings.

8.3.15.2 Transponders. The CP potential of onshore pipelines is periodically checked by measuring the potential at test posts. These posts house above-ground points for connection of the voltmeter to wires that are welded or brazed to the buried pipeline. On submarine pipelines, test posts are not practicable, but they can be simulated by battery-driven transponders that are electrically connected to the pipeline and include a reference electrode. The transponders are interrogated from the surface. Divers or ROVs replace the batteries or the complete transponder. A new

battery-free technology developed for land pipelines is claimed to be suitable also for submarine pipelines. In this system, the electrical power to drive the electronics is derived from the CP system itself.

8.3.15.3 Close interval potential survey (CIPS). A transponder can only give information on the local potential. A CIPS measures the potential of the complete line.[40] The simplest system uses a drum of wire. One end of the wire is connected to the pipeline at a suitable point (e.g., at the landfall or at an offshore riser) and the other end is connected to a voltmeter. A boat tows a reference electrode mounted on a submersible device that allows the reference electrode to be trailed close to the pipeline along the pipeline route. The wire is fed out and floats on the surface of the sea as the boat follows the pipeline route. The potential is logged, plotted, and analyzed. Inadequate potential indicates where the coating is damaged and corrosion may be occurring.

The trailing wire technique is rapid and has been used extensively for isolated pipelines, but it has limitations. It becomes less reliable as the length of the pipeline increases; wire failures occur, and there is a drift in potential as the resistance of the return path increases. In some cases the potential can fluctuate as geomagnetic potentials and currents (called telluric currents) build up in the wire and/or the pipeline. The potential of a cluster of pipelines that are electrically connected will be a blend of the potentials of all the pipelines, resulting in small defect areas being missed.

Several inaccuracies result from the inevitable variations in the geometrical relationship of the reference electrode to the pipe. The measured voltage changes as the distance between reference electrode and pipe or the environmental resistivity changes. This effect is most pronounced for buried pipelines because of the higher resistivity of seabed sediments and because the nature of the sediment can alter along the pipeline route. To an extent, these problems can be overcome by later calculation; and modern computers make this a simple procedure, but only if the relevant distance and resistivity factors are known. The distance between pipelines can be measured from coincident side scan sonar, but the local resistivity must be estimated from prior knowledge.

8.3.15.4 Current flux technique. For long pipelines and clusters of pipelines, it is more common to use a current flux technique. The current flowing from the anodes through the seawater to bare areas of the pipeline where the coating has failed creates potential gradients in the seawater. The potential gradients can be measured by a combination (termed an array) of reference electrodes. The potential gradient, the geometrical relationship between the reference electrode array and the pipeline, and the local resistivities are used to calculate pipeline potential and current flux from the anodes to the pipeline. Because the measurement of potential is indirect, the calculations require a more precise location of the reference electrode arrays with respect to the pipeline than does the trailing wire technique. A precise location can be obtained by mounting the reference electrode arrays on a submersible that can be flown along the pipeline, but this location method increases the cost of the survey. A current flux survey can be used to evaluate both the severity of coating defects and to estimate the output (and, therefore, the life expectancy) of the sacrificial anodes. Obviously, the accuracy of the estimate depends on the quality of the computer model and the accuracy of the geometrical and resistivity information. Reference electrode arrays are small, and it may be possible for the CP survey to be done in conjunction with an annual pipeline survey for span detection.

There are three basic types of current flux systems:

- An array of fixed reference electrodes
- Rotating twin reference electrodes
- A fixed reference electrode and a remote reference electrode

All systems measure the local potential gradient as the reference electrodes are transported along the pipeline. The reference electrodes measure the voltage differences in the seawater over a volume of space, and a computer calculates the current flux through the seawater from the distance apart of the electrodes, the distance of the array from the pipeline, and the seawater resistivity. The method is less accurate for buried pipelines. The rotating system generates an AC signal with the signal amplitude directly related to the local current flux.

8.3.16 Interactions between CP systems

Connected CP systems can interact, resulting in a higher rate of consumption of some anodes. The most common case occurs when a pipeline is electrically connected to an uncoated structure such as a jacket, which represents a large current sink compared to the pipeline. The extent of depletion of the pipeline anodes depends on the relative areas of jacket and pipeline; but usually the pipeline anodes are affected for a distance of 1–2 km. One option to prevent pipeline anode depletion is to insulate the pipeline from the jacket. Submarine isolation joints between the pipeline and riser are not used; the pipeline is isolated using an isolation flange or monobloc between the riser and the topside pipework. The riser clamps are also required to be designed to ensure isolation between the riser and the jacket.[41] Pipeline isolation was often done in the past, but is no longer common practice. It is more reliable to provide additional sacrificial anodes on the section of pipeline closest to the structure.

A more significant interference can occur when a pipeline that is protected by sacrificial anodes is electrically connected to a structure that is protected by an impressed current CP system. The ICCP system will have a greater driving voltage, and current will flow to the sacrificially protected pipeline. The sacrificial anodes will be cathodically protected and two kinds of interference can result. In the first case, tenacious oxides and hydroxide films form and passivate the sacrificial anodes. In this case, the anodes may not reactivate and corrode as required in the future. In the second case, corrosion of the sacrificial anodes is enhanced because alkali is generated on the anode surface and at high pH amphoteric dissolution reactions can occur. In the first case aluminum alloy anodes are affected, in the second case both aluminum and zinc are depleted.

8.3.17 Isolating flanges and monobloc joints

Pipelines can be electrically isolated from each other or from structures using non-conductive flanges or joints (monoblocs). Isolation flanges are the more common. In an isolation flange, a non-conductive gasket (e.g., FRP) is fitted between the bolted flanges. The bolts may bridge the flanges and must also be isolated using non-conductive washers and sleeves fitted through the flange holes. Movement of a pipeline can fracture these sleeves; therefore, it is a good practice to ensure that the flanges are hot boltable so that sleeves can be replaced while the pipeline is in service.

A monobloc, or isolating joint, is fabricated from lengths of pipe of different diameters so that one fits within the other. The pipe section ends have overlapping flanges within the annular space to transfer the longitudinal stress between the pipes. These flanges are separated by insulating gaskets, and the joint is sealed with epoxy cement. The completed joint is welded into the pipeline. Monobloc joints are available for all pipeline diameters but are most widely used for larger diameters (above 14 in.). If the transported fluids will have a high water cut, it is a good practice to coat the inner pipe within the monobloc to prevent stray currents bridging across the isolation. Inspection and maintenance are easier if the monobloc is located above the waterline.

Zener diode arrays are used to prevent sparking across isolation joints and to allow current to pass if a high voltage pulse occurs across the flange. Voltage pulses occur from lightning discharges, earth faults, and telluric currents. Telluric currents are irregular and short-lived currents induced into long pipelines by the rotation of the pipeline in the earth's magnetic field. They are particularly severe during periods of high sunspot activity.

For a riser to be isolated from topside pipework on a platform or a jetty, the riser clamps must also be designed to ensure the riser remains isolated from the structure. One approach is to line the clamps internally with polychloroprene (neoprene).

Pipelines and risers attached to or entering reinforced concrete structures may require electrical isolation from the reinforcement in the concrete. Reinforcing steel is cathodic to external steel because the alkaline environment around the steel, produced by the concrete, establishes a galvanic cell between embedded and exposed steel. There is limited oxygen diffusion into the concrete and the cathodic activity per unit area will be low, but there is a large area of reinforcing steel so that the overall current drain may be high. There will be an additional drain on the sacrificial anodes on the pipeline if the pipeline is electrically connected to the embedded steel. Typical design values for reinforced concrete are in the range 1–3 mA/m^2 of reinforced concrete surface. The alternative to electrical isolation is to attach additional anodes to the pipeline adjacent to the structure.

Electrical isolation flanges and joints must be checked periodically to ensure that they are providing isolation. Most failures result from breakage of non-conductive washers or sleeves around flange bolts and failure of the Zener diodes fitted across the isolation flange or joint. Monobloc joints have low failure rates. Measurement of isolation may be done by measuring the potential of the riser or pipeline on the two sides of the isolation unit. One side of the joint will be at the potential of the jacket and the other at the potential of the pipeline. The potential difference is usually greater than 50 mV. It is possible to use a resistance meter to check the isolation; however, the current flow across an isolation flange or joint can be very high, and this method is less reliable than potential measurements.

8.4 Concrete Weight Coatings

8.4.1 Coating application

Many submarine pipelines have to be overcoated with a concrete weight coating to provide negative buoyancy and ensure stability on the seabed. Stability is discussed in chapter 11. The concrete weight coating also provides mechanical protection against dropped objects and impact by trawl boards. Concrete weight coatings cannot be used on pipelines laid by the reel method.

The concrete weight coating is applied to the coated pipe by impingement, extrusion, or slip-forming. The minimum thickness for a concrete coating applied by impingement is about 40 mm, and the minimum thickness applicable by the extrusion process is 35 mm. Figure 8–5 illustrates an impingement process.

A typical concrete density is 2400 kg/m^3, but the density can be increased by incorporating a heavier aggregate such as iron ore, blast furnace slag, or barytes. It is important that the aggregate used is tolerant to the sulphate in seawater. Some carbonate materials react with sulphate and swell, resulting in a crumbling of the concrete coating.

An intercoat may need over thin coatings such as FBE to protect the thin epoxy coating from damage by the high velocity impinged concrete or the concrete extrusion process. Research by British Gas indicated that concern about coating damage by impinged concrete may be unfounded, as damage to coating without an intercoat was below 2%.

During S-lay the tensioners grip the concrete weight coating. Slip bands may be required for FBE, polyethylene, and polypropylene coatings to prevent the pipeline sliding out from within the weight coating. Slip bands are 1 m wide bands of an epoxy coating containing abrasive or angular sand applied at each end of the pipe before the concrete weight coating is applied.

The concrete coating will crack during installation because the pipeline bends over the stinger and reverse bends on the seabed. The cracks generally close up when the pipeline is finally at rest on the seabed. However, very thick weight coatings may need to be provided with slots to provide flexibility during installation. Concrete itself has inadequate strength to resist bending and must be reinforced. For large diameter pipe, a cage of welded steel often provides the reinforcement but, increasingly, and for the smaller diameter pipelines, a zinc-coated mesh is used to provide the reinforcement.

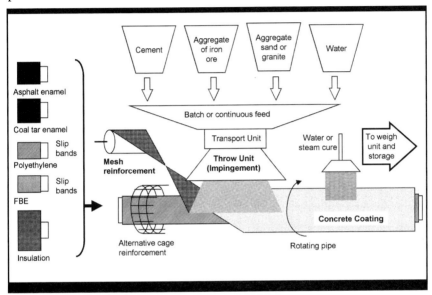

Fig. 8–5 *Concrete Impingement*

Studies on reinforced concrete indicate that exfoliation of the concrete can occur with consequent loss of stability if the steel reinforcement corrodes. Rusted steel has from 10 to 15 times the volume of the original steel from which it was formed, and this volume increase can cause cracking and detachment of the overlying concrete. Mesh formed from small-diameter wire has insufficient steel section to produce enough rust to cause cracking. The zinc coating on the steel mesh corrodes to produce a soft voluminous slime that blocks the cracks and pores in the concrete and reduces the rate of diffusion of corrodants to the reinforcement steel.

It is important that the reinforcement does not make electrical contact with the pipe steel or with the anodes because the reinforcement could then act as a Faraday cage or drain the cathodic protection current intended for the pipeline. Some contractors allow for a notional drain of cathodic protection to the reinforcement of 1 mA/m^2 of pipeline surface, but there is no technical evidence to support this allowance.

8.4.2 Anode installation on concrete-coated pipelines

The sacrificial anodes used on pipelines are cast as two half-shells, termed bracelets, and normally installed at the center of a pipe length over the anti-corrosion coating. It is common practice to coat the inside faces of the anodes with a coal tar epoxy or enamel to prevent anode dissolution in the gap between the pipe and the anode. Excessive dissolution would lead to an expansive stress on the anode and possible cracking. The anodes may be pre-installed before the concrete weight coating is applied, or the weight coating may be cut out before the concrete sets and the anode post-installed. For pre-installation the anode must be adequately protected from the impinged concrete, and it is vital that the protection is removed before the pipe joint is shipped offshore.

Anodes for use on hot pipelines may be installed over the insulation coating, or the anode may have an insulating (e.g., elastomeric) coating applied on the inner surface. These approaches reduce the temperature of the anode and increase the anode efficiency.

8.4.3 Field joints

The ends of each pipe section are left bare of anti-corrosion coating because the pipe sections are to be welded together. These bare sections represent around 4% of the total area of the pipe and need to be coated to reduce the drain on the CP system. If factory-prepared double-jointed pipe is to be used to minimize offshore welding, the weld is dressed and the field joints are finished in the coating yard before shipment of the pipe.

On the lay barge the welded field joint is cleaned to minimum SA 2½ and covered. The cover options are adhesive tape wrap, shrink sleeve wrap, or a powder epoxy coating applied to a pre-heated pipe. Tape is adhesive backed heavy-duty plastic tape that is applied as a double wrap around the pipe.[42] Shrink sleeves, originally developed for making water-proof electrical connections, are a specially formulated plastic that contracts when heated. The material is formed by heating and irradiating the film, stretching it, and allowing it to cool while stretched. On reheating, the tape returns to its original size. The shrink-wrap is applied around the cleaned, bare heated steel and additional heat is carefully applied with a torch to trigger the shrinkage onto the pipe. An FBE coating is applied in a very similar manner to the parent coating. The joint is cleaned, heated using an induction coil, and the epoxy powder flocked onto the heated area. Field joint protection procedures are illustrated in Figure 8–6.

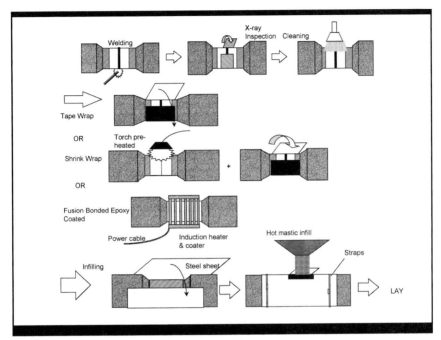

Fig. 8–6 *Field Joint Corrosion Protection and Infill*

An FBE field joint sometimes blisters as hydrogen is evolved from the weld area. The rate of hydrogen diffusion out of the weldment is accelerated by the heating of the steel by the induction coil. The area exposed at the blisters is very small and would be protected by the CP system.

For concrete-coated pipelines, the field joint has to be filled so that the pipeline has a constant outside diameter. The tracks or rollers that grip the weight coating might damage the weight coating if the field joints were not filled. The infill also provides mechanical protection to the field joint. The majority of infills used are formed from mastic applied when molten. A steel sheet is wrapped around the field joint and secured by large circle clips. The steel sheet includes a flap cut into the top. With the flap lifted, the annular space is filled with the molten mastic, which is then chilled and set using a water quench.

Alternate infills are formed from foamed plastics or rapid-setting concrete set with boiling water. Both these methods are applied using a temporary mould to contain the foam or concrete during the curing stage. Most foam infills appear, however, to confer little or no impact resistance; the foam becomes water saturated and may fail explosively when impacted because of the transfer of the hydraulic force through the foam. Foams with high solids content provide some mechanical protection. Rapid set concrete provides additional weight for stability, and simulation tests indicated a very high resistance to mechanical impact. Using metal or fibre filaments to provide concrete reinforcement can further increase the concrete impact strength.

Fishermen have complained about the steel sheets left on the joints infilled with hot mastic. Over time, the retaining bands corrode, and the sheets spread apart and damage trawl nets. To overcome this, either mouldless infill processes have to be used or the sheets must be consumable. Rapid corroding magnesium sheets were developed but are expensive. Sheet metal joints that are scored to encourage corrosion and fragmentation have also been suggested. Formless joints produced using foam or rapid-set cement are presently the favored alternatives to hot mastic.

8.5 Thermal Insulation Coatings

It is sometimes necessary, for flow assurance, to maintain the temperature of the fluids in a pipeline to prevent the formation of hydrates, to reduce wax deposition, to reduce the pressure drop by reducing the viscosity of heavy crudes, or to aid subsequent processing. Insulating coatings have seen much development recently.

There is a conflict between the needs for low thermal conductivity and for good mechanical properties. Good thermal properties imply a low-density open structure of foam or powder, but that inevitably makes the material mechanically weak. Solid elastomers and polymers are mechanically strong but are not good thermal insulators. Some fine powdery materials apply the microporous principle: if the pore size is less than the mean free path of the gas molecules within the pore, the thermal conductivity is smaller than that of the gas alone; and if the pores are small enough, this effect can occur even at atmospheric pressure. Mean free path is inversely proportional to pressure, so the thermal conductivity can be further reduced by reducing the pressure.

Multilayer coatings are the most usual form of coating, and coating applicators have developed many proprietary systems. The pipe is coated with an anti-corrosion coating able to tolerate the temperature, then a layer of insulating foam is applied, and finally the foam is protected with an outer wrap. Among the systems available are the following:

- FBE primer + high density polyurethane foam + HDPE sheath
- FBE primer + PVC foam + PE sheath
- EPDM corrosion coating + PVC foam + EPDM outer sheath
- FBE primer + polypropylene coating + PP foam + PP outer sheath

Foams exposed to hydrostatic pressure are progressively crushed, sometimes immediately and sometimes slowly in response to creep. A compromise option is a syntactic material in which small hollow glass microspheres are included within a polymer matrix.

The requirement for high mechanical strength and resistance to hydrostatic pressure can be eliminated by a pipe-in-pipe scheme. The transported fluid is carried by an internal line, which lies within a larger external pipe, usually supported on spacers. The inner line carries the operating pressure. The outer line carries the external hydrostatic pressure and resists bending and external concentrated and impact loads. The insulant partly or completely fills the annular space between the two, and does not need more mechanical strength than is required for installation within the annulus. The annulus can be partially evacuated and can contain another gas such as nitrogen or argon. Pipe-in-pipe systems can be fabricated on a lay barge, but doing so is a slow process and much more commonly the pipe-in-pipe is fabricated onshore then installed by reeling or by one of the tow methods. Both are described in chapter 12. A pipe-in-pipe may be relatively expensive because of the need for a second external pipe and because fabrication is much more complex than for a single pipe.

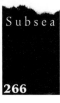
References

1   Ashworth, V., C. Booker, eds. (1986) *Cathodic Protection Fundamentals, Cathodic Protection—Theory and Practice.* Chichester, UK: Ellis Horwood Ltd.

2   King, R. A., J. D. A. Miller, D. S. Wakerley. (1973) "Corrosion of Mild Steel in Cultures of Sulphate-Reducing Bacteria: The Effect of Changing the Soluble Iron Concentration During Growth." *British Corrosion Journal.* 8, 89-93.

3   Chalke, P. (1992) "The Practicalities of Basing Your Corrosion Protection Budget on the Estimated Lifetime of a Pipeline." Conference on Cost-Effective Corrosion Protection of Offshore Pipelines. London: Institute International Research.

4   NACE Recommended Practice 0675–88: "Control of External Corrosion on Offshore Steel Pipelines." National Association of Corrosion Engineers.

5   DNV-RP-F106: "Factory Applied External Pipeline Coatings for Corrosion Control, 2000." Norway: Det Norsk Veritas.

6   SSPC PS10.01: "Hot Applied Coal Tar Enamel Painting System." USA: Structural Steel Painting Council.

7   NACE Recommended Practice 0185: "Extruded Polyolefin Resin Coating Systems for Underground or Submerged Pipe." National Association of Corrosion Engineers.

8   NACE TM0170: "Visual Standard for Surfaces of New Steel Air Blast Cleaned with Sand Abrasive." National Association of Corrosion Engineers.

9   NACE TM0175: "Visual Standard for Surfaces of New Steel Centrifugally Blast Cleaned with Steel Grit and Shot." National Association of Corrosion Engineers.

10   ISO 8502: "Visual Standard for Surface Preparation for Painting Steel Surfaces." International Organization for Standardization.

11   NACE Recommended Practice 0287: "Field Measurement of Surface Profile of Abrasive Blast Cleaned Steel Surfaces using a Replica Tape." National Association of Corrosion Engineers.

[12] ASTM G12: "Non-Destructive Measurement of Film Thickness of Pipeline Coatings on Steel." American Society for Testing and Materials.

[13] NACE Recommended Practice 0186: "Discontinuity (Holiday) Testing of Protective Coatings." National Association of Corrosion Engineers.

[14] ASTM G62: "Holiday Detection in Pipeline Coatings." American Society for Testing and Materials.

[15] NACE Recommended Practice 0274: High Voltage Electrical Inspection of Pipeline Coatings Prior to Installation." National Association of Corrosion Engineers

[16] ASTM G8: "Cathodic Disbonding of Pipeline Coatings." American Society for Testing and Materials.

[17] ASTM G80: "Specific Cathodic Disbonding of Pipeline Coatings." American Society for Testing and Materials.

[18] ASTM G42: "Cathodic Disbonding of Pipeline Coatings Subjected to Elevated Temperatures." American Society for Testing and Materials.

[19] ASTM G89: "Cathodic Disbonding of Pipeline Coatings Subjected to Cyclic Temperatures." American Society for Testing and Materials.

[20] ASTM G13/G14: "Impact Resistance of Pipeline Coatings (Limestone/Falling Weight)." American Society for Testing and Materials.

[21] ASTM G17: "Penetration Resistance of Pipeline Coatings (Blunt Rod)." American Society for Testing and Materials.

[22] Marine Technology Directorate. (1990) *Design and Operational Guidance on Cathodic Protection of Offshore Structures, Subsea Installations and Pipelines.* UK: MTD Publications.

[23] Parker, M. E., E. G. Peattie. (1988) *Pipeline Corrosion and Cathodic Protection.* Houston: Gulf Publishing.

[24] Baeckman, W. V., W. Schwenk. (1975) *Handbook of Cathodic Protection.* Translated by E. Molesley. UK: Portcullis Press.

[25] Morgan, J. (1993) *Cathodic Protection.* Houston: NACE.

26 Elliason, S. (1992) "Experience with Cathodic Protection Design in Norwegian Waters." Redefining International Standards and Practices in the Oil and Gas Industry Conference. London: Institute International Research.

27 Ffrench-Mullen, T., R. Jacob. (1985) "Pipelines Undersea." *Cathodic Protection—Theory and Practice*. eds. V. Ashworth, C. Booker. Chichester: Ellis Horwood Ltd.

28 Oganowski, C. H. (1985) "Studies on the Marine Corrosion of Cathodically Protected Steel by the Sulphate-Reducing Bacteria." (MSc dissertation) University of Manchester Institute of Science and Technology (UMIST).

29 King, R. A. (1980) "Prediction of Corrosion of Seabed Sediments." *Materials Performance* 19 (1). 39–43.

30 King, R. A., J. D. A. Miller. (1989) "Cathodic Protection and Sulphate-Reducing Bacteria." *Corrosion*. 1. 1–15.

31 DNV RP B401-1996: "Cathodic Protection Design." (Also released as DNV RP F-103, 2002) Norway: Det Norsk Veritas.

32 NACE RP 0169-92: "Control of External Corrosion on Underground or Submerged Metallic Piping Systems." National Association of Corrosion Engineers.

33 Harvey, D. (1992) "Cathodic Protection Standards for Land and Marine Applications." Presented at the Conference on Redefining International Standards and Practices in the Oil and Gas Industry. London: Institute International Research.

34 NACE TM0190-90: "Impressed Current Test Method for Laboratory Testing of Aluminium Anodes." National Association of Corrosion Engineers.

35 NACE Recommended Practice 0492-92: "Metallurgical and Inspection Requirement for Offshore Pipeline Bracelet Anodes." NACE RP0387-90: "Metallurgical and Inspection Requirements for Cast Sacrificial Anodes for Offshore Applications." National Association of Corrosion Engineers.

36 ASTM B418: "Cast and Wrought Galvanic Zinc Anodes." American Society for Testing and Materials.

37 D. Efird. (1992) "Overview of NACE Offshore Cathodic Protection Standards." Presented at the Conference on Redefining International Standards and Practices in the Oil and Gas Industry. London: Institute International Research.

38 S. N. Smith. (1993) "Analysis of Cathodic Protection on an Under-protected Offshore Pipeline." *Materials Performance*, 23(4). National Association of Corrosion Engineers.

39 DNV-OSS-301: "Certification and Verification of Pipelines." (2000) Norway: Det Norsk Veritas.

40 Backhouse, G. H. (1985) "Equipment for Offshore Measurements." *Cathodic Protection—Theory and Practice*. V. Ashworth and C. Booker, eds. Chichester: Ellis Horwood Ltd.

41 NACE Recommended Practice 0386-86: "The Electrical Isolation of Cathodically Protected Pipelines." National Association of Corrosion Engineers.

42 NACE MR 0274: "Material Requirements in Prefabricated Plastic Films for Pipeline Coating." National Association of Corrosion Engineers.

9 Pipeline Hydraulics

9.1 Introduction

The purpose of a pipeline is to carry a fluid from one point to another. A primary design objective, therefore, must be to ensure that the required flow can be driven by the pressure available, and that the line is correctly designed to optimize the balance between construction cost (which usually increases as the line diameter increases) and operating cost (which decreases as the line diameter increases, because of the reduced pressure drop and pumping losses). Hydraulic design interacts with calculation of temperature profile. If the temperature is too low, wax disposition and hydrate formation occurs and restarting the flow is difficult. If the velocity is too high, internal erosion and noise occur. Irregular flow and vibration can occur if the flow is multiphase. This is a huge subject, and this chapter can provide only a brief introduction. Multiphase flow is a very active area of research.

9.2 Single-Phase Flow of Newtonian Fluids

Single-phase means that the whole cross-section of the pipe is filled with either a gas or a liquid but not part gas and part liquid. A flow where the pipe contains some gas and some liquid is called *two-phase* and so is a flow with two separate liquids, such as oil and water. A *three-phase* flow has one gas and two liquid phases.

Newtonian means that the shear rate is linearly proportional to the shear stress, and a single number describes the relation between them called the *viscosity*. Gases, water, and most kinds of oil behave as Newtonian fluids. Some heavy oils are non-Newtonian, and a discussion of this complication follows.

An example to consider is a short length ds of a horizontal pipeline with uniform internal diameter D (Fig. 9–1) and the forces acting on an element of fluid flowing within it at a mean velocity U. Distance along the pipeline is measured by s that increases in the direction of flow. At the left end of the element, the pressure is p; at the right end, the pressure is $p+dp$. The pressure difference is balanced by the shear stress τ across the pipe wall. This shear stress occurs because the fluid has a finite viscosity and because there is a velocity gradient at the wall.

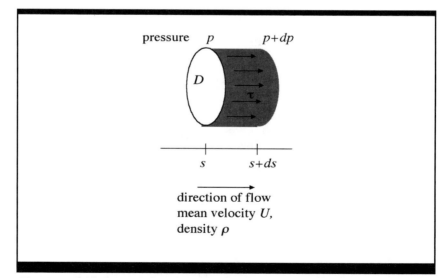

Fig. 9–1 *Fluid Element in Steady Horizontal Flow*

Since the fluid is moving at a uniform velocity, the resultant force on it must be zero: the force related to the pressure difference balances the force related to the wall shear stress t, and so

$$0 = \left(\frac{\pi D^2}{4}\right) dp + \tau(\pi D ds) \tag{9.1}$$

which rearranges into

$$\frac{dp}{ds} = -\frac{4\tau}{D} \tag{9.2}$$

The pressure decreases in the direction of flow, as expected since τ is positive. The pressure gradient dp/ds depends on the wall shear stress and the diameter. The shear stress is related to the mean velocity by

$$\tau = (1/2)f\rho U^2 \tag{9.3}$$

where

f is a skin friction coefficient

ρ is the fluid density

U is the mean velocity, equal to the volumetric flow rate q divided by the internal cross-section $\pi D^2/4$

Substituting from Equation (9.3) into Equation (9.2)

$$\frac{dp}{ds} = -\frac{2\rho f U^2}{D} \tag{9.4}$$

which is the fundamental equation that governs pressure drop.

In pipeline hydraulics, f is called the Fanning friction factor. Unfortunately, hydraulics also uses another friction factor, the Moody friction factor, here denoted m but often given other symbols, related to f by

$$m = 4f \tag{9.5}$$

It is clearly important not to confuse the two.

In single-phase flow of Newtonian fluids, the friction factor f is a function of Reynolds number Re, defined by

$$Re = \frac{\rho U D}{\mu} \tag{9.6}$$

where

μ is the viscosity.

The friction factor also depends on the relative roughness of the pipe, described by a ratio k/D, where k is a roughness height that describes the internal surface. For new steel pipe, k is about 0.05 mm, but it can be much increased by corrosion, erosion, and wax and asphaltene deposition. Many experiments have measured pressure drops in pipes, and the relation between f, Re, and k/D can be summarized in charts (Fig. 9–2). Also, f can be derived theoretically. In laminar flow (Hagen-Poiseuille flow), it is the following:

$$f = \frac{16}{\text{Re}} \qquad (9.7)$$

However, this formula only applies to Reynolds numbers up to about 2000. Most flows encountered in practice have Reynolds numbers much higher than that. Laminar flow breaks down and is succeeded by turbulent flow characterized by a turbulent boundary layer at the pipe wall. Most pipelines operate in the turbulent regime unless the flow is very slow or the fluid very viscous. For very smooth pipes in the turbulent regime, the relation between f and Re is the Kármán-Nikuradse implicit formula

$$\frac{1}{\sqrt{f}} = 4\log_{10}(\text{Re}\sqrt{f}) - 0.4 \qquad (9.8)$$

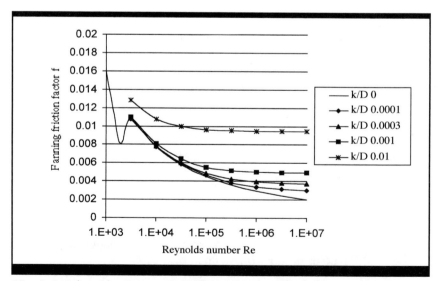

Fig. 9–2 *Relationship Between Reynolds Number, Friction Factor and Roughness*

Theoretical analysis of turbulent flow in rough pipes is described in standard texts on flow.[1,2] Thus far the pipe has been taken as horizontal. Equation (9.1) can be generalized by considering an element of fluid in a pipeline along a non-horizontal profile (Fig. 9–3):

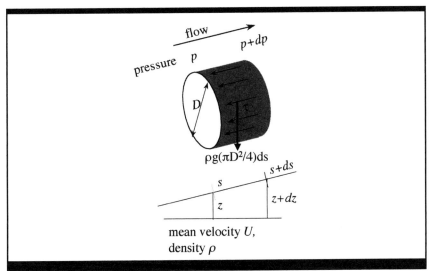

Fig. 9–3 *Fluid Element in Steady Inclined Flow*

Where

s is distance along the pipe, increasing in the flow direction

z is height measured upwards from a datum

ρg is the unit weight of the fluid

Again the flow is steady, so the resultant force on the element in the direction of the pipe axis must be zero.

$$0 = \left(\frac{\pi D^2}{4}\right) dp + \tau(\pi D ds) + \rho g\left(\frac{\pi D^2}{4}\right) ds\frac{dz}{ds} \qquad (9.9)$$

which rearranges into

$$\frac{dp}{ds} = -\frac{4\tau}{D} - \rho g\frac{dz}{ds} \qquad (9.10)$$

Again substituting from Equation (9.3)

$$\frac{dp}{ds} = -\frac{2\rho f U^2}{D} - \rho g\frac{dz}{ds} \qquad (9.11)$$

Equation (9.11) gives the pressure gradient in terms of the diameter, the velocity, the gravitational acceleration, the density, and the friction factor (in turn a function of the viscosity). If, and only if, all these quantities are constant along the length of the pipeline, Equation (9.11) can be integrated along the length of the pipeline to find the pressure drop from the upstream end 1 to the downstream end 2.

$$p_2 - p_1 = - \frac{2\rho f U^2 (s_2 - s_1)}{D} - \rho g(z_2 - z_1) \qquad (9.12)$$

where

$s_2\text{-}s_1$ is the length of the pipeline

$z_2\text{-}z_1$ is the height difference between the ends.

This formula can be used to find pressure drops in liquid flows where the temperature changes are small enough to have little effect on the viscosity and in gas flows where the pressure changes are small enough to have little effect on the density.

In many flows, Equation (9.11) applies, but the integrated form Equation (9.12) cannot be used because the properties that control f, ρ and U change significantly along the length. The following describe these factors:

- In many oil and gas flows, the temperature changes significantly along the length of the pipe because of heat transfer to or from the surroundings. The rate of heat transfer depends on the thermal conductivity of the pipe, on its insulation method, whether it is trenched and buried, and on the temperature of the surroundings. Oil viscosity μ is strongly dependent on temperature, and, therefore, temperature alters Re defined by equation (9.6)and therefore f.

- In all gas flows, density ρ is a function of pressure, and therefore changes from point to point: both pressure changes related to flow and pressure changes related to height play a part in this.

- In steady gas flows along a pipe of uniform diameter, the mass flow rate is constant along the length, but since the density changes because of pressure changes, the velocity is not constant along the length.

- While the second case applies to all gas flows, a further complication arises in most hydrocarbon gas flows: the gas cannot be idealized as a perfect gas because the operating temperature is not high compared to the critical temperature. This is taken into account by a compressibility factor Z, which is a function of gas properties and is particularly important for gases with a critical temperature close to the operating temperature, such as propane and ethylene.[2]

- Yet another complication in flows of real gases (as opposed to ideal perfect gases) is that enthalpy is a function of pressure as well as temperature. Therefore the temperature falls as the pressure decreases, even in a fully insulated pipeline. This phenomenon (Joule-Thomson cooling) can reduce the pipeline temperature below the temperature of the surroundings and lead to ice or permafrost formation. The Joule-Thomson coefficient $(\partial T / \partial p)_h$ expresses the rate at which temperature changes with pressure at constant enthalpy and can be determined from the relation between compressibility factor Z, pressure and temperature using thermodynamic relationships.[2, 3]

Because of all these factors, the integrated form is not applicable. The correct procedure is to integrate the fundamental Equation (9.11) in increments from the upstream end to the downstream end, taking into account:

- Heat transfer to the surroundings (through a heat transfer model)
- Variation in physical properties of the fluid, through a model of properties such as density and viscosity as a function of temperature and pressure
- Changes in height through the pipeline profile

This integration method is extremely laborious to do by hand but can be accomplished rapidly by a simple program. The program takes the flow rate and upstream conditions as input and outputs the downstream pressure and complete pressure and temperature profiles. The inverse problem of finding the flow rate given both the upstream and downstream conditions is best solved by systematic repeated trial.

There are simple approximate formula for gas flows, such as the Panhandle A formula used for transmission lines at relatively low pressures up to 7 MPa.[2] In American units

$$Q = 435.87 \, E \left(\frac{T_0}{p_0} \right)^{1.07881} \left(\frac{p_1^2 - p_2^2}{LT} \right)^{0.5394} G^{-0.4606} D^{2.6182} \quad (9.13)$$

where

Q is the flow rate in cu ft/day, measured at reference temperature T_0 deg R (deg F + 460) and reference pressure $p0$ psia

E is an efficiency factor, typically 0.92

P_1 is pressure at the upstream end, in psia

P_2 is pressure at the downstream end, in psia

L is the length in miles

T is the mean flowing temperature, in deg R

G is the gas gravity (air = 1)

D is the diameter in inches

The corresponding form in metric units is the following:

$$Q = 438 \times 10^{-3} E \left(\frac{T_0}{p_0}\right)^{1.07881} \left(\frac{P_1^2 - P_2^2}{LT}\right)^{0.5394} G^{-0.4606} D^{2.6182} \quad (9.14)$$

where

Q is the flow rate in m³/day, measured at reference temperature T_0 K (deg C + 273.16) and reference pressure p_0 mPa

E is an efficiency factor, typically 0.92

P_1 is pressure at the upstream end, in Pa

P_2 is pressure at the downstream end, in MPa

L is the length in km

T is the mean flowing temperature, in K

G is the gas gravity (air = 1)

D is the diameter in mm

Formulas of this kind are useful for preliminary sizing, but are much less used than they once were, because of the availability of user-friendly programs that do the job better and take more factors into account.

The discussion so far has considered steady flow at a uniform rate. Unsteady flow is also important, particularly in gas pipelines where line pack may be used to increase the storage capacity of the complete system. Numerical methods are used to solve unsteady flow problems.

Several hydraulics problems can occur in unfavorable conditions. In crude oil lines, mixtures of water and some crudes can form emulsions that have a much higher viscosity than oil alone and increase the pressure drop. In heavy waxy crudes, wax and asphaltenes drop out as a solid phase at low temperatures and accumulate on the pipe wall. The resulting loss of internal diameter has a dramatic effect on pressure drop. The volumetric flow rate q is related to the mean velocity U as illustrated by the following equation:

$$q = \frac{\pi}{4} D^2 U \tag{9.15}$$

and so, solving for U and substituting into Equation (9.4) neglecting gradient,

$$\frac{dp}{ds} = -\frac{32}{\pi_2} \frac{\rho f q^2}{D^5} \tag{9.16}$$

Since f only varies slowly with D (Fig. 9–2), for a given flow rate the pressure gradient is nearly inversely proportional to the fifth power of D, and a 10% reduction in D multiplies the pressure drop not by 1.1 but by $0.9^{-5} = 1.7$.

Wax deposition is controlled by keeping the temperature as high as possible, by pigging, and sometimes by adding inhibitors.

If the flow velocity in an oil pipeline is high, the wall of the pipe can be eroded, and the flow can be noisy (if that is significant). The allowable maximum velocity at which erosion will not occur is given by a semi-empirical formula

$$U_{max} = \frac{122}{\sqrt{\rho}} \tag{9.17}$$

where

U_{max} is the maximum velocity in m/s

ρ is the density in kg/m^3

If the flow velocity is low and the flow contains a significant fraction of water, water can drop out and flow as a second phase along the bottom of the pipe. If the water contains CO_2 or H_2S, preferential corrosion occurs at the 6 o'clock position. In any case, water will drop out if the flow stops and will tend to accumulate at low points in the profile. Combined erosion and corrosion are discussed further in section 7.4.3.

Some heavy crudes at low temperatures behave as non-Newtonian, non-linear fluids that have a yield stress; therefore, a finite shear stress is needed to produce any movement at all. If the yield stress in shear is denoted Y, the pressure difference needed to initiate movement over a length L in a line of diameter D can be obtained immediately from the fundamental Equation (9.2), and is

$$p_2 - p_1 = \frac{4YL}{D} \qquad (9.18)$$

So if the yield stress is 0.1 kPa (20 lb/ft^2), about the strength of tomato ketchup, the pressure difference required to start movement over a 50 km length of a 24-in. line is 33 mPa (4800 psi), which is substantial.

Finally some crudes form emulsions with water, and the emulsion can be much more viscous than either the crude or the water. Some of these problems can be controlled by maintaining the temperature and by chemical additives, but it is prudent to proceed cautiously and to seek specialist advice.

9.3 Calculation Example

A good example to study is a hypothetical oil pipeline carrying a maximum 100,000 barrels a day (b/d), reached 1 year after start-up, with a flow rate that progressively decreases and falls to 40,000 b/d after 6 years and 10,000 b/d after 10 years. At the same time, the amount of water produced to shore will increase from 0 after 2 years to 30,000 b/d after 10 years. Initially, the oil will be *sweet*, so the H_2S content will increase substantially. The task is to:

- Choose a diameter for the pipeline, assuming the following conditions:

  Oil density   850 kg/m^3

  Viscosity   0.01 Pa s

Maximum pressure at field   5 MPa

Minimum pressure at shore   1 MPa

- Consider problems that might be anticipated in the later stages of the life of the pipeline

The maximum flow rate is 100,000 b/d at the start. A starting point is a useful rule of thumb that says that the optimal diameter for an oil pipeline is roughly given by

$$D = \sqrt{\frac{q}{500}} \qquad (9.19)$$

where

q is the flow rate in b/d

D is the diameter in inches

However, note that this formula does not takes into account oil viscosity or density. It is based on a typical trade-off between capital and operating expenditure. The metric version of this formula is

$$D = \quad 840\sqrt{q} \qquad (9.20)$$

where

D is the diameter in mm and q the flow rate in m^3/s.

Rearranging (9.18) and putting in the numbers,

$$\sqrt{\frac{100000}{500}} = 14.1$$

Try a nominal 12-in. (actually 12.75 in. = 12.75x25.4 mm = 323.85 mm) outside diameter. Guess a 12.7 mm wall thickness for the purposes of hydraulic design. The velocity U is

$$\frac{100000 \ b \ / \ day \times 0.159 \ m^3 \ / \ b}{\frac{24 \times 3600 \ s/day}{\frac{\pi}{4} \ (0.32385 - 2 \times 0.0127)^2 \ m^2}} = 2.63 \ m/s$$

This is acceptable because it is less than the value given by Equation (9.16).

$$\frac{122}{\sqrt{\text{density}}} = \frac{122}{\sqrt{850}} = 4.2 \ m/s$$

for the velocity at which erosion will occur. Reynold's number is the following:

$$\frac{\rho U D}{\mu} = \frac{850 \times 2.63 \times 0.2985}{0.01} = 6.7 \times 10^4$$

From the friction factor chart (Fig. 9–2) for pipe with a relative roughness 0.001, the friction factor f is 0.0059. Therefore, the pressure drop is defined by the following equation:

$$\frac{2\rho f U^2 L}{D} = \frac{2 \times 850 \text{kg/m}^3 \times 0.0059 \times 2.63^2 \ (\text{m/s})^2 \times 28200 \ m}{0.2985 \ m}$$
$$= 6.554 \times 10^6 \ Pa = 6.554 \ MPa$$

which is too large.

If the outside diameter is increased to 14 in. and the wall thickness remains the same, D becomes $(14-1)(0.0254) = 0.3302$ mm. Then

$$U = 2.63 \times \left(\frac{0.2985}{0.3302}\right)^2 = 2.15 \text{m/s}$$
$$\frac{\rho U D}{\mu} = \frac{850 \times 2.15 \times 0.3302}{0.01} = 6.0 \times 10^4$$

From the friction factor chart for pipe with a relative roughness 0.001, the friction factor f is 0.0060, (only a little different from the previously calculated value because f varies only slowly with Re. The pressure drop is now:

$$\frac{2\rho f U^2 L}{D} = \frac{2 \times 850 \text{kg/m}^3 \times 0.0060 \times 2.15^2 \ (\text{m/s})^2 \times 28200 \ m}{0.3302 \ m}$$
$$= 4.03 \ MPa$$

which is acceptable.

The selected outside diameter is 14 in. (355.6 mm). In the latter part of the life of the field, the mean velocity will be much lower. After 10 years, it will be

$$2.15 \times \frac{40000}{100000} = 0.86 \ m/s$$

Water will begin to drop out and flow along the bottom of the pipe. It may be saturated with H_2S. Extensive corrosion may occur. Moreover, the oil is relatively heavy, and there may be wax deposition.

9.4 Heat Transfer and Flow Temperature

Heat can be transferred between a pipeline and its surroundings. This is important to the flow regime in various contexts, and it can also have significant effects on the surroundings. Consider for example, the common situation in which oil flows from a reservoir up well tubing to a subsea wellhead, then in a subsea flowline to a platform riser. The oil in the reservoir is at a high reservoir temperature, typically between 100 °C and 200 °C. The temperature falls in the tubing by conduction to the surrounding formations, but that kind of heat transfer is comparatively inefficient, so the oil reaches the seabed still at a high temperature. Heat is then transferred to the sea by conduction and by free and forced convection. The temperature of the oil falls between the wellhead end and the platform end. Because viscosity decreases with decreasing temperature, the viscosity increases towards the platform. This increase affects the pressure gradient (through the effect of Re on f).

If the flow stops during a well shutdown, the oil cools towards the sea temperature, and when flow restarts, the governing viscosity is the viscosity at the sea temperature. Moreover, if the oil cools sufficiently, wax may begin to separate and to form on the walls of the pipe. If water is present and the temperature is low and the pressure high, solid hydrates may form. If the arrival temperature at the platform is too high or too low, there may be process problems in equipment such as separators. If the pipe is hot near the platform, there may be large thermal expansion movements. If the pipe is hot and thermal expansion is constrained by friction against the seabed or by anchors, large axial compressive forces develop, and the pipe may buckle. Lastly, the temperature influences the rate at which the chemical reactions involved in corrosion take place.

For all these reasons, the engineer needs to know the distribution of temperature along the pipeline. Most hydraulic calculations include a parallel calculation of the temperature. The two calculations are coupled because the temperature affects the flow calculation (through the factors mentioned in section 9.2), while the flow calculation affects the temperature (through the effect of the flow velocity on the heat transfer coefficient).

Computer programs that generate flow calculations invariably calculate temperature as well. The program has to incorporate a heat transfer model, which determines the heat transfer rate to the surroundings as a function of the temperature of the pipeline contents. The heat transfer model covers both conductive heat transfer (through the pipe wall, its coatings, and the soil around the pipeline if the line is buried) and convective heat transfer (at the internal boundary between the fluid and the inner wall of the pipe, at the boundary between the pipe and the sea, and within the soil or rock cover if the pipe is buried). Convection can be important if the pipe is buried in rock or gravel because these materials are highly permeable to water and allow the formation of convection cells within the rock.

Many underwater pipelines are thermally insulated. The choice of insulation system is difficult. Foamed polymeric materials such as polyurethane foam are good insulators but mechanically weak because their strength has to come from the thin walls between the cells of the foam. If the foam is to retain its insulation properties, it has to be strong enough not to be crushed by the pressure of the surrounding water, either immediately or by slow creep. Solid elastomers and polymers are mechanically much stronger and tougher than foam but are less efficient insulators. An alternative is a compromise material such as Carizite™, which has hollow glass microspheres in a polypropylene matrix that keeps most of the mechanical strength of the polymer while cutting the thermal conductivity by the incorporation of the microspheres. Another alternative is a microporous material such as Microtherm™ or Wacker WDS™, which have extremely small nanopores that are smaller than the mean free path of the gas in the pores and give low thermal conductivities, particularly at low gas pressures.[4]

Still another alternative is a pipe-in-pipe scheme: the pipeline is enclosed in a second outer pipe, called a *carrier*, and the annular space between them is evacuated or filled with an inert gas. This filling reduces or eliminates conductive and convective heat transfer, and radiative heat transfer is reduced by incorporating reflective foils that radiate poorly.

9.5 Hydrates

Hydrates are solid compounds of hydrocarbons and water, rather like wet snow, formed when the components are in contact at low temperatures and high pressures.[2, 5, 6] They can be troublesome in hydrocarbon pipelines because once they have formed, they partially or completely block the flow. Hydrates are also important in other contexts: methane hydrates occur naturally and may be an important source of fuel in the future. However, drilling into undersea hydrates, particularly with water based muds, can lead to hazardous gas releases and well control problems, and dissociation induced by global warming may create positive feedback that accentuates climate change processes because methane is itself a greenhouse gas.

Figure 9–4 is a phase equilibrium diagram that expresses the conditions under which hydrates can form in a water/methane and in a water/95% methane/5% ethane system. The presence of higher hydrocarbons such as ethane (C_2H_6, C2) and propane (C_3H_8, C3) shifts the phase equilibrium towards higher temperatures and lower pressures, therefore making hydrate formation more likely for higher gas gravities. Katz and Edmonds give extensive data on the effects of higher hydrocarbons, and Katz gives a calculation method for phase equilibrium of gas mixtures.[2, 5]

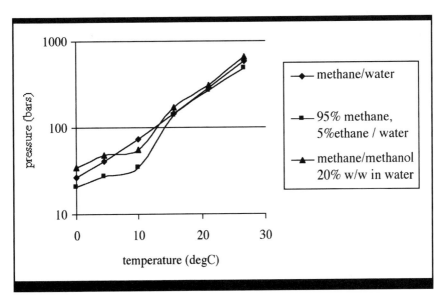

Fig. 9–4 *Phase Equilibrium of Hydrates*

Thermodynamic equilibrium is not the whole story because there is a nucleation barrier—similar to the barrier that allows pure water to be supercooled below the freezing point—that may prevent hydrates from beginning to form even though they could be stable if they had formed. However, once a hydrate plug has formed, it may be necessary to raise the temperature and/or lower the pressure substantially before the plug dissociates.

Hydrates are an important factor in marine pipeline design. One strategy is to keep the temperature as high as possible using one of the insulation strategies described in section 9.4. However, if the flow stops, the contents of the pipeline begin to cool and if hydrates form and obstruct the flow, it becomes difficult to raise the temperature to a level at which the hydrate dissociates. An active heat tracing system can counter this difficulty: Heat tracing is feasible, but relatively costly, and requires a power source.

Another strategy is to add a substance to the flow that shifts the phase equilibrium, as shown in Figure 9–4, toward higher pressures and lower temperatures. These additives are called *thermodynamic inhibitors*, and methanol and glycol are the most widely used. Figure 9–4 shows the effect of 20% w/w methanol in the water phase.[5] Inhibitors of this kind have to be added in relatively large quantities, a fact that usually requires a separate pipeline, Also the additive must be recovered at the downstream end. Salt (NaCl) is an inhibitor that often occurs naturally in produced water.

A third strategy is to add a different kind of inhibitor, a *kinetic inhibitor* (threshold hydrate inhibitor (THI) and low dose hydrate inhibitor (LDHI)) that alters the rate of nucleation, growth, or clumping together of the hydrate crystals. A kinetic inhibitor does not prevent hydrate formation, but it may delay it so that it is not troublesome; or it may switch it into a different form that does not block the pipeline. Polymers such as polyvinylpyrrolidone and polyvinylcaprolactam are THIs. They are much more expensive per unit volume than simple chemicals such as methanol, but the dosage rates are very much lower, so the total cost may be competitive.

9.6 Multiphase Flow

Multiphase flow occurs in many hydrocarbon pipeline systems where the total fluid flow can include a gas phase, one or more liquid phases (oil and water), and sometimes a solid phase (sand). The hydrodynamics of multiphase flow are much more complex than single-phase flow because the phases have very different densities and different mechanical properties and because several different flow regimes can occur. Multiphase flow is a specialist subject that could easily be the subject of several books and is a major research topic.

In a single-phase flow, the mean velocity is defined by the volumetric flow rate of the single phase divided by the cross-sectional area. If there is more than one phase, there needs to be a way of describing the velocity of each phase. The most convenient way is the phase superficial velocity, which is the velocity the phase would have if it were flowing in the pipeline on its own without the other phase or phases. Considering the case of two phases, gas and oil with volumetric flow rates q_g and q_l, the corresponding superficial velocities U_g and U_l, are the following:

$$U_g = \frac{q_g}{(\pi/4)D^2} \tag{9.20}$$

$$U_l = \frac{q_l}{(\pi/4)D^2} \tag{9.21}$$

Note that superficial velocity is not the same as the velocity at which the phase moves because superficial velocity reflects the relative flow rates. This can be understood by thinking of the case in which most of the cross-section is liquid, but a few gas bubbles move with the liquid at the same velocity. The superficial velocity of the gas is then much smaller than the superficial velocity of the liquid even though their phase velocities are the same. It should be noted that the ratio between the superficial velocities is not the same as the ratio between the fractions of the cross-section occupied by each phase. A case in point is a pipeline for which half the cross-section is liquid and half the cross-section is gas, but the gas is moving 10 times faster than the liquid.

The first complication compared to single-phase flow is that several different flow modes can occur. The modes can be distinguished on a flow map, as illustrated in Figure 9–5, which has gas superficial velocity on one axis and liquid superficial velocity on the other. The bottom left of the map corresponds to both phases moving slowly. The bottom right corresponds to the gas moving fast or in a large quantity and the liquid moving slowly or in a small quantity. The inset sketches show a short section of pipe with a schematic of the flow mode, with liquid (l) and gas (g) phases identified.

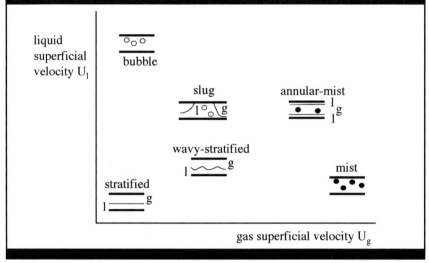

Fig. 9–5 *Flow Regime Map*

Much can be learned by thinking of everyday examples of two-phase flow. For example, water being poured slowly through the neck of a bottle. The water is the liquid phase, and the air is the gas phase. The unit weight of the water is about 800 times the unit weight of the air. The flow is stratified: the heavier water occupies the lower part of the neck, and the lighter air the upper part. Unsurprisingly, then, if both superficial velocities are very low, the flow is stratified, with liquid at the bottom and gas at the top, as shown in the bottom left-hand corner of Figure 9–5.

Now should the gas superficial velocity increase, the gas is moving faster than the liquid. That is what happens when wind blows across the surface of the sea: Unless the relatively velocity is very small, waves form on the surface. The stratified flow becomes a wavy-stratified flow.

When the gas velocity is higher still, the effect is loosely corresponds to a high wind at sea. The wind picks up some of the water and carries it along as spray: this is a mist flow in which the gas phase is continuous, and the liquid is carried as separate droplets. If there is more liquid, the wall of the pipe may be covered with liquid blown along by the gas, and this is an annular-mist flow.

Another case to consider is one that occurs when the liquid superficial velocity is high and the gas superficial velocity low. The gas forms separate bubbles in a continuous liquid flow; the bubbles tend to flow along the upper part of the pipe because they are lighter than the liquid.

A final case to consider is a flow that starts as wavy-stratified, but the superficial velocities increase. The waves grow until they fill the whole cross-section of the pipe. The flow becomes a slug flow in which the pipe is alternately filled with bubbly liquid (a *slug*) and with gas plus little or no liquid. Slug flow can be very troublesome because any treatment process at the downstream end of the flow has to be able to accept the irregular and uneven arrival of large volumes of liquid with little gas and then large volumes of gas with little liquid. The presence of slug flow may require that the flow first be taken to a large separator (*slug catcher*). Much of the research effort given to multiphase flow is devoted to predicting and dealing with slug flows.

Until recently, it was difficult to make reliable predictions of pressure drops in multiphase systems, and equally difficult to predict liquid hold up, a measure of how much liquid is in the pipeline. Although research has been conducted using several available correlations, it had a mixed empirical and theoretical basis and was based on small-scale experiments in low-pressure air/water systems. A typical test was on a 2-in. pipe containing an air/water mixture at atmospheric pressure, giving a density ratio of 800, whereas a full-scale hydrocarbon system might have a pipe diameter of 20 in. and a density ratio of 20. Since the scaling laws were imperfectly understood, tests results could not be reliably scaled up. Multiphase flow research has become a very active field, and the ability to predict pressure drops and hold ups and to deal with slug flow has much improved. There are several new test loops, which allow hydrocarbon systems to be tested on a reasonably large scale. Much more operating data from real pipelines have been made available and correlated systematically.

Flow analysis is complicated by several factors that have much more effect in multiphase flows than in single-phase. Whereas changes of level have little effect in single-phase, except through the hydrostatic height component in Equation (9.9), they have a major effect on stratified flows (with the denser liquid phase at the bottom and the lighter gas phase above it). The flow regime on a downslope of 0.002 (2 m per km) can be completely different from the flow on an upslope of 0.002, even though everything else is the same. On the down slope, gravity helps to drive the liquid flow so that the liquid runs down the slope on its own. This effect produces unexpected results. For example, when the pumps driving the flow at the upstream end are stopped, the flow at the downstream end continues for some time (for two hours in one instance of a pipeline across the Alps).

On an upslope, on the other hand, gravity holds the liquid back, and it can be carried forward only by shear forces transmitted from the faster-moving gas flow. The slower-moving liquid tends to accumulate and forms liquid slugs filling the entire cross-section. The dynamic pressure that occurs when the gas flow is cut off by the slugs drives the slugs violently forward.

An additional complication is that as the pressure and temperature change, the compositions of the phases change. In rapidly changing flow conditions in slug flows, there may not be time to reach phase equilibrium.

Several correlations have been incorporated in computer programs. The systems of correlations are self-contained and internally consistent, but they often give different answers. Much work has been done to compare the correlations with each other and with field measurements, and there are accepted views about which correlation is best for which prediction in which situation. It is wise to use a correlation that has been validated by comparison with a similar situation for which real measurements exist.

Complex situations occur in unsteady two-phase flows. They have important practical consequences: For instance, a slug catcher must be designed both for the slug volume occurs during quasi-steady flow and for the larger slugs that form when flow restarts. The behavior of the flow depends on the details of the restart procedure.

References

1   Kay, J. M. (1957) *Fluid Mechanics and Heat Transfer*. Cambridge, England: Cambridge University Press.

2   Katz, D. L. et al. (1959) *Handbook of Natural Gas Engineering*. New York: McGraw-Hill.

3   Pippard, A. B. (1964) *Classical Thermodynamics*. Cambridge, England: Cambridge University Press.

4   Fricke, J. (1993) "Materials Research for the Optimisation of Thermal Insulations." *High-Temperature High Pressures*, 25,. 379–390.

5   Edmonds, B., R. A. S. Moorwood, R. Szczepanski. (2002) "Controlling, Remediation of Fluid Hydrates in Deepwater Drilling Operations." *Ultradeep*. March 2002. 7–11.

6   Sloan, E. Dendy. (1990) *Clathrate Hydrates of Natural Gases*. New York: Marcel Dekker, New York.

10 Strength

10.1 Introduction

A pipeline clearly has to be strong enough to withstand all the loads that will be applied to it, both during its construction and in operation. During construction, it will be bent and pulled and twisted. When it goes into operation, it will be loaded by internal pressure from the fluid it carries, by external pressure from the sea, and by stresses induced by temperature changes. Sometimes it will be loaded by external impacts from anchors and fishing gear.

This chapter considers design to resist internal pressure, design to resist external pressure, longitudinal stress, bending, indentation, and impact. It focuses on the underlying structural mechanics and simple analytical approaches that provide solutions to most problems that arise in practice. More complex situations are nowadays analyzed by finite-element analysis, applying large-deformation program suits such as ABAQUS.

10.2 Design to Resist Internal Pressure

Internal pressure from the contained fluid is the most important loading a pipeline has to carry. It is easy to forget how large the forces generated by pressure are. Figure 10–1(a) is a cross-section of a typical large-diameter gas pipeline, a 30-in. pipeline, carrying an internal pressure of 15 MPa. To determine the tension set up in the pipe walls in each metre of pipe, one must consider the equilibrium of everything below the sectioning plane drawn in the Figure 10-1(a).

Half the pipe and half the contents are redrawn in Figure 10–1(b). Pressing downwards on a 1 m length of this system is the pressure of the gas multiplied by the area over which it acts, a rectangle 0.75 m broad and 1 m long. Pulling upwards on the same 1 m length is the tension in the pipe walls. There are no other vertical forces in a straight section of pipe. Since the system is in equilibrium, the tension in the pipe walls must be

$$(15 \text{ MPa}) \times (0.75 \text{ m}^2) = 11.25 \text{ MN}$$

This value is 1100 tonnes, a substantial force, more than the take-off weight of three 747 aircraft. The same force is carried by each and every meter of the pipeline. It is useful to remember how large the forces are that are interfered with if a pipeline is dented or gouged or if it is perforated intentionally or accidentally.

The hoop stress generated by internal pressure is statically determinate, so that no significant stress redistribution can occur; and the stress is not modified or relieved by plastic yield. If the hoop stress is too large, the pipeline can yield circumferentially, and continued yielding will lead to thinning of the pipe wall and ultimately to a rupture.

Going on to consider stress rather than force, the hoop stress can be calculated from equilibrium by essentially the same argument put into algebraic terms. For example, a pipeline with outside diameter D_0, inside diameter D_i, wall thickness t, internal pressure p_i, and external pressure p_0. The following Figure 10–2(a) shows a cross-section.

Now consider the equilibrium of everything within the circumscribing rectangle shown in (b): the rectangle is bounded by the diameter, two tangents at the point where the diameter intersects the outside surface, and a tangent parallel to the diameter (b) shows the stress components that act across the boundaries of different parts of the rectangle. s_H is the circumferential stress (hoop stress).

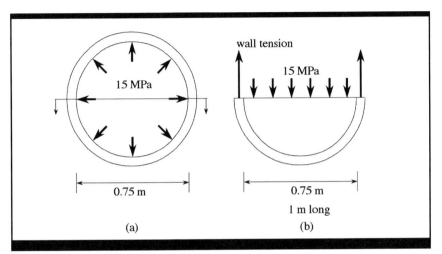

Fig. 10–1 *Forces in Pressurized Pipeline*

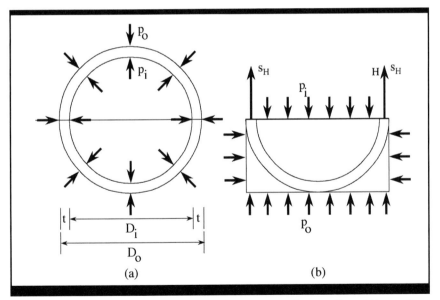

Fig. 10–2 *Circumferential Stress in a Pipeline Pressurized Internally and Externally*

The resultant force in the vertical direction must be zero. For unit length in the direction perpendicular to the paper:

$$0 = 2s_H t - p_i D_i + p_o D_o \qquad (10.1)$$

which rearranges into

$$s_H = \frac{p_i D_i - p_o D_o}{2t} \qquad (10.2)$$

This equation gives the mean circumferential stress exactly, whatever the diameter/ thickness (D/t) ratio.

A strong case can be made for basing all pipeline design for circumferential stress on Equation (10.2). However, for reasons of history and inertia, code committees have not come to that point; and various other versions of the formula are in use. The simplest and most widely used is the Barlow formula:

$$s_H = \frac{p_i D}{2t} \qquad (10.3)$$

This formula is derived by neglecting the external pressure term, $p_o D_o$, in Equation (10.3). D is normally taken as the outside diameter, which is obviously larger that the inside diameter. This can be interpreted as a rough-and-ready way of allowing for the small variation of hoop stress through the wall thickness. Comparison between Equations (10.3) and (10.2) shows that Equation (10.3) is always conservative in the sense that it gives a higher circumferential stress for the same internal pressure. It overestimates the maximum hoop stress, but many codes specify it use.

Clause 210 of the Det norske Veritas (DnV) 1996 rules requires the designer to use the following formula:[1]

$$s_H = (p_i - p_o) \frac{D_o - t}{2t} \qquad (10.4)$$

where

t is a minimum wall thickness that allows for manufacturing tolerance and for corrosion in operation

It can be interpreted as a compromise between Equations (10.2) and (10.3). The same formula is implicit in the treatment of internal pressure in clause 403 of the DnV 2000 rules, but those rules are based on a limit state approach that is discussed later in this book.[2]

The Equations (10.2), (10.3), and (10.4) are all derived from statics alone. An alternative approach is to carry out a full analysis, using the theory of elasticity, applying the full stress-strain relation and imposing the strain compatibility requirement and assuming the circumferential stress is not uniform. This analysis leads to the Lamé formula for the maximum hoop stress, which occurs on the inner surface and is

$$s_H = (p_i - p_o) \frac{D_o^2 + D_i^2}{D_o^2 - D_i^2} - p_o \qquad (10.5)$$

This formula calculates the maximum stress exactly, provided that the pipe remains elastic. The stress calculated from Equation (10.5) is always slightly higher than the stress calculated from Equation (10.2) but lower than the stress calculated from Equation (10.3) with D taken as the outside diameter. The differences increase as D/t decreases. Some codes allow the designer to use Equation (10.5).

Each of these formulae can be rewritten as a design formula for the nominal wall thickness required to keep the circumferential stress at or below a specified fraction f_1 of the yield stress Y, so that

$$s_H \leq f_1 Y \qquad (10.6)$$

f_1 is called the design factor.

Combining the Barlow formula, Equation (10.3), with Equation (10.6) results in the following equation:

$$\frac{p_i D}{2t} = s_H \leq f_1 Y \qquad (10.7)$$

This can be rewritten as a formula for the least wall thickness that satisfies Equation (10.6):

$$t \geq \frac{p_i D}{2 f_1 Y} \qquad (10.8)$$

It is often useful to add the second factor f_2, a manufacturing tolerance factor, which is intended to allow for variations in the actual minimum wall thickness below the nominal wall thickness. If, for example, a minimum wall thickness 12.5% below nominal is allowed, f_2 is 0.875. Incorporating this factor, Equation (10.8) becomes a formula for the minimum wall thickness

$$t \geq \frac{p_i D}{2 f_1 f_2 Y} \qquad (10.9)$$

Codes that impose a limit on hoop stress invariably specify a maximum value for the design factor (*usage factor* or *utilization factor*). In the past, the design factor was almost always taken as 0.72 for pipelines, and a lower value, 0.6 or lower, for risers and sections of pipelines close to platforms. Some codes still prescribe those values. However, it has been realized that the 0.72 factor was chosen more than 60 years ago, that it has been copied from code to code without much thought for a long time, and that it reflects the practice of a time when standards of pipe manufacture, welding, and construction were far inferior to that of today.[3, 4] Recent work has shown that the factor can safely be increased. How far it can be increased is the subject of lively debate in the industry. In the 1994 Canadian code, for example, the 0.72 design factor has been replaced by a higher factor of 0.80. The same change has been made in the 1995 Germanischer Lloyd rules. The DnV 1996 rules clause 204 imposes two conditions, one that limits hoop stress to a multiple η_s of specified minimum yield stress Y, and a second that limits hoop stress to a multiple η_u of specified minimum ultimate tensile strength T. The multiplier η_s takes the place of f_1 and depends on the safety class. Its values are listed in clause 204. For safety class *low*, the design factor is 0.83; for safety classes *normal* and *high*, the design factor is 0.77. The other multiplier η_u defined in clause 204 is 0.72 for safety class *low*, 0.67 for safety class *normal*, and 0.64 for safety class *high*.

Combining the DnV Equation (10.4) with the first of these conditions rearranges the equation into:

$$t \geq \frac{D}{\dfrac{2\eta_s Y}{p_i - p_o} + 1} \tag{10.10}$$

If, therefore, for a 30-in. pipeline

D	is 762 mm (30 in.)
p_i	is 20 MPa (200 bars, 2900 psi)
p_o	is 2 MPa (20 bars, 290.1 psi, corresponding to 200 m of water)
Y	is 413.7 MPa (N/mm^2) (60000 psi, X60)
η_s	is 0.83

the minimum wall thickness derived from equation (10.10) is 19.46 mm.

If the Barlow Equation (10.8) is applied with the same data, the minimum wall thickness is 22.19 mm. Most of the difference arises because Equation (10.8) leaves out the external pressure, which is relatively large in this example.

Alternatively, taking the *exact* Lamé Equation (10.6) as the basis for design, the minimum allowable thickness/outside diameter ratio t/D_o for the maximum hoop stress not to exceed $f1Y$ is given by the quadratic equation:

$$0 = \left(\frac{t}{D_o}\right)^2 - \left(\frac{t}{D_o}\right) + \frac{1}{\beta} \qquad (10.11)$$

where

$$\beta = 2\left(1 + \frac{f_1 Y + p_o}{p_i - p_o}\right) \qquad (10.12)$$

and the solution for Equation (10.11) is

$$t = \frac{1}{2}D_o\left(1 - \sqrt{1 - \frac{4}{\beta}}\right) \qquad (10.13)$$

Repeating the example above, the minimum wall thickness is 19.37 mm. The difference between the wall thicknesses derived from the DnV and Lamé formulas is normally small. The difference increases for smaller Do/t ratios, which correspond to higher internal pressures.

The DnV 2000 rules take a limit state approach to pressure containment design, based on partial factors.[2] The clause 402 requirement for pressure containment (bursting) requires that the difference between the local incidental pressure pli and the external pressure pe meet the condition

$$p_{li} - p_e \le \frac{p_b(t_1)}{\gamma_{sc}\gamma_m} \qquad (10.14)$$

where

γ_{sc} is a safety class reduction factor, listed in Table 5–5 of the rules, and for pressure containment taken as 1.046 for safety class *low*, 1.138 for safety class *normal*, and 1.308 for safety class *high*

γ_m is a material resistance factor, listed in Table 5–4 and taken as 1.15 for serviceability, ultimate and accidental limit states, and as 1.00 for the fatigue limit state

pb   is the pressure containment resistance, defined by

$$Pb \ = \ \text{Min} \left(\frac{2t}{D-t} f_y \ \frac{2}{\sqrt{3}}, \ \frac{2t}{D-t} \ \frac{f_u}{1.15} \ \frac{2}{\sqrt{3}} \right) \qquad (10.15)$$

in which the first term relates to the yielding limit state and the second to the bursting limit state.

In addition

t   is the wall thickness, allowing for fabrication tolerance and in the operational condition for a corrosion allowance (clause C300)

f_y is the characteristic yield strength, which is the specified minimum yield stress, where necessary derated to allow for a reduction due to increased temperature, multiplied by a material strength factor which takes into account the variability of properties (Table 5–2)

f_u is the characteristic tensile strength, which is the specified minimum tensile stress, where necessary derated to allow for a reduction due to increased temperature, multiplied by a material strength factor which takes into account the variability of properties (Table 5–2)

Rearranging Equations (10.14) and (10.15), the minimum wall thickness is

$$t \ = \ \cfrac{D}{1 + \cfrac{2}{\gamma_{SC} \gamma_m \ (P_{li} - P_e) \sqrt{3}} \ \cfrac{2}{} \ \text{Min} \left(f_y, \ \cfrac{f_u}{1.15} \right)} \qquad (10.16)$$

Continuing the example, and taking the following parameters in addition

γ_{SC} 1.138 for safety class normal,

γ_m 1.15,

f_u 413.7 MPa,

f_u at least 1.15 times 413.7 MPa, so that the yielding limit state governs,

the minimum wall thickness is 18.34 mm.

10.3 Design to Resist External Pressure

A large external pressure tends to make a pipeline *ovalize* (take on an oval shape) and collapse. A perfectly round pipeline loaded by a steadily increasing internal pressure would remain circular until the pressure reached the elastic critical pressure p_{ecr}, given by

$$P_{ecr} = \frac{E}{4(R/t)^3(1-v^2)} \qquad (10.17)$$

where

R is the mean radius (measured to half-way through the wall thickness)

t is the wall thickness

E is the elastic modulus

v is Poisson's ratio

and the pipeline would suddenly collapse. For most marine pipelines, the elastic critical pressure is quite high; for instance, for a 30-in. pipeline with a wall thickness of 22.2 mm, the elastic critical pressure is 12.5 MPa, corresponding to a water depth of 1250 m, which would generate a hoop stress of 208 MPa. Circumferential pressure yield is possible, but elastic collapse occurs first except for very thick pipelines.

Real pipelines are not perfectly circular but are always out-of-round to some extent. When an out-of-round pipeline is subject to external pressure, the out-of-roundness progressively increases and becomes very large at the pressure given by Equation (10.17). Before the elastic critical pressure is reached, the combination of hoop and circumferential bending stress reaches yield, and beyond that the pressure can only increase slightly before collapse occurs.

A simple analysis treats the out-of-roundness deviation of the cross-section from a circle as sinusoidally distributed around the circumference and treats the pipe as a thin-walled shell. The pressure p at which yield begins is the solution of a quadratic equation in p

$$p^2 - (p_Y + (1 + 1.5g(R/t))P_{ecr})p + p_Y p_{ecr} = 0 \qquad (10.18)$$

where

$$p_Y = \frac{Yt}{R} \tag{10.19}$$

which is the pressure at which the pipe would yield circumferentially if it were perfectly round and instability did not occur and if is a measure of out-of-roundness defined by

$$g = \frac{\text{maximum diameter} - \text{minimum diameter}}{\text{mean diameter}} \tag{10.20}$$

in the initial unloaded condition.

Equation (10.18) is found to give a safe estimate of collapse pressure although it is open to criticism. In particular, the analysis implicitly assumes that the pipe is stress-free in the initial condition. This is possible, but it is more likely that initial out-of-roundness is the result of some imperfection in the manufacturing process that has left the pipe with an initial circumferential bending moment and, therefore, not stress-free.

The DnV 2000 rules, clause D503, require the characteristic capacity for external pressure (collapse) to be calculated from a different formula which can be rewritten as

$$(p - p_{ecr})(p^2 - p_Y^2) = 2p_{ecr}p_Y g(R/t)p \tag{10.21}$$

in the present notation, but where

R is taken throughout as $D_0/2$, half the outside diameter, rather than half the mean diameter

g is denoted f_0 in the DnV rules, not to be taken smaller than 0.005

Equation (10.21) is a representation of the interaction between the three effects: circumferential yield, elastic instability, and circumferential collapse due to out-of-roundness that interact to lead to collapse. The equation is obviously based on idealizations of both the pipe geometry and the material behavior. It is surprisingly simple and can hardly represent the last word on the subject. Some of the details might be questioned, such as the selection of the thinness parameter as D/t based on the outside diameter D, rather than on the mean diameter $(D-t)/2$, which would be more consistent with thin-shell theory. However, Equation (10.21) is reasonably consistent with many tests, in particular a dataset of more than 300 tests on casing. One might guess that further work will produce a formula of the same type but with different coefficients.

The geometry is represented by the diameter, the wall thickness, and an out-of- roundness defined in terms of the maximum and the minimum diameters. A pipe with a cross-section of a Reulaux heptagon (the form of a British 50-pence coin) is obviously not round. It would have a collapse pressure lower than the collapse pressure of round pipe with the same diameter because the heptagon has corners and flatter sides, which together increase circumferential bending. However, under the DnV definition, its out-of-roundness is precisely zero because its maximum diameter and its minimum diameter are equal (defining diameter as the distance between parallel tangents on opposite sides). The curvature of the sides of the Reulaux heptagon is actually half as large as that of a circle with the same diameter, a fact that makes the Reulaux heptagon less able to resist external pressure.

The need for more sophisticated descriptions of geometric out-of-roundness has important implications for the description of tests, which have often been poorly reported. Similarly, the description of the geometry does not take into account the lengthwise variations in cross-section, which must be important, particularly if rapid changes in cross-section create additional flat spots.

The description of the plastic properties is based solely on yield stress. This is far from adequate. Yield stress as conventionally defined in the pipeline industry is, in any event, an incomplete description because it is measured at a strain of 0.005, which approaches the plastic range. Departures from linear-elastic behavior almost invariably occur at a stress substantially lower than the yield stress as conventionally defined; but specifications do not constrain this property, and the formula does not allow for it. In the same way, strain-hardening at strains beyond 0.005 varies between steels and pipe-making processes and influences the pressure at which the pipe buckles, but strain-hardening is not allowed for. A pipe made from a steel that yields at a low stress but strain-hardens rapidly would be expected to have a higher collapse pressure than one made from a steel that yields at a higher stress but strain-hardens only a little, but equations (10.20) and (10.21) predict the opposite.

Finally, one of the striking features of the formula is that it does not take into account the residual stresses left by manufacturing processes. Those stresses are very large, of the order of the yield stress at the temperature at which the pipe reached its final form. Residual stresses are only partially erased by hydro test and hydraulic or mechanical expansion and have a significant effect on the pipe's bending response to external pressure and on later instability.

Equation (10.20) can readily be rewritten as a equation whose solution is the required maximum D/t to withstand a given pressure p, as

$$\alpha^3 \beta(D/t)^5 - \alpha\beta(D/t)^3 - \alpha(\alpha+g)(D/t)^2 + 1 = 0 \qquad (10.22)$$

where

$$a = \frac{p}{2Y} \qquad (10.23)$$

$$b = \frac{Y(1-v^2)}{E}$$

which is a quintic in D/t and readily solved. It can be translated into a design chart. Figure 10–3 that follows plots D/t against α for three values of out-of-roundness and for β equal to 0.0389, which corresponds to X65 steel.

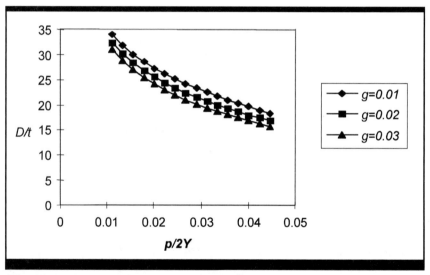

Fig. 10–3 *Minimum D/t Ratio as a Function of p/2Y*

Pipe out-of-roundness for deep-water pipelines is normally limited by the pipe specification, typically to 0.01 (1%). Figure 10–3 shows that reasonable values of round-of-roundness do not have a very large effect. Instances for which the out-of-roundness is great are in any case unacceptable for other reasons, such as line-up for welding (*hi-lo*).

Though the broad picture is clear, much more research on buckling under external pressure needs to be done. Selection of wall thickness is one of the most important design decisions to be made by the design engineer, and the wall thickness has a central influence on the cost and technical feasibility of a deep-water pipeline. In deep water, collapse under external pressure is usually found to drive the wall thickness decision. Deep water requires thick-walled pipe. In the design studies of the projected Oman-to-India pipeline in maximum depths of the order of 3000 m, for example, experimental and analytical studies indicate a minimum required wall thickness of well over 30 mm even for modest diameters of 20–26 in.[5, 6] These wall thicknesses burden the economic feasibility of projects that are attractive on other grounds and lead to many other difficulties, such as welding and upheaval buckling. However, a better approach is to construct the pipeline fully or partially liquid-filled, so that it never has to bear the full external hydrostatic pressure.[6, 7]

Figure 10–4 is a histogram of results of collapse tests carried out for the Oman-India pipeline and compares the measured collapse pressure with the collapse pressure calculated from Equation (10.21). The equation is slightly conservative, but not overly conservative. Though a more sophisticated analysis might lead to slightly lower wall thickness, big reductions are not to be hoped for. Instead, analysis can be based on finite-element analysis, which can take more complete account of the geometry and the material properties, not only strain-hardening and residual stress but also the departures from idealizations. One such idealization is the von Mises yield condition, which is known to be an incomplete representation of the behavior of real materials.

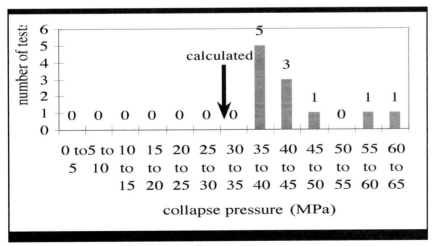

Fig. 10–4 *Histogram of Measured Collapse Pressures*

The analysis previously described applies to cylindrical tubes with a radius/thickness (*R/t*) ratio in the range 10 to 40, typical of pipelines. It does not apply to very thick tubes that collapse by a more complex interaction of yield and ovalization, or to very thin tubes, which collapse elastically. Thin tubes under external pressure are very sensitive to geometric imperfections, and there is a vast amount of literature describing tests.

10.4 Longitudinal Stress

A pipeline in operation carries longitudinal stress as well as hoop stress. Longitudinal stresses arise primarily from two effects. The first is the Poisson effect: a bar of metal loaded in tension extends in the tension direction and contracts transversely. If transverse contraction is prevented, a transverse tensile stress is set up. The analysis in section 10.2 shows that internal pressure induces circumferential tensile stress. If there were only circumferential stress and no longitudinal stress, the pipe would extend circumferentially (so that its diameter would increase) but would contract longitudinally (so that it would get shorter). If friction against the seabed or attachment to fixed objects such as platforms prevents longitudinal contraction, a longitudinal tensile stress occurs.

The second effect that tends to introduce longitudinal stress is temperature. If the temperature of a pipeline is increased and if the pipeline is free to expand in all directions, it expands both circumferentially and axially. Circumferential expansion is usually completely unconstrained, but longitudinal expansion is constrained by seabed friction and attachments. It follows that if expansion is prevented, a longitudinal compressive stress will be induced in the pipe.

Expansion stresses can be very large. The stress required to suppress uniaxial expansion completely under a temperature difference θ is $E\alpha\theta$, where E is the elastic modulus, and α is the linear thermal expansion coefficient. For steel, E_α is 2.4 N/mm^2 deg C, so that a temperature increase of 100° in a constrained situation induces a longitudinal stress of 240 N/mm^2, which is almost 60% of yield for X60 steel.

Calculations of longitudinal stress easily go wrong if they are not tackled systematically. The calculations are best carried out using a thin-walled tube idealization; it is scarcely ever worthwhile to use thick-walled analysis, though it is not particularly difficult to do so if required. In all the rest of this section, stress and strain are counted as positive when they are tensile. This convention is consistent with solid mechanics generally and is strongly recommended because otherwise it is easy to make mistakes with signs.

The circumferential stress is statically determinate and is given by the Barlow formula:

$$s_H = \frac{pR}{t} \tag{10.24}$$

with D replaced by twice the mean radius R. If the absolute value of the stress is required, p should be taken as the difference between the internal and external pressure. If the change of stress from the installation condition to the operational condition is required, p should be taken as the operating pressure.

The longitudinal strain ε_L is given by the stress-strain relation for a linear elastic isotropic material:

$$\varepsilon_L = \frac{1}{E}\left(-vs_H + s_L\right) + \alpha\theta \tag{10.25}$$

where

ε_L	is longitudinal strain
sL	is longitudinal stress
s_H	is circumferential stress
E	is Young's modulus
v	is Poisson's ratio
α	is linear thermal expansion coefficient
θ	is change of temperature (increase positive)

The effect of the radial stress in the pipe wall has been neglected (which is consistent with the thin-walled shell idealization).

The longitudinal stress s_L is not statically determinate but depends on the extent to which longitudinal movement is constrained. If there is complete axial constraint

$$\varepsilon_L = 0 \tag{10.26}$$

and Equations (10.25), (10.25), and (10.26) together give the longitudinal stress:

$$s_L = \frac{\nu PR}{t} - E\alpha\theta \tag{10.27}$$

Longitudinal stress, therefore, has two components, the first related to pressure and the second to temperature. The pressure component is positive (tensile), and the temperature component is usually negative (compressive). Whether the resultant is compressive or tensile depends on the relative magnitudes of the pressure and the temperature increase. Again for the 30-in. outside diameter pipeline described earlier, if the wall thickness is 20.8 mm(the next API standard size up from the calculated minimum), the internal pressure is 18 MPa, and the temperature increase is 90 °C, $E\alpha$ is 2.4 N/mm^2 °C and ν is 0.3, the pressure component of longitudinal stress is +96 N/mm^2, and the temperature component is –216 N/mm^2. Therefore, the resultant longitudinal stress is 96 – 216 = –120 N/mm^2. Neither component is negligible, and both are substantial fractions of the yield stress.

A different situation occurs when the pipeline is completely unconstrained; for instance, if it is close to a right-angled bend at the base of a riser. Both the circumferential stress and the longitudinal stress are statically determinate, and the longitudinal stress is

$$s_L = \frac{1}{2} \frac{p_i R}{t} \tag{10.28}$$

Close to a platform or to an expansion loop, the pipeline is partially free to expand, and an intermediate situation occurs. The longitudinal strain ε_L is positive, and the longitudinal stress is more tensile than the fully constrained stress calculation from Equation (10.28). Although that situation is not strictly statically determinate, it proves it is possible to carry out a straightforward analysis and to derive formulae for movements in partially constrained segments and for stress distributions.[9]

The pipe wall is in a state of biaxial stress, with both circumferential and longitudinal stress components. Strictly, in fact, the state of stress is triaxial. The third principal stress in the radial direction changes from (–internal pressure) at the inside to (–external pressure) at the outside. The third principal stress is usually ignored, and this is consistent with the thin-walled shell idealization, which neglects stresses of order p by comparison with stresses of order pR/t.

Calculating an equivalent stress determines how near to yield the combined stress comes. The most commonly used equivalent stress is the von Mises equivalent stress s_{eq}, given by

$$s_{eqvM} = \sqrt{s_H^2 - s_H s_L + s_L^2} \qquad (10.29)$$

for the case in which torsion is negligible and the radial third principal stress is negligible.

Continuing the earlier example after Equation (10.27)

$$s_H = 321 \quad \text{N/mm}^2$$

$$s_L = -120 \quad \text{N/mm}^2$$

therefore

$$s_{eqvM} = \sqrt{(321)^2 - (321)(-120) + (-120)^2} = 395 \text{ N/mm}^2$$

If the equivalent stress reaches yield, the pipe begins to deform plastically. In the usual case, when the circumferential stress is tensile and the longitudinal stress is compressive, the circumferential plastic strain is tensile and the longitudinal plastic strain is compressive. Under the conditions of full axial constraint, development of a longitudinal tensile plastic strain is accompanied by an equal and opposite longitudinal compressive elastic strain, so that the total strain remains zero. These strains add a plastic strain component to Equation (10.26). Its effect is to make the longitudinal stress less compressive and partially to relieve the axial stress. Elastic-plastic interaction, therefore, allows a degree of stress redistribution. It is self-limiting, and the pipe cannot continue to deform.

Older codes put limits on equivalent stress. More modern codes such as the 1988 Dutch NEN3650 rules, BS8010 Part 3: 1993 and the 1996 and 2000 DnV rules allow and sometimes encourage strain-based design. Strain-based design replaces limits on equivalent stress by limits on strain, though only in conditions where the deformation is restricted, for example by continuous vertical support. Strain-based design also has many advantages but can lead to a large reduction in bending stiffness, which needs to be carefully considered if the pipe is not fully constrained. BS8010 Part 3: 1993 permits allowable strain design for pipelines and includes the following clause:

The limit on equivalent stress imposed by clause 4.2.5.4 may be replaced by a limit on allowable strain, provided that all the following conditions are met:

1 *under the maximum operating temperature and pressure, the plastic component of the equivalent strain does not exceed 0.001 (0.1 per cent): the reference state for zero strain is the as-built state (after pressure test). The plastic component of the equivalent strain is an equivalent uniaxial tensile strain defined as*

$$\varepsilon p = \sqrt{\frac{2}{3}} \sqrt{\varepsilon_{pH}^2 + \varepsilon_{pL}^2 + \varepsilon_{pR}^2} \qquad (10.30)$$

where

ε_p *is the equivalent plastic strain*

ε_{pL} *is the principal longitudinal plastic strain*

ε_{pH} *is the principal circumferential (hoop) strain)*

ε_{pR} *is the radial plastic strain*

2 *any plastic deformation occurs only when the pipeline is first raised to its maximum operating pressure and temperature, but not during subsequent cycles of depressurization, reduction in temperature to the minimum operating temperature, or return to the maximum operating pressure and temperature;*

3 *the D^o/t ratio does not exceed 60*

4 *welds have adequate ductility to accept plastic deformation*

Plastic deformation reduces pipeline flexural rigidity. This effect may reduce resistance to upheaval buckling and should be checked if upheaval buckling is a possibility. Note. This clause does not affect the limits on allowable hoop stress.[10]

Similar conditions appear in clauses D507 and D508 of section 5 of the DnV 2000 rules, though the allowable strain is defined differently, and the D/t ratio is limited to 45. The rules permit larger strains if D/t is less than 20, provided that tests or experience indicate that there is still a safety margin.

The condition represented by these clauses were intended to be based on the results of an analysis that idealizes the pipeline material as an ideal elastic/perfectly-plastic von Mises material. Almost invariably, this is the idealization used in pipeline stress analysis. However, experiments show that plastic deformation begins before the yield condition that corresponds to the yield stress is reached. This is to be expected because the nominal yield stress normally corresponds to a strain of 0.005, whereas the elastic component of strain at the nominal yield stress, Y/E, is smaller, about 0.002. Repeated cyclic loading leads to small plastic strains, even though the nominal yield condition is not reached. There have been concerns that these strains might be cumulative and lead to unacceptable deformations. Although these concerns seem to be exaggerated, further research to clarify this question is in progress.

10.5 Bending

Pipelines are often bent during construction. In lay barge construction (Figure 12–2 in chapter 12), a pipe is bent first in one direction in the overbend and then in the opposite direction in the sagbend. In reel construction, a pipeline is bent well into the elastic range when it is wound onto the reel and is later straightened by reversed plastic bending, then bent again in the sagbend. If the pipeline is laid on an uneven seabed, it bends to conform to the bottom profile, and it may be bent yet again when it is trenched.

Figure 10–5 illustrates the relationship between bending moment and curvature in a pipeline bent into the plastic range. At a small curvature, the pipe bends elastically, and the ratio of the moment to the curvature is the flexural rigidity F. When the curvature is increased beyond the yield curvature at which the points furthest from the neutral axis begin to yield plastically, the relationship begins to curve over.

Under further increase of curvature, the bending moment continues to increase but does so much more slowly at a rate controlled by the interaction between strain-hardening (which tends to increase the bending moment) and ovalization (which tends to reduce the bending moment). If at that stage the curvature is reduced, the bending moment decreases linearly, and when the moment is zero, there is a residual curvature.

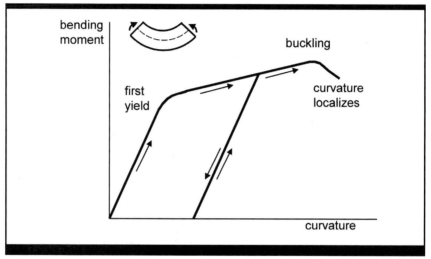

Fig. 10–5 *Relation Between Curvature and Bending Moment*

If the curvature is continuously increased, the bending process ultimately becomes unstable, and the pipe begins to wrinkle on the compression side. A buckle develops, and the bending moment decreases. The curvature is no longer uniform, and it localizes at the buckle, forming a kink.

The following analysis treats a pipeline as a thin-walled cylindrical shell made from an elastic/perfectly-plastic material. This approach has the advantage of simplicity and has been found satisfactory in a large number of problems of large deformation of pipelines, among them buckle propagation, bending, buckling, and denting. The effect of additional factors such as strain hardening and more complex thick-walled cylinder theory is invariably to increase the calculated resistance, though generally it will be only an amount of the order of 10%. The thin-walled elastic shell approach is, therefore, conservative. The strengthening effect of concrete weight coating is also neglected in the analysis for normal coating thickness, it is again generally of the order of 10% (though somewhat larger for concentrated loads).

At small curvatures, the pipe deforms elastically with a flexural rigidity F given by

$$F = \pi R^3 t E \qquad (10.31)$$

where

R is the mean radius,

t is the wall thickness, and

E is Young's modulus.

At large curvatures, the bending moment is the full plastic moment

$$M_0 \;=\; 4R^2 t Y \qquad (10.32)$$

where Y is the yield stress.

There has been extensive theoretical research conducted on plastic bending of tubes, many model tests, and a few full-scale tests. At relatively small curvatures, up to three times the yield curvature, the behavior is adequately described by a very simple model, which applies elementary beam theory and adopts the Euler hypothesis that plane sections remain plane and that the changes in cross-section consequent on bending have a negligible effect. At larger curvatures, change in cross-section becomes significant and has been considered theoretically by Calladine and in experiments by Reddy, Murphy, Kyriakides and Corona, Wilhoit and Merwin, Bouwkamp and many others.[11-17]

Bending buckling begins at a curvature that depends primarily on the radius/thickness ratio and to a lesser extent on strain hardening. Figure 10–6 plots experimental data of the dimensionless buckling curvature $\kappa_b R$ against the radius/thickness ratio R/t. Buckling curvature is defined as the curvature at which the bending moment is at maximum. $\kappa_b R$ is the nominal bending strain.

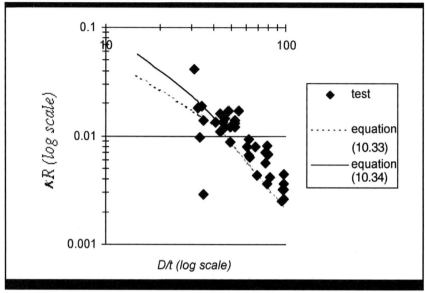

Fig. 10–6 *Relation Between D/t and Buckling Strain*

The following empirical relationship is found to give a safe estimate for steel pipelines:

$$\kappa_b R = \min\left(\frac{1}{4}\frac{t}{R}, 5\left(\frac{t}{R}\right)^2\right) \tag{10.33}$$

The DnV 2000 rules (section 5, clause D507) modify this as part of a limit-state approach. The design compressive strain for the case where the internal pressure is greater than or equal to the external pressure is expressed in the following equation:

$$\varepsilon_d \leq \frac{\varepsilon_c}{\gamma_c}$$

$$\varepsilon_c = 0.78\left(\frac{t}{D} - 0.01\right)\left(1 + 5\frac{s_H}{f_y}\right)\alpha_h^{-1.5}\alpha_{gw} \quad D/t < 45 \tag{10.34}$$

where

γ_c is a resistance strain factor

s_H is circumferential stress, calculated from equation (10.4)

f_y is the characteristic yield stress defined earlier

α_{gw} is a girth weld reduction factor

α_h is the maximum allowable yield to tensile ratio

The $s_H f_y$ term reflects the stabilizing effect of internal pressure and is only taken into account when the internal pressure is greater than the external pressure. The relationships Equations (10.33) and (10.34) are plotted in Figure 10–6. In Equation (10.34), the girth weld factor has been taken as 1, the maximum allowable yield-to-tensile ratio at 0.85, and the circumferential stress as zero. The design compressive strain ε_c has been equated to the nominal bending strain κR, the strain that would have developed in the pipe under the same curvature if the pipe had not ovalized.

The relationships are unexpectedly simple. They are almost purely geometric except for the circumferential stress term in Equation (10.34), whereas they would be expected to include material properties, such as the ratios between strain-hardening tangent modulus and yield stress, and between yield stress and elastic modulus. The reasons why simplified empirical formulas do, in fact, work are not fully understood, but Calladine explains some of the reasons.[11]

In a typical large-diameter pipeline, D/t is 50 (based on outside diameter), and R/t is 24.5 and the nominal buckling strain of 0.0083 from Equation (10.33) and 0.0100 from Equation (10.34), both more than four times the yield strain for typical pipeline steel. Small-diameter flowlines generally have a smaller D/t ratio. This is because they either have a high operating pressure or because they have been designed to be installable by reel ship, and they can be bent even further into the plastic range.

A limited amount of work has been done on the post-buckling behavior of tubes, as part of research on collision resistance. A simple way of accounting for it is to assume that the material that forms part of the wrinkle contributes nothing to the bending moment.

External pressure reduces the buckling curvature. The DnV 2000 rules say that if the external pressure is greater than the internal pressure, the design compressive strain should meet the condition

$$\left(\frac{\varepsilon_d}{\varepsilon_c \gamma_c}\right)^{0.8} + \frac{P_e}{P_e / \gamma_{sc}\gamma_m} \le 1 \tag{10.35}$$

where

$$\varepsilon_c = 0.78\left(\frac{t}{D} - 0.01\right)\alpha_h^{-1.5}\alpha_{gw}D/t < 45$$

The effect of external pressure occurs because an incipient wrinkle bends inward, rather than outward, and external pressure helps to drive the inward movement. Internal pressure, on the other hand, counteracts the tendency for the pipe to ovalize as it bends, and, therefore, tends to increase the allowable strain. The sequence of loading is significant.

Corona and Kyriakides show that plastic buckling occurs at a lower combination of external pressure and curvature if the pressure is applied first and the curvature second than if the curvature is applied first and the pressure second.[15] Radial loading, in which pressure and curvature are increased together, leads to buckling at a still lower combination.

Just as external pressure reduces the buckling curvature, so internal pressure increases it. High internal pressure also reduces the bending moment at which the pipe yields in bending. It has been suggested that pipelines could be made to conform to uneven seabed by raising the internal pressure to a high level and allowing the pipeline to flex under its own weight, in this way preventing the formation of long spans. This effect may happen unintentionally during hydrostatic testing.

Pressure and axial force interact with resistance to bending. They modify the distributions of longitudinal and hoop stress so that yield begins at a smaller nominal bending stress than it would otherwise. This effect can be important in span and upheaval buckling analysis.

If a pipeline is bent far into the plastic range under external pressure, the mode of buckling can change. The buckle is no longer a local wrinkle on the compression side, but it spreads out and moves along the pipe, flattening it into a dumb-bell cross-section (rather like a tube of toothpaste crushed between finger and thumb). This process is called buckle propagation. It only starts if the external pressure is greater than an initiation pressure that depends on the size of the bending buckle. Once it has started, however, a buckle can continue to propagate, even into lengths of pipe that are not being bent. However this action occurs only if the external pressure is greater than the propagation pressure p_{pr}, which is smaller than the initial pressure and is given by the Battelle formula

$$p_{pr} = \frac{6Y}{(R/t)^{5/2}} \tag{10.36}$$

The DnV 2000 rules give almost the same formula, with 6.19 in place of 6 and based on an outside radius, rather than a mean radius.[2]

Buckle propagation is a serious potential threat to submarine pipelines because of the risk that a buckle might run along a great length of pipe and destroy it. Some lines are designed so that the maximum external pressure difference during construction is greater than the propagation pressure. It is then prudent to protect the line against the possibility of propagation by installing buckle arresters at regular intervals, typically about every 15 joints (180 m). If a buckle is initiated and propagates, it runs to the arresters and stops, and only the length between the arresters is lost. The arresters are needed only during construction.

Once the line is in service, the internal pressure is almost always greater than the external pressure, and propagation cannot occur. There are semi-empirical design rules for buckle arresters, and designs can be checked against large-displacement analysis by ABAQUS.

The arrester most commonly used is a short length of heavier pipe (integral arrester), welded into the pipeline. An alternative is the sleeve arrester, a separate short length installed around the pipeline with the gap filled with grout. The integral arrester uses steel slightly less efficiently but has the advantages of eliminating the need for additional welding. Also with an integral arrester, the outside diameter can match the outside diameter of concrete weight coating so that the pipe has a uniform outside diameter. Another alternative is a short heavy ring.

A pipe becomes slightly ovalized when it is bent. In the elastic range, *ovalization* is negligible, but it may become significant in the plastic range and can in a few circumstances be large enough to reduce resistance to external pressure and cause trouble if the pipe needs to be perfectly round. Figure 10–7 illustrates why ovalization occurs by showing in (a) a length of bent pipe and within it two smaller elements 1 and 2. Element 1 is on the tension side below the neutral surface. The forces on either end of the element are perpendicular to the local cross-section. Because the pipe is curved, the forces are not quite in line and, therefore, exert a resultant upward force on the element. Transferred to a cross-section (b), the upward forces on elements like 1 induce an upward force on the lower half of the cross-section.

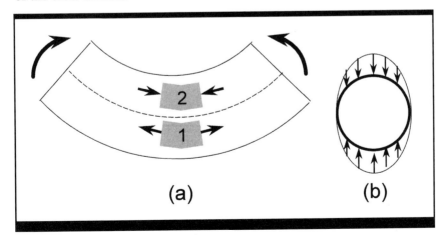

Fig. 10–7 *Ovalization During Bending*
   (a) Forces on Elements Below and Above Neutral Surface
   (b) Resultant Forces Transferred to Cross-Section

Similarly, the forces on the ends of element 2 have a downward resultant, and transferred to the cross-section, they exert a downward force on the upper half of the cross-section. Now looking at the cross-section (b), the forces create circumferential bending and deform the original circular cross-section into a roughly elliptical section with its major axis perpendicular to the plane of bending. This is a nonlinear effect, induced by the interaction between the longitudinal stresses induced by bending and the bending curvature itself.

Ovalization under a bending curvature K can be predicted by

$$z = (1 - v^2)\left(\frac{\mathcal{K}R^2}{t}\right)^2 \tag{10.37}$$

where

z is change in diameter/original diameter, and

v is Poisson's ratio.

This relationship was derived in the classical work of Brazier on elastic tubes, but measurements by Reddy show it to be a safe estimate in the plastic range. A less conservative empirical relationship is due to Murphey and Langner as shown in the following expression:[13]

$$z = 0.24 \left(1 + \frac{1}{60}\frac{R}{t}\right)\left(\frac{\mathcal{K}R^2}{t}\right)^2 \tag{10.38}$$

Pipelines are sometimes bent into the plastic range and then restraightened. In the reel ship method of construction, the pipe is bent into conformity with the reel hub, straightened as it is unwound off the drum, curved again on the aligner at the top of the ramp, and bent in the reverse direction in the straightener. Cyclic loading tests by Kyriakides show that when a tube is bent and restraightened, about three-quarters of the maximum ovalization is recovered after straightening. However, a cyclic process of bending, straightening, bending in the opposite direction, straightening, bending in the first direction, and so on, leads to a progressive increase of ovalization and can ultimately lead to collapse. Therefore, cyclic plastic bending repeated many times ought to be avoided but is unlikely to occur to a real pipeline.

10.6 Indentation

Various concentrated loads are applied to pipelines while they are being constructed. For example, a pipeline leaving a lay barge over a stinger has to withstand concentrated forces from the stinger rollers. Close to the departure point, the pipe will lift off the rollers and may move back into contact with them as the barge pitches in a seaway; and the loads are then applied dynamically. In operation, trawl boards moving at up to 3.5 m/s (7 knots) can strike a pipeline and induce a high local impact stress.

Concentrated loads can generate various kinds of deformation, depending on the load application method and on the shape and stiffness of the object applying the load. In critical cases, it may be necessary to carry out an elastic analysis by a finite-element method. In many other cases, it will be enough to carry out a plastic analysis to find the load under which the pipe wall will collapse then to compare the applied load with the collapse load.

Analysis of denting is based on a theoretical analysis developed by Wierzbicki and Suh.[19] Their analysis corresponds to a sharp-edged line indenter, which is aligned in a plane orthogonal to the pipe axis and advances towards the pipe. The local circumference is transformed from a circle into a form that consists of a straight line in contact with the indenter, two relatively sharp circular arcs forming knuckles, and a larger-radius arc within which the cross-section radius increases slightly. Moving plastic hinges form at the ends of the arcs.

The deflection u_d at the indenter is related to the indenter force P by

$$u_d = \frac{3}{32\pi} \frac{P^2}{Y^2 t^3} \tag{10.39}$$

Though Suh and Wierzbicki's analysis includes some bold simplifications, it agrees well with measurements in tests. Figure 10–8 compares the predictions of Equation (10.39) with measurements made when a section of 812.8 mm (32-in.) outside diameter 19.05 mm wall concrete-coated X65 pipe was indented by a transverse small-diameter roller. The agreement is close enough for any practical purpose. The graph also shows that a thick-walled pipe is quite strong against external loads: a load of 1 MN (100 tonnes), larger than one would want to apply deliberately, induces a deflection of 20 mm.

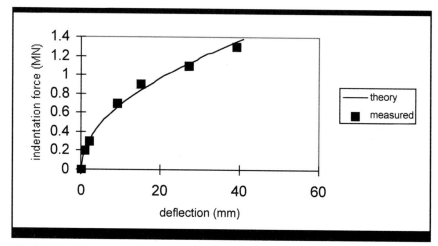

Fig. 10–8 *Indentation of 32-in. Pipeline*

An alternative analysis is more appropriate to point loads applied over a roughly circular area, small by comparison with the pipe radius. Morris and Calladine considered an elastic-plastic tube radially loaded through an integral rigid boss and based their analysis on a collapse mechanism of inward-rotating segments.[20] They found the relationship between load P and deflection ud under the load to have the form

$$\frac{P}{Yt^2} = \Phi\left(\frac{u_d}{t}, \frac{a}{\sqrt{Rt}}\right) \tag{10.40}$$

where

α is the radius of the loaded area

Φ is a function plotted in reference 19

If an object such as an anchor hooks a pipeline, severe bending occurs at a point where the longitudinal compression side of the pipeline is also subjected to a large inward-acting radial force. Since both inward force and axial longitudinal compression induces similar modes of deformation, a severe interaction is to be expected. Unfortunately, this condition has not been investigated experimentally.

10.7 Impact

Impact loads occur as a result of accidents if, for instance, something falls off a platform and onto a pipeline or if an explosion generates missiles. Impact resistance to missiles moving at high velocities is a complicated phenomenon that is not fully understood theoretically, but there is a substantial body of experimental evidence. What is most important is to know whether the pipe is perforated so that the contents can escape. Nielsen gives an empirical correlation for the kinetic energy of a missile that perforates a pipe:[21]

$$U = Ct^{1.7}a^{1.8}D^{-0.5} \tag{10.41}$$

where

U	is the minimum missile kinetic energy at which perforation occurs
C	is an empirically determined constant, 8×10^9 in SI units
t	is the pipe wall thickness
a	is the missile diameter
D	is the pipe diameter

This correlation was developed for relatively high missile velocities of the order of 100 m/s, but it has been shown to agree well with test results at much lower velocities.[22] In almost all impact calculations, there is no practical need for very high accuracy: the objective is to assess risk of perforation by a comparison between the pipeline's ability to resist perforation and the kinetic energy of a potential missile.

References

1   DNV. (1996) *Rules for Submarine Pipeline Systems*. Norway: Det norske Veritas.

2   Offshore Standard OS-F101: "Submarine Pipeline Systems." Det norske Veritas, Høvik, Norway (2000).

3   Palmer, A.C. (1996) "The Limits of Reliability Theory and the Reliability of Limit-State Theory Applied to Pipelines." (OTC8218) *Proceedings of the 28th Annual Offshore Technology Conference*, 619–626. Houston.

4   Palmer, A.C., N.S.T. Curson. (1996) "The Role of Pipeline Engineering Design Codes." *Pipes and Pipelines International*. 41, 18–21.

5   McKeehan, D. (1996) "Oman India Gas Pipeline Project: Technology Development." *Proceedings of the Offshore Pipeline Technology Conference*. Amsterdam.

6   Tam, C., et al. (1996) "Oman India Pipeline: Development of Design Methods for Hydrostatic Collapse in Deep Water." Proceedings of the Offshore Pipeline Technology Conference. Amsterdam.

7   Palmer, A.C. (1997) "Pipelines in Deep Water: Interaction Between Design And Construction." *Proceedings of the COPPE-FURJ Workshop on Submarine Pipelines*. Federal University of Rio de Janeiro. 157–166.

8   Palmer, A. C. "A Radical Alternative Approach To Design and Construction of Pipelines in Deep Water." (OTC8670) Proceedings of the Offshore Technology Conference, Houston, TX.

9   Palmer, A.C., Ling, M.T.S. (1981) "Movements of Submarine Pipelines Close to Platforms." *Proceedings of the 13th Annual Offshore Technology Conference*. Houston, TX.

10 BS 8010 Part III. British Standards Institution, London (1983).

11 Calladine, C.R. (1983) *Theory of Shell Structures*. Cambridge, England: Cambridge University Press.

12 Reddy, B.D. (1979) "An Experimental Study of the Plastic Buckling of Circular Cylinders in Pure Bending." *International Journal of Solids and Structures*. 15: 669-683.

13 Murphey, C. E., C. G. Langner. (1985) "Ultimate Pipe Strength under Bending Collapse and Fatigue." *Proceedings of the International Symposium on Offshore Mechanics and Arctic Engineering*. Dallas. 467–477.

14 Kyriakides, S. P. K. Shaw. (1985) "Inelastic Buckling of Tubes Under Cyclic Bending." (Report EMRL 85.4) Engineering Mechanics Research Laboratory. University of Texas at Austin.

15 Corona, E., S. Kyriakides. (1988) "Collapse of Inelastic Pipelines Under Combined Bending and External Pressure." *Proceeding of the International Conference on Behaviour of Offshore Structures*. Trondheim. 3. 953-964

16 Corona, E., S. Kyriakides. (1988) "On the Collapse of Inelastic Tubes Under Combined Bending Pressure." *International Journal of Solids and Structures*. 24, 505–535.

17 Wilhoit, J. C., J. E. Merwin. (1973) "Critical Plastic Buckling Parameters for Tubing in Bending Under Axial Tension." (OTC 1874) *Proceedings of the 5th Annual Offshore Technology Conference*. Houston, TX.

18 Boukamp, J. G., R. M. Stephen. (1973) "Large Diameter Pipe Under Combined Loading." *Transportation Engineering Journal*. ASCE, 99 TE3, 521–536.

19 Wierzbicki, T., Myung Sung Suh (1986) "Denting Analysis of Tube Under Combined Loadings." (Report MITSG 86-5) *MIT Sea Grant College Program. Massachusetts Institute of Technology*.

20 Morris, A. J., C. R. Calladine. (1971) "Simple Upper-Bound Calculations for Indentation of Cylindrical Shells." *International Journal of Mechanical Sciences*. 13. 331–343.

21 Neilson, A. J., W. D. Howe, G. P. Garton. (1987) "Impact Resistance of Mild Steel Pipes." (Report AEEW-R 2125).UKAEA, Winfrith.

22 Palmer, A. C., A. Neilson, S. Sivadasan. (2004) "Impact Resistance of Pipelines and the Loss-of-Containment Limit State." *Journal of Pipeline Integrity*. In press.

11 Stability

11.1 Introduction

A pipeline has to be stable on the seabed. If it is too light, it will slide sideways under the action of currents and waves. On the other hand, if it is very heavy, it will be difficult and expensive to construct.

Designers can increase the weight of the pipeline by adding an external concrete weight coating that also gives mechanical protection to the anti-corrosion coating. Alternatively, they can increase the submerged weight by increasing the wall thickness of the pipe, though this is a relatively costly option, particularly if the pipe is a corrosion-resistant alloy. They can also reduce hydro-dynamic forces and increase stability by trenching the pipeline into the seabed or add weight by adding bolt-on weights or mattresses. To eliminate the possibility of instability, their designs can call for burying the line in the seabed or covering it with rock.

The first step in design against hydrodynamic forces induced by current and wave is to determine how large the design-steady current and the design wave ought to be.

The design currents section 11.2 provides an introduction to this topic for currents, and section 11.3 for waves. Both are extremely extensive areas of oceanographic research, and these sections are only introductions. Section 11.4 discusses the calculation of the hydrodynamic forces. Security against wave forces requires there to be enough resistance to lateral movement to withstand the hydrodynamic force, and section 11.5 describes the analysis of lateral resistance.

The conventional approach to design is to determine the submerged weight required so that the lateral resistance is large enough to hold the pipe in equilibrium against the combination of weight and hydrodynamic force. Section 11.6 describes how the ideas developed in the afore-referenced sections are brought together in the design process.

There are good grounds for thinking that the conventionally accepted design method is in fact irrational and incorrect in principle. That method wrongly assumes that the seabed itself is stable. In reality the seabed usually becomes unstable and mobile before the design conditions for a pipeline are reached. Section 11.7 describes how this happens and suggests alternative approaches for stabilizing the pipeline. Research is in progress.

11.2 Design Currents

The largest component of quasi-steady current is usually due to tide. Tides in most places are semi-diurnal so that there are two roughly equal high tides and two roughly equal low tides per day. In other places there are diurnal tides, or large differences between the heights of the highs and lows. The tidal cycle does not repeat itself exactly from tide to tide because of the varying relative positions and distances of the moon and the sun in relation to the earth. In a lunar month, the height of the tide and the intensity of tidal currents occur twice in a cycle from high spring tides (just after full moon and new moon) to lower neap tides.[1]

Information about tidal currents is available in most coastal areas from charts, almanacs, and from previous current surveys stored in databases. This information may be enough for conceptual design and, sometimes, for final design. Tidal heights and currents can be computed numerically, and specialist research groups routinely carry out calculations of this kind.

In a location that has not been studied before, it will be wise to check published or calculated tidal currents against measurements at sea. A current-meter survey can be carried out on the proposed pipeline route. It is usual to install a string of at least three recording current meters, immediately above the bottom, at mid-depth, and near the surface, so as to generate additional data for construction planning. The meters are left in place for one lunar month, and they record magnitude and direction of the current at 10-minute intervals. The record is then analyzed to determine the maximum current and the statistical distribution of magnitude and direction. A design extreme 10-year or 100-year current can be estimated by extrapolating the observed distribution using extremal statistics. Although this is often done, it is open to some logical objections for a deterministic phenomenon like tide, and in practice probably tends to overestimate extreme currents.

Tidal currents vary through the depth of the sea. Currents measured at one depth can be used to estimate currents at a different depth by the van Veen power law, which states that the current is proportional to the 1/7th power of height above bottom. But this formula cannot be relied on over substantial depth differences.

Not all quasi-steady currents are tide-related. In estuaries, river flow sometimes has a significant effect. In shallow seas such as the North Sea, storm surges induced by deep cyclonic weather patterns occasionally generate storm surges that may be several meters high, accompanied by surge-related currents of 1 m/s or more. In the open sea, there are occasional tidal whirls and loop currents linked to meteorological effects, density currents, and internal waves. In the Gulf of Mexico, for example, loop currents can reach 2 m/s (4 knots). They reach to surface and can be detected by satellite measurements of surface temperature, but at present their development cannot be reliably forecast.

11.3 Design Waves

How the design wave should be selected depends on the data available. In a few instances, extensive wave data sets are available, either from instrumental measurements by wave-rider buoys or wave staffs on platforms or from much less reliable visual measurements from lightships or merchant ships. NOAA (National Oceans and Aeronautics Administration, NOS (National Ocean Service), NIMA (National Imagery and Mapping Agency) in the United States and the

Meteorological Office and Marine Information Advisory Service (MIAS) in the UK hold catalogues of data of this kind, obtained by parallel organizations in individual countries. Oil companies and research groups also hold additional data.

Wave data sets are useful for only extreme wave prediction if they extend for at least five years, preferably more. Many of them do not meet this criterion. Direct extrapolation from data covering one or two years is so unreliable as to be dangerously misleading.

When satisfactory wave data are available, the methods for predicting extreme waves are based on the standard methods of extremalstatistics.[2, 3] When there are no adequate wave data or when an independent check is warranted, the design has to be based on wind data. Numerical modeling of waves surges and currents has reached a level of accuracy at which hindcast data (obtained by numerical modeling of past events) are probably a more reliable guide to future extreme events than measured data. The primary value of measured data is to validate numerical models.

Wind data are more widely available than wave data and have several advantages:

- Wind speed and direction are easier to describe and characterize than wave height and are easier to measure, and measurement techniques are better standardized.

- Wind records are kept by meteorological stations and airfields and often extend over 30 years or more, whereas there are no continuous records of waves over such long periods.

- The horizontal variation of wind is less rapid than that of waves, whereas the extreme wave climate in a bay will be quite different from the extreme wave climate at an offshore rock lighthouse 10km away. The extreme wind climate will be nearly the same, though in coastal areas speed and direction can change substantially over short distances.

- The statistical population associated with extreme events is easier to identify from wind records, interpreted in conjunction with meteorological records, than from wave records.

If there are no direct records of wind, it can be reliably hindcast from synoptic charts.

Extreme waves are generated by extreme storms, and the preferred procedure is to identify a series of major storms over a period of at least

10 years, to calculate the corresponding extreme wave height, and then to input the results into a procedure for estimating the extreme design wave corresponding to an acceptable encounter probability. It is important for the set of storms to include all the storms that occurred.

Brink-Kjaer describes systematic procedures to identify extreme storms.[4] In areas that have been extensively studied, such as the North Sea, it will often be possible to take advantage of earlier work and escape having to repeat the identification.

Methods for predicting waves from wind are described in standard references.[1] It is recommended that numerical methods be used for wave prediction for final design of important pipeline projects, though approximate techniques such as the Sverdrup-Munk-Bretschneider method remain extremely useful for preliminary design.

A wave record may consist of a continuous record of surface height, of a continuous record for part of an interval in which the sea state remains roughly constant (such as a 10-minute record every 3 hours), or of an estimate of the maximum wave observed in each recording interval. The analysis can be based on either a maximum wave in each recording interval or on a significant wave. If it is based on the maximum wave height, extremal statistics can be applied to extrapolate the maximum wave height statistics to a maximum wave corresponding to a design return period. Then the corresponding maximum wave height can be estimated by applying a known relation between the maximum wave and the significant wave.

The design maximum wave is related to the design return period T_R and an encounter probability E. The return period is the average time interval between successive events in which the design wave is exceeded. The exceedance probability is the probability that the design wave is exceeded during the design lifetime L, and is

$$E = 1 - exp\left(-\frac{L}{T_R}\right) \qquad (11.1)$$

When the design lifetime of a pipeline is 60 years, for instance, and the engineer is prepared to accept a 10% chance that the design wave is exceeded once during the design lifetime, the return period is the solution of

$$0.1 = 1 - exp\left(-\frac{60}{T_R}\right) \qquad (11.2)$$

which is 570 years. It must be added that it is not customary practice to carry out this calculation, and a more usual approach is to assert that a 50- or 100-year return period is enough. The exceedance probability is then quite large, particularly when the overall uncertainties in the estimate are

taken into account. The designer cannot escape from the need to make a rational choice of design return period in conjunction with an examination of the consequences of exceeding the values used in design.

The methods of extremal statistics applied in extreme wave prediction are used in other areas of water engineering, such as the estimation of extreme floods. Sarpkaya describes their application to extreme waves.[2]

Statistical theory shows that the extremal distribution has to take one of three asymptotic forms. The method has the following steps:

1. The observed or estimated wave height maxima in each recording interval are ranked in order of height, so that H_1 is the highest and H_i the i-th highest.

2. Each maximum is associated with an exceedance probability $1-p(H)$.

3. An extremal distribution is derived from the heights H_i and the corresponding probabilities $1-p(H_i)$. This is often done by plotting $1-p(H_i)$ against H_i on a diagram on which the selected extremal distribution ought to plot as a straight line.

4. The extremal distribution is extrapolated to the design exceedance probability r/T_R, where r is the recording interval and T_R the design return period.

There are several alternatives, particularly in steps 2 and 3. In step 2, there is more than one way of assigning $p(H)$ to each maximum. If the total sample contains N wave height maxima, the natural choice is to assign to the i-th height maximum H_i and exceedance probability i/N, or perhaps $i/(N+1)$ or $(i-2)/N$. Each alternative is known to introduce a systematic error, and Sarpkaya suggests a general plotting formula.

In step 3, several extremal distributions are possible. There has been much argument about which is the best distribution. Likewise there has been discussion on whether it matters that the type I and type III distributions assign a small but finite probability to negative values of H and that all but type III_U allow infinitely-high values. These issues may be merely academic because all that is required is to fit data and extrapolate.

The extremal distribution can be estimated by standard statistical methods of moments or maximum likelihood, and the method of moments can be used to determine the extremal distribution that best fits the data. Statistical methods such as the chi-square test can also be applied to check closeness of fit. Statistics is not magic, and extrapolating from limited data inescapably leads to substantial uncertainties. It is a sobering exercise to estimate confidence limits on design wave heights calculated by the method previously described. St. Denis and Borgmann made these estimates.[5,6] Confidence limits ought always to be estimated in practice, but, regrettably, often are not, possibly because the results are so discouraging. The further the extrapolation, the wider apart are the confidence limits and the more uncertain the estimate, a fact that is important to remember in subsequent calculations.

The extrapolation process is purely statistical and, of course, does not check that the design wave is physically realizable. The height may be limited by factors such as breaking, either at the location of interest or at a place where a wave crosses a shallower bank further offshore.

Analysis requires a design wave period as well as a design wave height, but the problem of choosing the design period has been given much less attention. One approach is to estimate a most probably maximum steepness from analysis of a scatter diagram. The period T_{max} associated with the maximum height H_{max} lies between $(6H_{max})^{1/2}$ and $(15H_{max})^{1/2}$, where H is measured in m and T in seconds. A more ambitious approach is to attempt to determine a joint probability distribution of height and period. It is not always clear whether a long-period wave or a short-period wave is the more critical for the stability of the pipeline, and it may be necessary to repeat the analysis for a range of periods.

Much remains to be learned about extreme waves. It is thought that many unexplained losses of ships may be ascribed to *superwaves* brought about by instantaneous coincidence of two or more large waves moving at different speeds, but little is known about this.

A wave acts on a pipeline through the oscillatory currents it creates, so the velocities have to be calculated from the height and period. Doing so is an exercise in fluid mechanics. A number of theoretical approaches exist: they are not in conflict, but represent alternative idealizations adopted in attempts to resolve mathematical difficulties and to reach a balance between accuracy and complexity. Wave theories are the subject of an enormous amount of literature, and for detailed discussion of the complex issues involved, the reader is referred to the standard texts such as Sarpkaya.[2]

The simplest approach is Airy small-amplitude theory, which corresponds to a limiting case in which the wave height H is small compared to the wavelength L and can be thought of as the first in a series of perturbation theories for increasing wave steepness H/L. It neglects all terms involving velocity[2], and applies the free-surface conditions not at the free surface but at the mean water level. Because of its simplicity, Airy theory is widely used, even outside its strict range of validity. Most analysis of shoaling and refraction, for instance, applies Airy theory. Caution is needed because the errors can be substantial, above all when forces are calculated because the drag component of force is proportional to velocity squared and a 20% error in velocity creates a 44% error in force.

Airy theory is not exact but is a limiting case for very small waves because of the approximation of the boundary condition at the free surface. The errors become more significant as the steepness H/L and the ratio H/d of wave height to water depth increases. Stokes and others developed a series of higher-order theories, which are more accurate for steeper waves and are readily available as computer programs.

Much attention has been given to the question of how to decide when each wave theory should be applied. The question can have no single answer, and a decision will, in any case, depend on the purpose for which the results are required and on the degree of accuracy necessary (and justified by the reliability of the underlying data). Moreover, an assessment of the optimal theory for one objective might be irrelevant to another objective. Wave theories are commonly compared by how well they satisfy the boundary condition at the free surface, whereas in pipeline design, the most important results are the velocity and acceleration just above the bottom. There is evidence that Airy theory does in fact predict near-bottom velocities well, even in conditions where it is a poor predictor of surface velocities.

Le Mehauté, quoted in Sarpkaya, presented a plot that shows the approximate range of validity of different wave theories. The axes are wave height and water depth, each non-dimensionalized by division by gT^2. Airy theory is adequate if the wave is not too steep. Steeper deep-water waves ($H/gT^2 > 0.001$) are better represented by higher-order Stokes theories. In shallower water, either stream function theory or cnoidal theory is more suitable.

The availability of computers has reduced the need for extensive assessment of different wave theories. If the engineer has available a computer that can output numerical results from one of the more sophisticated wave theories, such as stream function theory that is known to have a very wide range of application, it would seem sensible to use that theory universally.

11.4 Hydrodynamic Forces

A current across a pipeline creates a hydrodynamic force. Figure 11–1(a) is a schematic. A high-pressure region occurs on the lower part of the upstream side. Across the top of the pipeline, the velocity is higher than the free-stream current. The flow separates at a position that depends on the velocity. The flow is somewhat unsteady in the mixing zone downstream of the separation point, and a series of vortices is shed. A low-pressure wake develops on the downstream side. If the pipeline is slightly above the bottom (Fig. 11–1(b)), the flow is substantially modified. Then there is a high-velocity flow under the pipe, created by the pressure difference between the upstream and downstream sides. If the bottom is sediment, the high velocity will erode the bottom and enlarge the gap between the pipeline and the sea floor, then will weaken as the gap grows until a stable scour depth is reached.[1] If the pipeline is in a trench (Fig. 11–1(c)), the flow may separate from the upstream edge, and the pipeline will lie partly in the wake created by the side of the trench.

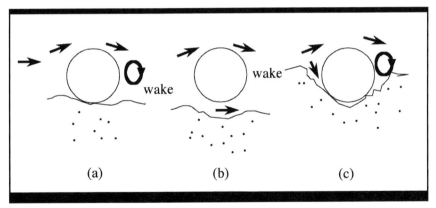

Fig. 11–1 *Water Flow Across Seabed Pipelines Pipeline on (a) Bottom (b) Pipeline Above Bottom (c) Pipeline in Trench*

In all these cases, the pressure differences around the pipeline create a hydrodynamic force. The hydrodynamic force can be divided into a horizontal component in the downstream direction (*drag*) and a vertical component, usually upwards (*lift*). Even in a steady flow, there are small but measurable variations in the force because of unsteady flow in the turbulent mixing zone. In an unsteady flow, there are very large fluctuations in force, especially if the flow reverses.

From the viewpoint of fluid dynamics theory, the pipeline is a blunt body. A boundary layer separates from the top of the pipeline as water flows across it, and a wake forms on the downstream side. Classical potential theory cannot adequately describe the flow. Computational fluid dynamics has made much progress in the analysis of flow over blunt bodies at high Reynolds numbers of this type, but its methods have not been refined to a level suitable for routine design.

In the case of steady flow, the horizontal and vertical force components can be related to the current and the pipe dimensions by

$$F_x = \tfrac{1}{2}\rho C_D D U^2 \tag{11.3}$$

$$F_y = \tfrac{1}{2}\rho C_L D U^2 \tag{11.4}$$

where

F_x	is the horizontal force per unit length of pipeline
F_y	is the vertical force per unit length of pipeline
ρ	is the density of the water
C_D	is a drag coefficient
C_L	is a lift coefficient
D	is the outside diameter of the pipeline
U	is the velocity of the water normal to the pipe axis

The pipeline is usually in a bottom boundary layer so that the velocity varies with height above the bottom. It is important to use a consistent reference height for the velocity U in the definition of the coefficients C_D and C_L. This is a frequent source of confusion because different authors have used different reference velocities.

C_D and C_L are functions of Reynolds number UD/ν, where ν is the kinematic viscosity. C_D and C_L also depend on the roughness of the pipeline because roughness influences the position at which flow separation occurs. Roughness can be described by a ratio k/D, where k is a roughness height. The coefficients also depend on the velocity profile, on the natural turbulence intensity within the flow, on the roughness of the seabed upstream, and on the level of sediment transport.

The most accessible and reliable source is a series of measurements carried out by Hydraulics Research Station (HRS) on instrumented sections of 305 mm, 610 mm and 915 mm pipe in a strong tidal current in the Severn estuary.[7] The site has a very large tidal range, which allows an instrumented section of pipe to be accessible above water at low tide and for the forces to be measured a few hours later in a 1.3 m/s ebb current with some 6 m of water over the pipe so that the forces are not influenced by blockage effects. The bottom is a marl ledge.

The diameters tested covered much of the range significant to pipeline design. The tests examined both rough pipes (*k/D* 0.018, corresponding to very rough concrete) and smooth pipes (*k/D* 0.00024, corresponding to a smooth material like fusion-bonded epoxy), and found significant differences, particularly in the lift coefficients. Two of the curves from the report are redrawn in simplified form as Figures 11–2 and 11–3.

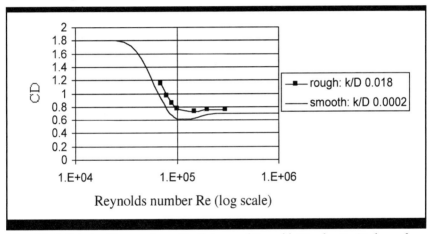

Fig. 11–2 *Relation Between Drag Coefficient and Reynolds Number Re (Redrawn from Reference 7)*

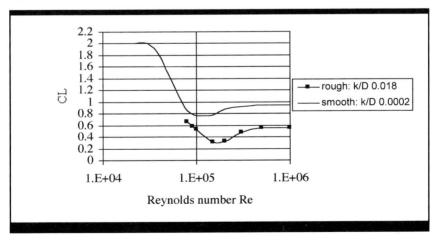

Fig. 11–3 *Relation Between Lift Coefficient and Reynolds number Re (Redrawn from Reference 7)*

A significant drop in lift and drag coefficients at a Reynolds number of about 10^5 is related to a transition from subcritical flow to supercritical flow, which delays separation to a point further downstream. This narrows the wake and the consequent reduction in form drag more than outweighs the increase in drag due to skin friction.

The coefficients found in the HRS test series are generally lower than those reported by earlier research workers. One explanation is that some earlier work had too small a flume in relation to the pipe diameter so that the flow was obstructed, a wave formed over the pipe, and the hydrodynamic force was artificially increased. Another part of the explanation of the discrepancy may be the presence of transported sediment in the Severn estuary. Since sediment transport is also likely to be present in field conditions with high currents, this would appear to be a positive reason to use the HRS results.

A pipeline is often partially embedded in the seabed. The HRS research program included measurements of lift and drag coefficients for this case. Both C_D and C_L fall with increasing embedment, as would be expected; and the lift approaches zero when the pipe is half-buried. A drag coefficient defined by reference to the projected area above the bottom falls slightly in the supercritical Re > 10^5 region. An assumption that the drag force is proportional to the projected area above the seabed is therefore slightly conservative.

Some pipelines are trenched for shelter against currents and for protection against fishing gear. Many pipelines in shore crossings are pulled along an open trench during construction. The condition during the pull can be critical to the design, even if the pipeline is buried later. At least two test programs have investigated forces on pipes in trenches. A typical test shows that if a pipe is in a steep-sided triangular trench half a diameter deep with sides at 30° to the horizontal so that half the pipe projects above the natural bottom, the drag coefficient is cut to about half the value of that for a pipe on the bottom. If the trench is one diameter deep, so that the top of the pipe is level with the bottom, the drag coefficient is reduced further, to about one-tenth its value for a pipe sitting directly on the bottom. If the pipe if partially embedded in the bottom of the trench or if a spoil furrow lies along the sides of the trench, the drag is further reduced.

If the flow does not cross the pipeline at right angles, the hydrodynamic flow is reduced. The HRS program included measurements of forces on pipes at angles of incidence of 15°, 30°, and 45° (between the flow and a normal to the pipe). A simple idealization is to assume that the force is determined by the transverse component $U \cos\theta$ of the velocity U (where θ is the angle between the flow direction and a normal to the pipeline axis) and is independent of any longitudinal component. Drag and lift coefficients for angled flow can then be defined by replacing U by $U \cos\theta$ in Equations (11.3) and (11.4). If the simple idealization applied, the coefficients would be independent of θ. The HRS tests showed that these coefficients in fact decrease slightly for increasing θ so that in the supercritical region, C_D falls from 0.7 at normal incidence and becomes 0.6 for 15°, 0.48 for 30°, and 0.3 for 45°. It is conservative to assume that Equations (11.3) and (11.4) apply with the perpendicular velocity component and the same coefficients as for perpendicular flow.

Two or more pipelines are sometimes installed together. Flow across the two pipes creates complex interference effects that can have surprising consequences. For instance, intuitively it would seem almost certain that if water flowed across two parallel cylinders in line with the flow, the upstream cylinder would shelter the downstream one and that the hydrodynamic force on it would be smaller than the force on the upstream cylinder. However, experiments demonstrate that the force on the downstream cylinder is sometimes the larger. Similarly, it seems reasonable to suppose that if the cylinders were more than, say, three diameters apart, their interaction would be negligible. However that supposition too is not always the case. The possibility of interference ought always to be considered when two or more pipes run parallel. Zdravkovich, Sarpkaya, and Bearman review the subject and give references to the specific cases that have been examined.[8, 9, 10]

Unsteady flow is more complicated than steady flow. Unsteady flow occurs under waves where a pipeline is in an oscillatory wave-induced current, possibly superimposed on a steady current component associated with tide, storm, and ocean circulation. Most analysis methods are based on the Morison equations, which are almost universally used in the offshore industry:[2]

$$F_x = \frac{1}{2}\rho C_D D u |u| + \left(\frac{\pi}{4}\right)\rho D^2 C_M\left(\frac{du}{dt}\right) \qquad (11.5)$$

$$F_y = \frac{1}{2}\rho C_L D u^2 \qquad (11.6)$$

where

F_x is the horizontal force per unit length of pipeline

F_y is the vertical force per unit length of pipeline

ρ is the density of the water

C_D is a drag coefficient

C_L is a lift coefficient

C_M is an inertia coefficient

D is the outside diameter of the pipeline

u is the instantaneous velocity of the water

du/dt is the instantaneous horizontal acceleration of the water

$|u|$ is the absolute value of u

The first terms in the equations are similar to those in Equations (11.3) and (11.4) but with minor differences. The first term in the horizontal force equation has $u|u|$, rather than u^2, so that the sign of the force changes when the sign of u changes. A current to the right induces a drag to the right, and a current to the left induces a drag to the left. The first term in the lift Equation (11.6) has $u2$ because the lift is upwards regardless of the way the current flows.

The second *inertia* term in Equation (11.5) accounts for the acceleration of the fluid and has two components. Any body in an accelerating fluid is subject to a force equal to the mass of displaced fluid multiplied by the acceleration. This force is the Froude-Krylov force. If the analysis outlined above were complete, C_M would be 1; but in addition, there is a second force component associated with additional accelerations in the fluid that exist because the fluid has to accelerate around the pipe. Therefore, C_M is greater than 1. Care should be taken in using published results because some researchers call the parameter in the Morison equation C_I and define a different C_M equal to C_I-1.

The horizontal acceleration du/dt in a plane flow is strictly given by

$$\frac{du}{dt} = \frac{\partial u}{\partial t} + u\left(\frac{\partial u}{\partial x}\right) + v\left(\frac{\partial u}{\partial y}\right) \tag{11.7}$$

where

x and y are horizontal and vertical coordinates

v is the vertical velocity

In applications, it is usual to ignore the convective terms $u\partial u/\partial x$ and $v\partial u/\partial y$. Sarpkaya examines the question in detail and argues that there is some justification for neglecting the convective terms. Leaving them out will, in most instances, lead to a slight overestimate of maximum forces. If the near-bed velocity corresponds to a progressive wave and if v is negligible, $\partial u/\partial t$ and $u\ \partial u/\partial x$ are 90° out of phase; and the ration of their instantaneous maximum values is $U_m T/L$, where U_m is maximum velocity, T is wave period, and L is wavelength. That group is normally small so that the convective acceleration component has little effect.

There is no inertia term in Equation (11.6) because the water has no vertical acceleration at the bottom and because the vertical acceleration at the level of the pipe is small.

Comparisons have been made between measured forces on a pipe in a water tunnel and forces calculated from the Morison equation after the coefficients have been optimized to obtain the best possible fit. Agreement is disappointing, but no one has yet been able to come up with an acceptable, simple alternative to the Morison equation. The limitations of the equation, which originally was not intended to be applied in this way, are discussed at length in texts on hydrodynamics. Nothing is to be gained by tinkering with the coefficients, and a radically new approach is needed if genuine progress is to be made. A difficulty of principle is that the flow is far too complex to be completely described by instantaneous values of u and du/dt. Flow-visualization experiments and computations show that as the flow moves backwards and forwards across the pipe, vortices are shed irregularly from the upper surface, move downstream, interact with each other (sometimes cancelling each other and sometimes coalescing) and are then swept back across the pipe when the flow reverses. Because of the influences of the vortices and the separated boundary layer, the local flow around the pipeline is not the same from one wave cycle to the next. Bryndum and other authors have found that large lift forces occur when the free-stream velocity is almost zero because of phase-leading wake reversal. This is plainly inconsistent with Equation (11.6).[10]

The Morison coefficients can again be determined by tests. In oscillating flow, they depend on Keulegan-Carpenter number UmT/D and to a lesser extent on Reynolds number. Keulegan-Carpenter number is a measure of the ratio between the distance moved by a water particle between its extreme positions in oscillating flow and the diameter of the pipeline. If the Keulegan-Carpenter number is large, the water moves much more than the diameter of the pipe in each oscillation. The drag term in Equation (11.5) is then larger than the inertia term, and conditions during a cycle begin to approach those in steady flow. If the Keulegan-Carpenter number is small, the inertia term becomes more important.

The importance of the inertia term can be seen by calculating the ratio between the maximum value of the inertia term and the maximum value of the drag term, and taking into account that the maxima occur at different times. For a sinusoidal oscillating flow, the ratio between the maxima of the drag and inertia terms is

$$(inertia)_m/(drag)_m = \pi^2(C_M/C_D)/(U_mT/D)\ 0 \qquad (11.8)$$

and, therefore, inversely proportional to the Keulegan-Carpenter number.

It is generally agreed that the coefficients depend primarily on the Keulegan-Carpenter number. There must also be some dependence on the other factors known to be important in steady flow, particularly Reynolds number and roughness; but it has not been possible to disentangle those factors from experimental scatter. Many pipelines rapidly become heavily covered with marine growth, but the effect of the growth on pipelines on the bottom seems not to have been investigated. Parallel work on growth-covered cylinders in a free stream has shown that it has a significant influence on hydrodynamic forces.

Most research results have been derived from experiments on oscillating flows in laboratory water tunnels. The flow is reasonably well defined, but the Reynolds number is usually rather lower than in that seen in the field. In an extensive series of tests by Hydraulics Research, a section of 273-mm (10-in.) pipe was mounted across a water tunnel in which forced oscillatory motion of a piston generated a movement that was a sinusoidal function of time.[11] Water pumped through a bypass to the piston generated a steady velocity component, which was superimposed on the oscillatory component generated by the piston motion. The forces on the center section of the pipe were measured by a load cell system, and the measured variation of force with time was matched to the Morison equation through a harmonic analysis. Bryndum carried out a separate investigation and found coefficients somewhat higher than the coefficients found in the Hydraulics Research tests.[10] A second paper from the same group argued convincingly that much better agreement between observed and calculated forces can be achieved if the velocity in the Morison equation is the measured velocity close to the pipe, rather than the free-stream velocity.

A major investigation of pipeline stability was also carried out in Norway.[12, 13, 14, 15] It tackled the known inadequacies of the Morison equation and examined a modification of the equation that takes into account the response of the wake to flow reversal. When the flow begins to reverse under the action of the horizontal pressure gradient, the water in the wake behind the pipe is almost stationary; and the pressure gradient accelerates the water back across the top of the pipe, creating a region of high velocity and low pressure which exerts a significant effect on the lift and drag.

A few attempts have been made to measure forces on actual pipelines. A Hydraulics Research program installed an instrumented length of pipe on the seabed in 25 m of water in Perran Bay near Perranporth in southwest England and measured forces under waves.[16] Current meters close to the section of pipe measured wave-induced velocities. The Morison coefficients were calculated from the measured forces by a least-squares fitting procedure. The results show a very high level of scatter (as usually happens when measurements of wave forces at sea are correlated with theory) and are lower than those seen in the water tunnel and flume tests. The reason may be that the flow in the sea has a far more elaborate random turbulent structure than the flow in a tunnel, which is very organized. Evidently, in the sea there is a strong tendency for forces in one direction to be partially cancelled by forces in the opposite direction on a section of pipe located a short distance away.

Grace made measurements on sections of pipe under waves off Hawaii.[17, 18, 19, 20] In one test series, a section of 323.8 mm (11.75 m) pipe, 8.5 m long, was held on saddles in 5 m of water at an angle of 55° to the wave crests. Forces were measured on the central 1.2 m under waves 2.2 m high with periods between 12 s and 16 s. A current meter close to the pipe measured instantaneous velocities. The results were analyzed in eight distinct ways. The simplest *zero-kinematics* analysis calculated the drag coefficient from the force measured when the measured acceleration was zero, and the inertia coefficient from the force measured when the measured velocity was zero.

Using this analysis, C_D has an average value of 1.204, a standard deviation of 0.207, and a range from 0.5 to 1.4; C_M has an average of 1.646, a standard deviation of 0.370, and a range from 0.9 to 2.5, again demonstrating the limitations of the Morison equation, even when coefficients are referred to measured velocities, rather than to those calculated from a wave theory.

An improvement is achieved by least-squares regression analysis, using different techniques (sometimes weighting peak forces more heavily); but Grace found that there was an irreducible normalized prediction error of 10%. As he pointed out, this analysis gives an over-optimistic impression because the use of measured velocities introduced an element of circularity in the analysis, which allowed each wave its own force coefficient.

The analysis so far has treated the flow around the pipeline as two-dimensional, as if the transverse current were uniform along the length of the pipeline. That ideal situation only exists when the oscillatory current is generated by an infinitely long-crested wave exactly parallel to the pipeline. In reality, waves are short-crested and approach a pipeline obliquely, particularly in shallow water where the pipeline is roughly at right angles to the depth contours and wave refraction has brought the wave propagation velocity nearly into the same direction as the pipeline. The wave-induced bottom currents are not in phase along the length of the line and vary in amplitude. These effects reduce the maximum force on the pipeline.

Some research workers concerned with wave forces on pipelines, notably Grace, have preferred a different approach. Grace defines an extreme horizontal force coefficient C_{max} and an extreme vertical force coefficient K_{max}, so that

$$F_{xmax} = \frac{1}{2}C_{max}DU_{max}^2 0 \tag{11.9}$$

$$F_{ymax} = \frac{1}{2}K_{max}DU_{max}^2 0 \tag{11.10}$$

where quantities subscripted max are maximum values during the wave cycle, and other values are as defined earlier. The coefficients are reported as functions of $U_{max}^2/A_{max}D$, where A_{max} is the maximum acceleration. This ratio is a multiple of Keulegan-Carpenter number.

There are theoretical objections to this method. It gives maximum values alone and cannot describe the variations in force during a cycle nor identify the most unfavorable combination of drag and lift. Against that, it can be argued that the Morison equation is demonstrably insufficient to permit a genuine identification of the variation of force through the cycle, and that with the force coefficient method it should be assumed that the maximum lift and the maximum drag force occur at the same instant. In Grace's words, "This unsatisfactory non-predictable situation may best be guarded against by using ample structural factors of safety…" That wise remark applies equally to calculations based on the Morison equation (and to other areas of submarine pipeline engineering). It is sometimes thought that one method is more rational than the other, but that is not so.

11.5 Lateral Resistance

If a pipeline is to be stable, its weight must generate enough lateral resistance to withstand the hydrodynamic force. The reaction from the seabed onto the pipeline has a vertical component R and a horizontal component S. The limiting conditions that allow the pipeline to be stable can be described in an interaction diagram (Figure 11–4) whose axes are R and S. The limiting condition is described by two curves, one corresponding to incipient movement to the left and the other to incipient movement to the right. A combination of R and S represented by a point between the two curves is stable, but combinations outside the curves cannot be reached.

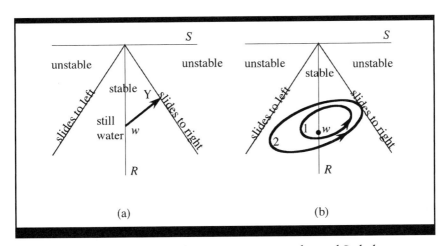

Fig. 11–4 *Horizontal and Vertical Reactions Between Pipeline and Seabed (a) Monotonically Increasing Water Velocity (b) Oscillatory Water Velocity*

Pipeline designers have generally adopted a frictional model of the limiting interaction between S and R and have taken the limiting condition as

$$S = \pm fR \qquad (11.11)$$

where

f is a coefficient of friction

The choice of f is largely based on experience and on a small number of experiments. Lyons carried out experiments on sand and found that small movements began at quite small ratios S/R, but that the threshold of large movements corresponded to a value of about 0.7 for f.[21] Behavior on clay is more complicated: f has a small value if the pipe is light enough not to sink into the clay significantly; but a heavier pipe sinks, and f and the lateral resistance then increase. In pipeline design, it is rare to encounter a combination of high currents and soft clay bottom because high currents would have eroded the clay. Lambrakos carried out field measurements of lateral resistance in an ambitious program that towed an instrumented sled across the seabed; but the results show a large degree of scatter, and the test could not be well controlled.[22] His coefficients are consistent with those found by Lyons.

Two recent research programs have investigated the limiting relation between R and S more thoroughly. A Norwegian Hydraulics Laboratory program carried out tests on pipe interaction with sand and clay. The principal results indicate that interaction with sand is consistent with the generally-accepted model if the degree of embedment into the bottom is small; but if there is significant embedment, the lateral resistance includes a significant passive-pressure component, generated by the weight and internal friction of the sand pushed in front of the pipe.[23, 24]

Danish Hydraulic Institute carried out tests on sand.[25] It investigated two kinds of loading history, the first with monotonically-increasing lateral force (representing a slow increase of a steady current across the pipe), and the second with an alternating force steadily increasing in magnitude (representing a cyclic increase of an alternating force generated by wave-induced currents that correspond to progressive growth of a storm). Very small movements occur at low values of S/R, but the pipe does not move far until S/R reaches about 0.5. The movement then increases rapidly and reaches about half a diameter of the pipeline when S/R is 0.7. Embedment into the bottom induces a much higher resistance, even if the embedment is only a small fraction of a diameter. Cyclic alternation of S leads the pipe to settle into the bottom and further increases its resistance to lateral movement later. The limiting value of S/R (measured when the movement has reached 0.2 diameters) is very much larger than the value observed under monotonically-increasing loading (but otherwise identical test conditions). Generally speaking, cyclic loading increases the limiting value of S/R by a factor of 2.

A pipeline in a trench has much more resistance to movement than a pipeline on a level seabed. It can become unstable either by sliding up the side of the trench or by deforming the soil on either side. In the first case, the coefficient f is increased to f', where

$$f' = \tan(\alpha + \arctan f) \qquad (11.12)$$

where

α is the side slope of the trench.

f' is much larger than f. Deformation of the side of a trench by hydrodynamic forces on a pipeline cannot occur unless the soil is very weak.

11.6   Stability Design

The process of stability design brings together the methods of wave and current prediction, hydrodynamic force calculations from currents, and lateral resistance analysis. A helpful way of considering the stability of unanchored pipelines is to think of the variation in the components R and S of the reaction between the pipeline and the seabed, defined in the lateral resistance section. For instance, in the case in which the current is steady and has no oscillatory component (Figure 11–4a) S is zero and R is equal to the submerged weight w per unit length in still water. When the current increases, S increases and R falls (because of the hydrodynamic lift). The pipeline becomes unstable at a current corresponding to point Y, and any further increase in current will sweep the pipeline away. The design problem is to make w large enough so that the combination of R and S remains within the stable region.

A similar diagram can be drawn for unsteady oscillatory flow under waves (Figure 11–4b). R and S then change through the wave cycle, and the point that represents them moves along a curve. In regular waves, the curve will be a closed loop. In random waves, it will be an irregular sequence of loops. If the oscillatory currents are small, loop 1, the pipeline will be stable. If the currents are larger, the pipeline will lose stability. Loop 2 represents a combined wave and current case in which the pipeline is just stable.

During the design procedure, the engineer has to confirm that the stability condition is satisfied. If it is not, weight has to be added to generate more lateral resistance. Adding weight externally (by increasing the thickness of a concrete coating, for instance) increases the diameter, and the hydrodynamic forces have to be recalculated. A simple computer program can carry out the design process very rapidly.

It is sometimes argued that it is unnecessarily conservative to design a pipeline to resist the maximum lateral force that can be applied because the maximum force cannot act simultaneously along the whole length of the line. A limitation of this argument is that sections of pipe that are less heavily loaded can only help to support more heavily loaded sections if shear forces are transmitted along the pipe. Shear forces can only be preset if the pipe bends horizontally. Moreover, there is a possibility of progressive instability, in which the pipe does not move bodily to one side but instead moves in one area at a time. The three-dimensional, time-domain analysis that would be required to take advantage of this effect in design has been the object of research by Danish Hydraulic Institute and Norwegian Hydrodynamic Laboratories. These organizations have developed computer models that allow the progressive motion of a pipeline to be followed in detail. In the present state of knowledge, it appears to be better practice to design a pipeline to be completely stable; but programs are available that allow the user to determine the movement of the pipeline and to design it so that the movement does not exceed a maximum acceptable limit. The results are extremely sensitive to the input assumptions.

Occasionally, it is argued that a pipeline should be designed to resist the maximum significant wave, rather than the maximum wave. That argument has little rational basis because if the maximum wave passes across the pipeline, it generates the corresponding bottom currents; and the pipeline has to resist the forces they induce. The fact that the existing sea state can be described by a statistical parameter called the *significant wave* does not alter this conclusion. However, it is possible to defend the significant wave method as a crude rough-and-ready way of accounting for various three-dimensional effects, such as wave directional spectrum spread, short-crested waves, and the variation of instantaneous force along the length of the pipeline. In the present state of knowledge, there is no simple way of incorporating these effects in analysis, though it is possible to calculate them by a full three-dimensional, finite-element analysis.

Veritec has put forward a revised method of stability analysis based on the results of an extensive dynamic analysis of various pipelines and has presented it through the use of a set of dimensionless parameters. The method is the basis of a Norwegian recommended practice.[26] The dynamic analysis is based on a time-domain solution of the pipeline stability and incorporates three-dimensional effects, surface wave spectra, and non-linear soil resistance.

The Generalized stability analysis is performed by comparing known dimensionless parameters against a set of parametric curves. The procedure does not give any indication of the physical processes involved nor of the magnitude of the hydrodynamic or soil resistance forces. It can be performed blind without knowledge of the physical procedures being modeled.

The analysis is based on the following dimensionless parameters:

Significant Keulegan-Carpenter number $K = U_s T_u/D$

Pipe weight parameter $\qquad\qquad L = W_s/(0.5\rho_w D U_s^2)$

Current to wave velocity ratio $\qquad M = U_c/U_s$

Relative soil weight (sand soil) $\qquad G = \rho_s/\rho_w - 1$

Shear strength parameter (clay soil) $\quad S = W_s/(DS_u)$

Time parameter $\qquad\qquad\qquad T = T_l/T_u$

Scaled lateral displacement $\qquad\quad \delta = Y/D$

where

U_s is the significant wave-induced seabed velocity,

T_u is the associated zero-crossing period,

U_c is the current velocity integrated over the diameter of the pipeline,

ρ_w is the mass density of sea water

ρ_s is the mass density of soil

S_u is the undrained shear strength of clay soil,

T_l is the duration of the sea state, and

Y is the allowable lateral displacement of the pipe.

If no information is available, the allowable displacement in sandy soils is recommended as 20 m in DnV Zone 1 (greater than 500 m away from a platform) or 0 m in DnV Zone 2 (less than 500 m from a platform).

In clay soils, no displacement is allowed. The wave-induced velocity U_s and period T_u are determined from a Jonswap surface wave spectrum. The standard includes a method that allows both values to be obtained directly from the surface peak period.

The required submerged weight for stability can be determined by comparing these parameters against dimensionless curves. A safety factor of 1.1 is recommended on the design submerged weight for clay soils.

The Generalized analysis is valid for a certain range of parameters, corresponding to the range for which the dynamic analysis validation was performed. This is

$4 < K < 40$

$0 < M < 0.8$

$0.7 < G < 1.0$ (for sand soil)

$0.05 < S < 8.0$ (for clay soil)

$D > 0.4$ m

Outside these ranges, Veritec recommends the Simplified Static stability analysis, based on a link between the traditional stability design procedure and the Generalized stability analysis. The results from the two distinct methods are made consistent with each other through the use of two calibration factors: one based on the soil conditions, and one based on the Keulegan-Carpenter number and ratio of wave to current velocity. The calibration factors ensure that the results of the Simplified analysis tie in with the Generalized analysis. Each stage in the Simplified analysis may not be truly representative of the actual process, but the results are corrected using the two calibration factors. The analysis is again based on pipelines designed with an allowable lateral displacement of up to 20 m in sandy soil and 0 m in clay soil.

The water particle velocities are obtained in the same way as in the Generalized analysis. The hydrodynamic forces on the pipeline are obtained from the Morison equation using coefficients of C_D 0.7, C_M 3.29 and C_L 0.9. (Note that these values are nominal, as correction is made through the calibration factor.) The soil resistance is modeled by a linear friction factor. Again, allowance for this assumption is made through the calibration factor. A safety factor of 1.1 is included in the results of the analysis for both clay and sandy soils. A paper compares the designs arrived at with those derived from a parallel program carried out by the American Gas Association.[27, 28]

11.7 Interaction with Instability of the Seabed

Traditional stability design treats the seabed as immovable. This assumption is not necessarily justified. In extreme wave and current conditions, the seabed too can begin to move in response to hydrodynamic forces so that pipeline movement and active sediment transport occur together. The stability of a pipeline cannot be considered in isolation from the stability of the seabed on which it rests.

If the pipeline is unstable, the stability of the seabed is likely to be marginal at best. If the seabed is unstable, the pipeline becomes unstable with it. Indeed, a possible interpretation of the success of the traditional approach to design is that it ensures that the pipeline is always heavy enough for the seabed to become unstable before the pipeline and that the pipeline then sinks into the moving seabed material.

If the pipeline is initially partially buried, and a top layer of the seabed begins to move, then:

1. The pipeline is exposed to hydrodynamic forces over a greater fraction of its diameter.
2. The fluid moving past the line has a density greater than water because of the presence of seabed particles.
3. A smaller fraction of the pipeline is embedded in stationary material, and the line's resistance to lateral movement is therefore reduced.

All three factors have an adverse effect and undermine (literally!) the validity of the traditional approach.

Palmer and Damgaard examine the issue of seabed instability in detail and describe two instances of pipelines off the coast of Australia where the seabed unquestionably becomes grossly unstable long before the extreme design conditions are reached.[29, 30] The conventional design calculations then become essentially irrelevant. They show that seabed mobility in extreme conditions is the rule rather than the exception, and that it applies over wide areas of the North Sea. Figure 11–5 shows the wave height at which seabed instability begins, as a function of particle size, for three water depths.

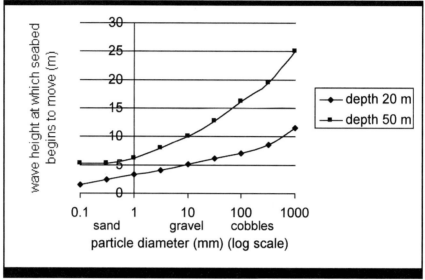

Fig. 11–5 *Onset of Seabed Instability*

The problem of determining the depth below mud line to which the seabed becomes mobile is only partly understood. That depth depends on the Sleath number:

$$S = \frac{U_0\omega}{g(s-1)} \tag{11.13}$$

where

U_0 is the amplitude of the orbital velocity

ω is the cyclic wave frequency

g is the gravitational acceleration

s is the particle specific gravity

The question is complicated by the increases of pore pressure that occurs as a result of cyclic shear stresses induced by waves. This is the subject of active research under the European Union (EU) Liquefaction Around Marine Structures (LIMAS) program. Within that program, tests on a pipeline resting on a fine-grained silty soil show that the soil partially liquifies and that the final level reached by the pipe depends on its mean specific gravity.[34]

Neglect of seabed instability is the central flaw in the conventional approach to stability design, and correcting that flaw is much more important that continued argument (or continued research) into the Morison equation coefficients or into more sophisticated hydrodynamic models. Future studies will have to take a more integrated view of a pipeline's interaction with seabed stability and with processes like self-burial and scour.

Having said that, the conventional design method does seem in practice to lead to satisfactory designs. There are several factors that make the conventional method conservative, among them the enhanced lateral resistance produced by embedment, the neglect of wave long-crestedness and three-dimensional effects generally, and the overestimate of seabed wave-induced velocities produced by idealizing the maximum wave as one of a series of regular waves all with the same height. Instability does occasionally occur, invariably as a result of design mistakes (particularly in estimates of maximum waves), construction mistakes, and loss of weight coating.

References

1   Komar, P.D. (1976) *Beach Processes and Sedimentation.* Prentice-Hall.

2   Sarpkaya, T., M. Isaacson. (1981) *Mechanics of Wave Forces on Offshore Structures.* New York: Van Nostrand Reinhold.

3   Embrechts, P., C. Kluppelberg, T. Mikosch. (1997) *Modelling Extreme Events.* Berlin: Springer-Verlag.

4   Brink-Kjaer, O., J. Knudsen, G. S. Roodenhuis, M. Rugbjerg. (1984) "Extreme Wave Conditions in the Central North Sea." *Proceedings of the 16th Annual Offshore Technology Conference.* Houston. 3. 283–293.

5   St. Denis, M. (1969) "On Wind-Generated Waves." *Topics in Ocean Engineering.* (ed. C. L. Bretschneider). Gulf Publishing. 37–41.

6   Borgmann, L. E. (1961) "The Frequency Distribution of Near Extremes." *Journal of Geophysical Research.* 66. 3295–3307.

7   Littlejohns, P. S. G. (1974) *Current-induced Forces on Submarine Pipelines.* (Report INT 138) Wallingford, England: Hydraulics Research Station.

8   Zdravkovich, M. M. (1985) "Forces on Pipe Clusters." *Proceedings of the International Symposium on Separated Flows.* Trondheim.

9   Bearman, P. W., M. M. Zdravkovich. (1978) "Flow Around a Circular Cylinder Near a Plane Boundary." *Journal of Fluid Mechanics.* 89. 33–47.

10 Bryndum, M. B. and V. Jacobsen. (1983) "Hydrodynamic Forces from Wave and Current Loads on Submarine Pipelines." *Proceedings of the 15th Annual Offshore Technology Conference.* Houston. 1. 95–102.

11 Wilkinson, R. H., A. C. Palmer, J. W. Ells, E. Seymour, N. Sanderson. (1988) *Stability of Pipelines in Trenches.* "Proceedings of the Offshore Oil and Gas Pipeline Technology Seminar." Stavanger.

12 Wolfram, W. R., J. R. Getz, R. L. P. Verley. (1987) "PIPESTAB Project: Improved Design Basis for Submarine Pipeline Stability." *Proceedings of the 19th Annual Offshore Technology Conference.* Houston. 3. 153–158.

13 Holte, K., T. Sotberg, J. C. Chao. (1987) "An Efficient Computer Model for Predicting Submarine Pipeline Response to Waves and Currents." *Proceedings of the 19th Annual Offshore Technology Conference*. Houston. 3. 159–169.

14 Verley, R. L. P., K. Reed. (1987) "Prediction of Hydrodynamic Forces on Seabed Pipelines." *Proceedings of the 19th Annual Offshore Technology Conference*. Houston. 3. 159–169.

15 Fyfe, A., J., D. Myrhaug, K. Reed. (1987) "Hydrodynamic Forces on Seabed Pipelines: Large-Scale Laboratory Experiments." *Proceedings of the 19th Annual Offshore Technology Conference*. Houston. 1. 125–134.

16 Wilkinson, R. H., A. C. Palmer. (1988) "Field Measurements of Wave Forces on Submarine Pipelines." *Proceedings of the 20th Annual Offshore Technology Conference*. Houston.

17 Grace, R. A. (1978) *Marine Pipeline Systems*. Englewood Cliffs, NJ: Prentice-Hall.

18 Grace, R. A., J. Castiel, A. T. Shak, G. T. Y. Zee. (1979) "Hawaii Ocean Test Pipe Project: Force Coefficients." *Proceedings of Civil Engineering in the Oceans IV*. American Society of Civil Engineers. 99–110.

19 Grace, R. A., S. A. Nicinski. (1976) "Wave Force Coefficients from Pipeline Research in the Ocean." (OTC 2676) *Proceedings of the 8th Annual Offshore Technology Conference*. Houston, TX.

20 Grace, R. A., J. M. Andres, E. K. S. Lee. (1987) "Forces Exerted by Shallow Ocean Waves on a Rigid Pipe Set at an Angle to the Flow." *Proceedings of the Institution of Civil Engineers*. 83. 43–59.

21 Lyons, C. G. (1973) "Soil Resistance to Lateral Sliding of Marine Pipelines." *Proceedings of the 5th Annual Offshore Technology Conference*. Houston. 2. 479–484.

22 Lambrakos, K. F. (1985) "Marine Pipeline Soil Friction Coefficients from In-Situ Testing." *Ocean Engineering*. 12. 131–150.

23 Brennoden, H., O. Sveggen, D. A. Wager, J. D. Murff. (1986) "Full-Scale Pipe-Soil Interaction Tests." *Proceedings of the 18th Annual Offshore Technology Conference*. Paper OTC 5338 presented at the 18th Annual Offshore Technology Conference. 4. 433–440.

24 Wagner, D.A., J. D. Murff, H. Brennoden, O. Sveggen. (1987) "Pipe-Soil Interaction Model." *Proceedings of the 19th Annual Offshore Technology Conference*. Paper OTC 5504 presented at the 19th Annual Offshore Technology Conference. 3. 181–190.

25 Palmer, A. C., J. Steenfelt, J. O. Steensen-Bach, V. Jacobsen. (1988) "Lateral Resistance of Marine Pipelines on Sand." (OTC5853) *Proceedings of the 20th Annual Offshore Technology Conference*, 4, 399–408.

26 RP E305: "On-Bottom Stability Design of Marine Pipelines." (1988) Veritec.

27 Hale, J. R., W. F. Lammert, D. W. Allen. (1991) "Pipeline On-Bottom Stability Calculations: Comparison of Two State-Of-The-Art Methods and Pipe-Soil Model Interaction." (OTC6761) *Proceedings of the 23rd Annual Offshore Technology Conference*. 4, 567–582.

28 AGA. (1991) "Submarine Pipeline On-Bottom Stability; vol. 1. Analysis and Design Guidelines." Final Report on Projects PR-178-516 and PR-178–717. American Gas Association.

29 Palmer, A. C. (1996) "A Flaw in the Conventional Approach to Stability Design of Pipelines." *Proceedings of the Offshore Pipeline Technology Conference*. Amsterdam.

30 Damgaard, J. S., A. C. Palmer. (2001) "Pipeline Stability on a Mobile and Liquefied Seabed: a Discussion of Magnitudes and Engineering Implications." *Proceedings of the 20th International Conference on Offshore Mechanics and Arctic Engineering*. Rio de Janeiro.

31 Sleath, J. F. (1994) "A. Sediment Transport in Oscillatory Flow." From *Sediment Transport Mechanisms in Coastal Environments and Rivers*. Belorgey, M., R. D. Rajaona, J. F. A. Sleath. (eds.), Singapore: World Scientific.

32 Sleath, J. F. A. (1998) "Depth of Erosion Under Storm Conditions." Proceedings of the 26th *Conference on Coastal Engineering*. ASCE. New York, 2968–2979.

33 Sassa, S., H. Sekiguchi. (1999) "Wave-induced Liquefaction of Beds of Sand in a Centrifuge." *Geotechnique* 4.: 621-638.

34 Teh, T. C., A. C. Palmer, M. D. Bolton, J. S. Damgaard. (2004) "Experimental Study of Marine Pipelines on Unstable and Liquefied Seabed." *Coastal Engineering*.

12 Marine Pipeline Construction

12.1 Introduction

This chapter examines the principal methods of construction. Most marine pipelines are constructed by the lay barge method described in section 12.2. Many small- and intermediate-diameter lines are constructed by the reel ship method described in section 12.3. Bundles and some other lines are installed by pull-and-tow techniques covered in section 12.4. Many pipelines have to be trenched, particularly those in shallow water, and some are buried: trenching and burial are described in section 12.5.

12.2 Lay Barge Construction

12.2.1 Introduction

Lay barge construction is by far the most frequently used technique for marine pipeline construction. It remains the method of choice for most pipelines. Lay barge construction is versatile, flexible, and self-contained. Though it may be expensive to mobilize a lay barge to a remote location, once the barge is in place it can start work and operate as efficiently with minimal support from the shore. It has little competition as a method for installing large-diameter single lines (though not bundles) and smaller-diameter lines and competes aggressively with the reel-and-tow techniques.

The lay barge system can be seen as a natural development from lowering-in methods of onshore pipeline construction. It was originally developed in shallow water in the near shore Gulf of Mexico in the 1940s and 1950s. The first North Sea pipelines were laid in the period from 1968 to 1975, among them West Sole, Leman, Forties (the first large-diameter line in more than 100 m), Frigg, Brent and Noordgastransport. All were constructed using the lay barge method. By today's standards, lay barge operations were prone to weather downtime and to mechanical breakdown, and productivity was extraordinarily low and costs very high. The first Forties line, 170 km long, took two lay barges two seasons each in 1973–74, and other lines took even more lay barge effort. In contrast, the second Forties line was laid by one barge in three months in the summer of 1990. Figure 12–1 shows the production rates on that project and the high lay rates that were achieved after a short time with few interruptions. Laying rates of several km/day are routine nowadays.

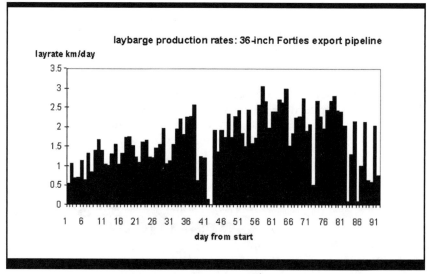

Fig. 12–1 *Lay Barge Production (from Reference 1)*[1]

Figure 12–2 shows the S-lay version of the lay barge system schematically and gives some of the terminology. The construction is based on a moored or dynamically positioned barge on which the pipeline is built on a ramp. Lengths of pipe are lined up at the upper end of the ramp and pass through a series of welding stations as the barge moves forward. Sometimes a separate welding line on the barge welds the pipe lengths together in twos (*double jointing*) before they join the main line.

Tensioners apply a force to the pipe near the stern end of the ramp. The pipe leaves the barge at the stern, and its configuration immediately beyond the stern is a convex-upward curved section, called the *overbend*, supported on rollers by a stinger structure. The stinger is a substantial structure, often nearly 100 m long, generally constructed as a single open framework rigidly fixed to the barge, but frequently it has one or more buoyant segments hinged to each other and to the barge.

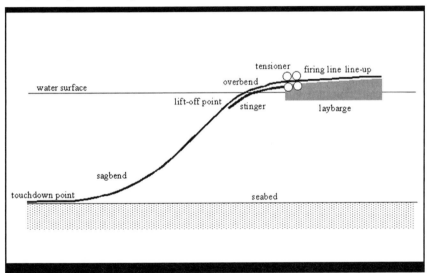

Fig. 12–2 *S-lay Lay Barge Pipelaying Schematic (Not to Scale)*

The pipe loses contact with the stinger at the lift-off point just above the end. It continues downward through the water as a long suspended span, a concave-upward curve, called the *sagbend*. It reaches the seabed tangentially at the touchdown point.

The shape of the pipe in the sagbend is controlled primarily by the interaction between the applied tension and the submerged weight of the pipe and, to a lesser extent, by the flexural rigidity of the pipe. If the applied tension is increased, the curvature of the pipe in the sagbend decreases so that the sagbend becomes longer and flatter. At the same time the touchdown point moves further from the barge, and the liftoff point moves up the stinger. If the applied tension is reduced, the sagbend curvature increases. If the tension is reduced too far, the bending curvature may become excessive, and the pipe may buckle.

On the other hand, the shape the pipe takes on in the overbend is controlled by the stinger geometry. Tension has almost no effect on the overbend configuration if the stinger is rigid and only a small effect if the stinger is segmented. The stinger has to be quite long because otherwise the stinger curvature will not match the suspended-span curvature and excessive bending will occur at the end of the stinger.

There are many lay barges of this type. Figure 12–3 depicts the Allseas lay barge, *Tog Mor*.

Fig. 12–3 *Tog Mor Lay Barge (Photograph by Courtesy of Allseas)*

The lay barge technique has proved itself resilient and flexible. There is no limitation on length and essentially no limit on the diameter of the pipe that it can accommodate. A barge can lay 42-in. pipe on one project, and within a few days lay 6-in. pipe on another project with minimal modifications to its systems. Many lay barges were constructed a long time ago: *LB200* went into service in 1975, *Semac* in 1977, and *Castoro 6* in 1979, so that their costs have been amortized.

There have been several new developments in lay barges. The most significant are the applications of dynamic positioning to lay barges and the development of the steep-ramp J-lay system.

12.2.2 Dynamic positioning

A lay barge has to be held precisely in position while it lays pipe. It cannot be allowed to drift sideways or to yaw away from the direction of the pipe. Either movement rapidly bends the pipe severely at the end of the stinger, and the pipe is easily buckled and kinked. Except in shallow water, a substantial tension has to be applied to the pipe to control the pipe curvature in the sagbend between the stinger and the seabed. That tension reacts on the barge, and the barge has to be held in position against it.

Until quite recently, lay barges were invariably positioned by mooring lines to anchors. The mooring lines were attached to winches, and the operator controlled the position of the barge by paying out or winding in the mooring lines. A typical anchored third-generation semisubmersible such as the Stolt Offshore *LB200* has 14 mooring lines, each 4200 m long. Not all of them need to be active at any one time, and only about eight are needed to hold the barge in position. The precise number depends on the current, the wind, the sea state, and the water depth. The anchors are progressively repositioned by anchor-handing tugs.

Positioning by mooring has several disadvantages, among them:

- The length of the mooring system often causes problems. Mooring lines have to radiate from the barge over a full circle of directions, and in confined or congested areas it becomes difficult to place the anchors so that the lines do not intersect features such as existing platforms, wellheads, pipelines, islands, or shipping lanes.

- The risk that an anchor might drag into a previously laid pipeline, a situation that would necessitate a major repair.

- The mechanical flexibility of the mooring system in deep water, which limits the precision with which the barge can be positioned.

- The need to repeatedly relocate anchors, action that becomes difficult and dangerous in rough seas and may limit the barge's progress. This factor has a greater and greater impact as lay rates increase because anchors must be relocated frequently. If there are 12 anchors in operation, each one has to be relocated once every 3000 m of progress and the barge is laying at 6000 m/day, then one anchor has to be relocated every hour.

It has been recognized for a long time that there would be advantages to applying the dynamic positioning (DP) principle that is routinely used for drill ships and diving support vessels to lay barges. The vessel is positioned by thrusters (shrouded propellers that can be rotated into the required direction) that are controlled by computer through a GPS (global positioning system) or an acoustic positioning system. Even in 1974, the designers of *Viking Piper* (now *LB200*) had considered adding part of a DP system to the barge thrusters.

Two factors had worked against DP for lay barges. The first was the reliability of the system and the severe consequences of buckling and barge damage that would rapidly follow any failure. The second was the power required to balance the applied tension. A typical tension is 1 MN (100 tons), and the corresponding thruster power requirement to offset it is roughly 7.5 MW (10,000 HP). The need for this steady force and for additional steady force to counteract current drag makes the system more difficult to control and leads to a significant fuel cost.

The Allseas lay barge *Lorelay* (Fig. 12–4) was the first to apply the dynamic positioning principle to pipe laying and has captured a significant segment of the North Sea and Gulf of Mexico pipe laying markets. The Allseas lay ship *Solitaire*, the McDermott *DB50*, the Saipem *S-7000*, and the Heerema *Balder* are all dynamically positioned. An integrated control system measures and controls the position of the barge and the configuration of the pipe relative to the barge. Today, all the current developments of deep-water lay barges rely on dynamic positioning, rather than anchors.

Fig. 12–4 *Lorelay Lay Barge (Photograph by Courtesy Of Allseas)*

12.2.23 J-lay

Conventional lay barges have an S-configuration (so-called after the shape taken on by the pipeline as it travels from the barge to the seabed). This is the version of the lay barge system depicted in Figure 12–2. The horizontal or near-horizontal ramp allows space for several welding stations, tensioners, an X-ray station, a field joint station, and a stinger of acceptable length to be combined with an acceptable tension level.

S-lay was used to lay the vast majority of pipelines in the North Sea, in depths up to 300 m in the Norwegian trench, and for deep-water pipelaying in the Gulf of Mexico. One of the most ambitious pipelaying projects so far completed, the crossing of the Strait of Sicily between Tunisia and Sicily, used the *Castoro VI* semi-submersible lay barge. That lay barge was built especially for that crossing. The barge has a relatively steep ramp at about 7° and a curved steep stinger with a departure angle of 39°.

S-lay encounters difficulties in very deep water. The problem is that the pipeline leaves the stern end of the barge almost horizontally, then goes over the stinger in an overbend (convex upward) until it leaves the stinger at the lift-off point. It then bends the other way in a suspended span forming a sagbend (convex downward). The weight of the pipe in the sagbend is supported by the applied tension. The tension has to be great enough for the slope of the pipe at the upper end of the suspended span to match the slope on the stinger otherwise the pipe will kink at the end of the stinger. The slope at the upper end of the suspended span depends on the applied tension; the larger the tension, the smaller the slope. In addition, the tension has to be great enough to keep the curvature in the sagbend within acceptable limits so that the pipe does not curve excessively.

In shallow water, it is possible to meet these conditions without either a long stinger or high tension. A very long stinger is undesirable because it is excessively vulnerable to wave and current forces. High tension is undesirable because of the risk of tensioner damage to the pipe coating and because the tension has to be balanced by the barge's mooring or dynamic positioning system.

In deep water, the tension has to be high, and the stinger has to be long. The J-lay concept offers an escape from these constraints. It gives up the idea that the pipe has to leave the barge horizontally. Instead, the pipe is made up on a steep ramp, typically at 75° to the horizontal, and leaves the barge steeply over a vestigial stinger. There is no overbend, and the entire suspended span has the form of an extended letter J. Figure 12–5 is a schematic of J-lay.

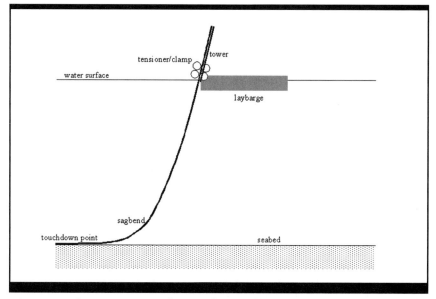

Fig. 12–5 *J-lay Lay Barge Pipelaying Schematic (Not to Scale)*

J-lay has many advantages, among them:

- Because the pipe leaves the barge steeply, the tension is dictated only by the need to limit bending in the sagbend, not by the need to match the suspended span to the stinger. The tension is usually much less that with other methods.

- Because the pipe leaves the barge steeply, it is not exposed to wave action except in a short region close to the surface.

- There is no stinger.

- Because the tension is reduced, the touchdown point is not as far behind the barge as occurs with the S-lay method; therefore, it is easier to position the barge so as to lay the pipe precisely in congested areas.

- Because the tension is smaller, spans are shorter.

- In severe weather, the barge can lower the pipe a short distance until it is just below the barge, then can weather vane around the pipe so as to head into the sea and minimize wave-induced movements.

Equally, there are clear disadvantages. In particular:

- Because most of the ramp is high above sea level, it is impracticable to position a series of operations such as line-up, welding, weld radiography, tension application, and field-joint infill at intervals up the ramp.

- The added weight high up on the barge has an adverse effect on its stability.

- If the barge has to continue laying pipe in shallow water, the ramp has to be lowered to a less steep angle, otherwise the pipe would be forced to bend to a sharp radius to reach the seabed horizontally.

The first factor indicates that all welding and radiography should be carried out at one station at the lower end of the ramp. That, in turn, suggests that the lengths of pipe to be welded on the ramp should be multiple joints, fabricated from several standard 12-m lengths welded together so that the barge can advance by the length of a multiple joint after each length is completed. It also suggests that there would be advantages in adopting a rapid one-shot welding system at the welding station so that each weld is completed more rapidly than by conventional automatic or hand-welding, or, alternatively, that welding be abandoned and replaced by an alternative joining system such as threaded connections.

It was recognized a long time ago that J-lay has compelling advantages in deep water. Small-diameter pipelines 3 or 4 in. in diameter were laid in the 1960s from drill ships in a vertical J-configuration with minimal horizontal tension. The first recent application of J-lay was the 20-in. Maui B to A pipeline, laid in 1991–92 in 105 m of water in the Tasman Sea to the west of the North Island of New Zealand. Special factors applied in that instance. Like the Strait of Georgia, New Zealand is remote from any center of the offshore industry and has little activity offshore, and no marine pipelaying equipment is maintained there. The line is too large for reel laying, and both bottom tow and lay barge construction were studied in detail. Both methods were shown to be practicable. Bottom tow was thought to have a higher risk.

While options were being considered, the Dutch contractor Heerema was awarded a contract to set jacket topsides at Maui B. Heerema pointed out that its semi-submersible crane vessel *Balder* could be adapted to lay pipe in a J-configuration, and that option would eliminate the need for a second vessel. That option was selected, and a J-lay tower was added to the vessel. It consists of a fixed steep ramp some 80 m high, a moveable transfer ramp, and a very short stinger that terminated about the sea surface.

Sextuple 72-m strings of internally clad pipe were made up on shore and transported out to the vessel by cargo barge, lifted onto the transfer ramp, and raised into alignment with the steep ramp. The pipe was held by a clamp at deck level, and all the welding was carried out at a single station.

The project was a technical success, but it encountered extensive delays. It had been anticipated that the 15 km would take two or three weeks to lay. It actually it took more than three months because of welding difficulties, which had been expected, and problems with the clamp, which had not been identified during the trials.[2]

The next application of the method was by McDermott's *DB50* derrick barge, which in 1993 laid the 12-in. Auger pipeline in 870 m of water in the Gulf of Mexico. The system is based on a self-contained tower that can be transferred between vessels and on a sophisticated ready-rack, pipe-handling system and transfer ramp. The tower is removed from the vessel between pipelaying projects.

Several contractors have now completed J-lay systems to take advantage of the deep-water market in the Gulf of Mexico, Brazil, and West Africa. Contractors that had already made heavy investments in S-lay argued in the past that S-lay could do anything that J-lay could do and that there was no need for J-lay. That position has shifted recently, and some contractors that formerly argued against J-lay have now committed to J-lay systems, sometimes with the enthusiasm of the recent convert. Saipem is one of the technical leaders in this field and has the *Saipem 7000* J-lay system. It has a diameter range from 4 to 32 in., a tower angle from 70° to 90°, one welding station, and one non-destructive testing station. The ramp is designed for quadruple 48-m joints. The system operates in either tensioner mode (in which an actively-controlled tensioner can pull the pipe in or pay the pipe out to maintain the tension within prescribed limits) or in clamp mode (in which the pipe is clamped to the barge while the weld is being completed and released only while the barge moves forward). In tensioner mode, the tensioner capacity is 5.1 MN (525 tons). In clamp mode the maximum holding capacity is 20 MN (2000 tons).

Heerema is another leading contractor. It has modified the *Balder* for deep-water J-lay, initially for the a large contract for some 400 km of pipelines and risers up to 28 in. (711.2 mm) in diameter for the Mardi Gras system in depths up to 2000 m (6500 feet) in the Gulf of Mexico. The tensioner capacity is 525 tons, and the standard configuration is for quadruple joints; but the tower is designed to carry 1050 tons and can be extended for six joints (hexjoints).

Figures 12–6, 12–7 and 12–8 show the lay barges *S-7000*, *Balder* and *Solitaire*. The first two apply J-lay, and the third S-lay. Whether J-lay or S-lay will dominate the deep-water market is highly controversial. Allseas has laid pipe in very deep water by S-lay, and its barge *Solitaire* is built in the S-lay configuration, though a very long stinger allows the pipe to reach a departure angle of 69° to the horizontal.

Fig. 12–6 S-7000 *Lay Barge (Photograph By Courtesy of Saipem)*

Fig. 12–7 Balder *Lay Barge (Photograph by Courtesy of Heerema)*

Fig. 12–8 Solitaire *Lay Barge (Photograph by Courtesy of Allseas)*

Most people take the view that J-lay will come to dominate the market for water depths greater than 1000 m and will compete vigorously in depths greater than 300 m.

12.3 Reel Construction

The first offshore pipelines in Europe were laid in 1944 in the Pipe Line Under The Ocean (PLUTO) project to build gas pipelines across the English Channel to supply the Allied armies in northern France. The development drew on oil industry experience, particularly from the Anglo-Iranian Oil Company, an ancestor of British Petroleum. The small group of engineers brought to the project an extraordinarily creative drive and imagination. Pushed by the pressure of war, they were a model and an inspiration to everyone concerned with pipelines.

PLUTO had two designs for 3-in. pipe, one a conventional steel tube and the other an armored lead tube made by cable technology. The pipe had no corrosion protection, because it would not be needed for long. Forty km lengths of pipe were wound onto large floating reels 15 m in diameter. The drums were towed by tugs, unwinding the pipe as they went and laying pipelines from England to France in as little as 10 hours. Asked why a 100 km marine pipeline cannot now be laid in 10 hours, the engineer can only ruefully reply that it must be because so much research and development have been carried out. Significantly, PLUTO encountered some technical problems such as buckle propagation, which were forgotten and rediscovered 30 years later.

The reel method was developed further in the 60s, initially in the Gulf of Mexico where Gurtler Hubert, a Louisiana contractor, converted a landing-ship hull into the reel barge *U-303* (later *RB-1*). It laid 6-in. pipe from a vertical-axis reel. The pipe was bent plastically in the horizontal plane and had to be straightened before it was laid otherwise the suspended span would bow sideways and form a series of kinks on the seabed. A second barge *RB-2* (later *Chickasaw*) was constructed in 1970, and it too laid pipe from a horizontal reel in up to 300 m.

Reel method technology was acquired first by Fluor and then by Santa Fe who went on to design and build the reel ship *Apache*. It went into service in 1979 and began to take an increasing share of the flowline market, taking advantage of the fact that the line could be made onshore, wound onto the reel, and unreeled on location in a few hours. McQuagge and Davey describe a typical project.[3] *Apache* is now operated by Technip-Coflexip.

Until recently, only Technip-Coflexip was active in reel ship pipelaying in the North Sea. More contractors have now entered the market. Global Industries operates the horizontal reel barges *Chickasaw* and *Hercules* in the Gulf of Mexico. *Hercules* is designed to lay pipe with a diameter up to 20 in. ETPM has laid small-diameter pipelines off the shore of West Africa from the *Norlift* reel ship.

DSND has installed a reel on the *Fennica*, which was built as a combined icebreaker and offshore construction vessel, to take advantages of the factor that Baltic icebreakers are busy in the winter and construction vessels are needed in the summer.[4] *Fennica's* reel hub is about 10 m in diameter, much smaller than the 16.4 m reel on *Apache*. The reel diameter is important because the maximum bending strain in the pipe is inversely proportional to the hub diameter. *Fennica* appears to be mainly aimed at the small-diameter market. It first went into service in September 1996 to recover the blocked 8-in. Staffa line for London and Scottish Marine Operations (LASMO), and then constructed twelve 10- and 6-in. pipelines for the Banff field.

Figures 12–9 and 12–10 show Global Industries' reel barges *Chickasaw* and *Hercules*. Figure 12–11 shows the Technip-Coflexip reel ship *Deep Blue*.

A number of enhancements of the reel ship method have been explored, but so far none has been actively promoted nor developed. The reel method also can be used to install pipeline bundles, with or without carriers.[5] A limitation has always been that reeled pipe could not be concrete-coated because the level of longitudinal bending strain (0.022 for 16-in. pipe on the *Apache* hub) would cause concrete to crack and break away. This has been a severe restriction of the applicability of the reel method. Because of this limitation, additional wall thickness had to be added to stabilize the pipe against currents and waves. That meant that the owners could not finalize the pipe specification and commence pipe procurement until they had committed to a lay method. Concrete able to be reeled would allow the two decisions to be made independently. A weight coating of polymer-modified concrete has now been developed that can be reeled, but unfortunately it has not been pursued by the operators.

Some people argue that reel laying is an inherently attractive technique that has never quite developed its full potential. It follows the sound principle that as much work as possible should be carried out onshore in a controlled factory environment, insensitive to weather and without the very large support costs involved in any offshore operation.

Fig. 12–9 *Chickasaw Reel Barge (Photograph by Courtesy of Global Industries)*

Fig. 12–10 *Hercules Reel Barge (Photograph by Courtesy of Global Industries)*

Fig. 12–11 *Deep Blue Reel Ship (Photograph by Courtesy of Technip-Coflexip)*

12.4 Pull and Tow

12.4.1 Introduction

Another method of pipeline construction is to construct the line onshore and drag or tow it into place on the surface, below the surface, or on the bottom. This technique has the advantage that the line can be inspected and tested before it is towed and can be used both for single lines and for bundles. Moreover, there is no limit on the size or complexity of a bundle.

Only controlled-depth tow and bottom tow have been used to install flowlines in the North Sea, but other kinds of tow have been applied elsewhere. They are considered separately.

12.4.2 Surface tow

In the surface tow method, the pipe bundle is made buoyant. The bundle is towed to location on the surface and is sunk in position, often by flooding pontoons or by drawing it down by tension. Surface tow has been used for short lengths of line in shallow water in protected environments such as the Gulf of Mexico and the Arabian Gulf and in the UK for outfalls.

Surface tow in a rough sea subjects the bundle to severe wave action and, for that reason, is not used in environments where rough seas are likely. For a 36-in. (914.4 mm) steel carrier in a 10-sec wave 5 m high (typical of a summer storm in the North Sea) the bending stress is at least 200 N/mm^2, a substantial proportion of the yield stress and enough to cause fatigue damage.

Lateral currents influence surface towing. In the final location of the bundle, north-easterly surface currents of, for example, 0.5 m/s (1 knot) are expected, and the corresponding lateral force on a bundle towed across the current is 100 N/m (7 lb/ft). Because of this, the bundle will not tow in a straight line behind the tug. Instead, it will follow a curve at an angle to the tug's track.

An additional difficulty with surface tow is that a towed pipe can be hydrodynamically unstable. This phenomenon was observed in development of the Dracone flexible barge, for which towed buoyant tubes were seen to undergo violent lateral oscillations, both in field trials and in model tests, at quite modest speeds. Instability, however, can be reduced by adding an unstreamlined connection sled to the trailing end, thereby increasing the tension, or by adding carefully designed tail fins.

A positively buoyant bundle in a carrier can be lowered to the bottom only by drawdown or by flooding the carrier, a task that is difficult to control. Because of these factors, one can see that towing at the surface almost always requires external buoyancy.

Surface tow generally has not been adopted as a method of construction of flowlines in deep water, though it is a useful method in shallow water. Once the complication of adding buoys has been accepted, it is preferable to suspend the bundle from the buoys, far enough below the surface so that the bundle is at a depth below substantial wave action. That method describes the technique of near-surface tow examined in section 12.4.3.

Another technique of surface tow is to wind the pipe into a floating spiral supported by floats, and then to unwind it onto the seabed. This method has been the subject of extensive studies and small-scale trials, but has not been applied to a project to date.

12.4.3 Near-surface tow

In this method, the pipe bundle is suspended below the surface from buoys, thereby reducing the direct action of waves. However, it is not possible to eliminate wave action entirely. To do so would require the bundle to be lowered beyond half the wavelength of the longest waves, typically are 100 m or more long. Submergence to that depth would prevent the line from being towed in shallow water and would lead to several difficulties when the buoys were released. Lowering to the tow depth can be staged so that the bundle is just below the surface in shallow water and further down in deep water where wave action is more severe.

Because the spar buoys also are subject to wave action and because the pipe is a long flexible system with independent buoy elements not rigidly connected to the pipeline, the dynamics of the tow are potentially highly complex. The issue of the buoy together with the difficulty of lowering the pipe are factors that have deterred potential users of surface and near-surface tow in exposed deep-water areas.

Near-surface tow for deep-water construction was tried in the 1960s as part of Gaz de France studies for a pipeline between Algeria and Spain. Similar methods were later used to construct a deep-water outfall at Cassidaigne and to lay a gas pipeline in Lake Geneva. A full-scale trial of near-surface tow in the North Sea followed in 1975, when a 1000-m length of 16-in. pipe was towed 900 km northeast of Scotland.

A second test was carried out two years later as part of a Compagnie Française des Pétroles (CFP)- Total/Elf Aquitaine joint program on the remorquage-aboutage-tension (RAT): towing, tie-in, tension method. Four strings of 20-in. pipe were made up, fitted with pontoons, and stored in a fjord before being towed in seas with waves up to 5 m and in 30-knot winds. Flexural stresses in the pipe strings were recorded as part of the program.

The most ambitious trial of the near-surface tow technique was conducted by a group of operators led by Mobil in 1980. The bundle was 600 m long, and consisted of three 50.8-mm (2-in.) ID through flowline (TFL) and several hydraulic control hoses, all bundled in a 324-mm (12-in.) carrier. The bundle was made up at an abandoned airfield on the shore of Matagorda Bay in the Gulf of Mexico.

During the tow, the bundle was supported by 29 buoys, of which 15 were spar buoys and 14 horizontal buoys. At launch and in shallow water, the spar buoys were lashed to the carrier in a horizontal position, causing the bundle to float at a depth of 3 m. Once the tow had crossed the intracoastal waterway ship channel, the end of each of the spar buoys was released from the carrier, causing the carrier to float at a depth of 12 m. Throughout the tow, the bundle was kept under tension, first between two spud barges in the shallow bay, then between two tugs during the open water tow, and finally between a semi-submersible drilling rig at one end and a derrick barge at the other. On arrival at site, the bundle was pulled down in an inverted catenary by tension applied at both ends. The buoys were released from the carrier while the bundle was under a 0.15 m/s transverse current so that it would lie sideways (as in the final stages of a surface tie-in). Originally, the plan was to release the buoys acoustically; but because nylon safety lines had been added for extra security during the buoy upending operation, a manned submersible had to release the buoys.

12.4.4 Mid-depth tow

In mid-depth tow, the pipeline or pipeline bundle is negatively buoyant, and it is suspended in a long, flat catenary between two tugs, a tug at each end. The tugs apply tension to the bundle, and the amount of tension determines how flat the catenary is. The lengths of the lines connecting the bundles to the tugs can be altered so that the bundle can be carried higher in shallow water and lower to escape wave action of deep water. The tugs themselves are influenced by waves. Surge movements are particularly significant because they induce relative longitudinal movement along the line of the pipe. That movement alters the tension. Tow speeds are from 2 to 6 knots (1 to 3 m/s).

In most applications of mid-depth tow, the bundle itself is positively buoyant, but hanging chains are added at intervals so that the net buoyancy is negative. Undertow, the motion of the chains through the water, generates hydrodynamic lift (as well as drag), creating an effect that helps to trim the towed bundle. When the line is lowered to the bottom, it floats just above the seabed, as in off-bottom tow. The chains also add substantially to the structural damping of the complete system and therefore help to stop oscillations.

The configuration of the catenary is critically dependent on the interaction between the tension, the submerged weight, and the hydrodynamic lift on the chains. Most towed bundles are long enough for their finite flexural stiffness to have a negligibly small effect on the configuration. Since the tension is limited by the available tug bollard pull, the submerged weight has to be accurately controlled to a low value for the method to be feasible. This in turn implies narrow tolerances on the weight of the carrier and its internals, on the coatings and the spacers, and on the outer diameter of the carrier. In the 1984 mid-depth tow of bundled flowlines for Cormorant, the submerged weight was 22.1 N/m (1.5 lb/ft). Submerged weights of this order are recommended by the Smit-Kestrel joint venture that carried out the tow.

The sag of the catenary is limited by the available water depth. The allowable maximum value of the submerged weight is inversely proportional to the square of the length of the bundle, and, therefore, the problem of achieving the correct submerged weight rapidly becomes more severe as the length increases. For this reason, the maximum length of a bundle for which mid-depth tow is thought to be about 8500 m. In the opinion of the contractors that practice the method in the North Sea, this length should not be regarded as a fixed upper limit.

The mid-depth tow method was pioneered by Conoco and Hamilton, but early experience in the North Sea was marred by the sinking and loss of two bundles for the Hamilton Argyll field., as a result of failure of the spiral-welded carrier pipe, probably a consequence of fatigue damage. The development was continued for Shell Expro and Occidental and has become widely and successfully used in the North Sea.

12.4.5 Off-bottom tow

In off-bottom tow, the pipe is buoyant, and it is held down to float 1 or 2 m clear of the bottom, by chains hanging from the pipe and dragging on the bottom. The pipe itself does not contact the bottom, and is not subject to abrasion. If the pipe encounters a rough bottom feature such as a ridge or a gulley, it does tend to conform to it, but less so than if it were in contact, because of the interaction between the flexibility of the pipe and the weight of the chains. If the tow route crosses a second pipeline, only the chains contact the line. It may be unnecessary to take any special measures at the crossing, if the coating of the second line can withstand the impact of the chain links, which a properly designed concrete weight coating often can. Alternatively, it may be necessary to place a bridge of mattresses or sandbags, or to seek out a crossing point at which the second line is buried.

There is only a narrow dividing line between off-bottom tow and mid-depth tow. If an off-bottom tow is towed rapidly enough or if a trailing tug creates enough tension, the tow lifts away from the bottom and becomes a mid-depth tow. Off-bottom tow is frequently combined with bottom tow. In that case, lengths at the ends of a bottom-towed pipeline, otherwise in continuous contact with the bottom, are buoyed and fitted with chains so that they float above the bottom. The ends are then laterally and vertically flexible and can be flexed as part of a primary alignment operation.

The installation of three flowline bundles in the Murchison field was originally planned as an off-bottom tow, but plans were later changed so that the bundles were towed at mid-depth.

12.4.6 Bottom tow

In bottom tow the pipeline rests directly on the seabed, and a tug pulls it. This technique is similar to bottom pull, which is the second most widely used construction method for submarine pipelines. It is used to build most shore crossings and also is used extensively to construct river crossings and crossings between shores. The pull is from fixed winches or from a pull barge. Pontoons may be added to reduce the submerged weight and make the line easier to pull. The longest single length of pipeline installed by the pull method is a 30-km pipeline in Iran.

Bottom tow has been widely applied in the Gulf of Mexico and in the Australia Bass Strait, principally but not exclusively for pipeline bundles. It has not been used in the North Sea since the installation of the Statfjord A loading line in 1978. The longest lines installed in this way are 16000 m long. There are several difficulties, among them:

- The abrasive effect of the seabed on the pipeline, which compels the use of a hard and tough coating

- The need to take special measures when one pipeline is dragged over a line already in place, requiring the first line to be trenched, buried, or protected by mattresses

- The need to find a tow route free of obstructions such as cliffs, wrecks, and boulder fields

- As with mid-depth tow, the need to find a long, straight, unobstructed make-up site from which the pipe can be launched, preferably into a area protected from rough seas and with no bottom obstructions

Because a pipeline is in contact with the bottom throughout the tow, it is more secure against currents and waves than it would be in other tow methods. If the sea becomes too rough for the tow to continue, the tug can simply abandon the towline and recover it later. Nothing else needs to be done.

Dragging the pipe across the bottom exposes the external coating to severe abrasion. The problem of finding a coating that can withstand the drag has been studied in a number of tests. In 1975, four sections of pipeline, 762 and 406 mm (30 in. and 16 in.) diameter, altogether 610 m long, were made up in Tananger in southern Norway. Some of the pipeline had concrete weight coating, and some had various kinds of thin-film coating. The string was towed along a winding route between the offshore skerries into the Norwegian Trench to a maximum depth of 350 m, into the shallower water to the west of the trench, through the trench a second time, and back to Tananger. It was found that the concrete weight coating had suffered little damage and that the loss of concrete by abrasion did not exceed a few mm. On the other hand, epoxy and polyethylene coatings were severely abraded.

A field trial investigated coatings for bottom tow in very deep water. The tests covered an extensive tow route in the Gulf of Mexico across sand and clay bottom. As a result of the tests, a silica-filled, liquid epoxy coating was chosen as the coating for subsequent projects. In the Placid Green Canyon project, a number of flowlines were made up on the beach of Matagorda Island, a barrier island in the Gulf of Mexico. They were launched sideways by a series of side boom tractors that drove along the beach carrying the pipe into the water. The flowlines were then towed into position and connected by a diverless deflection system in a depth of 600 m. During the tow, the longest bundle collided with a reef, partially floated to the surface, and then sank. It was recovered in sections, towed back to the island, welded together, and towed back, this time successfully. The incident emphasized the degree to which bottom towed pipelines interact with the bottom. It also points out the necessity of having detailed advanced knowledge of the seabed and careful navigation to follow the survey route. Since then, many other lines have been successfully towed from the Matagorda Island site.

12.5 Trenching

12.5.1 Objectives

Submarine pipeline trenching practice has developed in response to demand. There are two kinds of requirement, firstly for trenching over relatively short distances in shore crossings, and secondly for shallower but much longer trenches in the open sea. The first requirement has usually been met by some form of dredging technology, but sometimes by plows and jetting machines. The second requirement was met in the past by jetting equipment operated from jet barges, but in the last 10 years, plows and mechanical cutter systems have captured part of the market. This was at first a response to dissatisfaction with the high cost and limited protection given by trenches produced by jetting.

There is a resurgence of interest in jetting, in part due to the application of remotely operated vehicle (ROV) technology. Many contractors operate jetting spreads. It is estimated that jetting has about half the pipeline trenching market and a similar proportion of the cable trenching market.

In the early years of North Sea development, it was thought that all pipelines should be trenched; and government authorities made very stringent demands. When the Statfjord A loading line was designed in 1976, for instance, the Norwegian authorities initially asked for a trench depth of 3 m in hard clay with an undrained shear strength of 150 kPa. Only later did they agree to reduce the required trench depth to 2 m and then to 1.2 m. The early large-diameter lines in the central and northern areas of the UK sector were trenched to about 1.5 m, as were the large-diameter lines in the Netherlands and German sectors.

A move to less demanding trenching requirements was led by Shell Expro, which carried out an extensive program with the objective of demonstrating that the 36-in. Far-north Liquid And Gas System (FLAGS) pipeline did not need to be trenched. As a result of this program and others, the authorities in the UK reached a position in which they were prepared to allow some pipelines of 16-in. diameter and up not to be trenched, although each case is considered individually. This policy has now been in operation for several years, and the results have been satisfactory. In the Netherlands sector, there has been a similar reduction in trenching requirements, and trenching to 0.2 m below the natural bed level is thought satisfactory, although special requirements have been imposed in megaripple areas where there is large-scale movement of bed forms.[6] In the Norwegian sector, most of the Statpipe system was left untrenched.

Small-diameter pipelines are much more vulnerable to damage from fishing gear though less affected by waves and currents. They are generally trenched outside the immediate vicinity of platforms, though a few operators have opted not to trench. The 16-in. limit is somewhat arbitrary. A joint industry project has been carried out to determine if this limit can safely be reduced, but it did not arrive at a firm conclusion, though it did show that the level of impact forces is much influenced by the shape of the trawl board.

The risk is not only to the pipeline. If trawl gear suddenly snags a pipeline, the fishing vessel can be capsized by the sudden increase of tension in the towing warp. Awareness of this issue has been heightened by the *Westhaven* accident in the North Sea in March 1997 in which four lives were lost.

12.5.2 Trenching systems

12.5.2.1 Jet barge. In this system, a jet sled is pulled along the pipeline by a barge. It carries a vertical *claw* that consists of two tubes with jets, one on each side of the pipeline. Water is pumped from the barge down hoses to the jets and erodes the seabed, forming a slurry of water and soil. A jet eductor system ejects the slurry to one side. The sled carries instrumentation to monitor the forces between it and the pipe. A typical large jet barge has 24 MW (32000 hp) engines driving pumps supplying 7 m^3/min at 17 MPa. The barge is anchored and moves forward on its anchor cables in the same way as a lay barge. A typical rate of progress is 70 m/h in one pass. In some conditions, many passes are necessary, and in one notorious instance in the Netherlands sector, 13 passes were needed to lower the pipeline to the specified level, 2 m below the natural sea bottom.

Jetting was developed in the 1950s, and until about 15 years ago was the only system used in deep water. Early North Sea pipelines were all trenched by this technique. It had advantages of simplicity and security of the pipeline against damage. It can cope with a wide range of bottom soils, from sand to medium clay, but the shape of the trench is highly variable. In medium clay, jetting makes a neat rectangular trench, but in loose sand it makes a wide shallow trench with sides at less than 10° to the horizontal, which does little or nothing to protect the pipeline.

Enthusiasts for jetting argue that it has suffered from lack of investment and research. If the amount of investment that has been devoted to research and development on plowing and cutting had been matched by equal investment in jetting, jetting would be more competitive.

12.5.2.2 Jetting machines. This system is based on a self-contained machine, supplied with power from the surface by an electrical umbilical. Like the jet sled in the jet barge technique, the machine straddles the pipeline. A typical machine is the Land & Marine TM4, which has two jet pumps, each supplying 52 m^3/minute of water, and two sand pumps, each supplying 40 m^3/min. This machine has the following measurements:

length	6.6 m
width	5.8 m
height	8.2 m
air weight	85 tons
pipe diameters	0.7 to 1.1 m

The trenching speed is 100 m/h. The trench depth reached after one pass is about 2 m, depending on many factors, among them the type of soil and the weight and diameter of the pipeline. Machines of this type are often used to trench outfalls and shore crossings, and the TM4 machine trenched the shore crossings of the Forties and Frigg pipelines in the North Sea.

Modern jetting machines apply ROV technology and methods developed for deep-water burial of fiber-optic cable systems.

12.5.2.3 Mechanical cutting machines. Cutter machines cut the soil under the pipe by picks mounted on chains or cutter discs, entrain the material with a dredge pump system, and eject it to one side. There have been several programs to develop machines of this kind for North Sea service.

The Kvaerner-Myren trenching machine was developed in 1974–78 as part of efforts directed toward the projected construction of a pipeline from the Statfjord field to the Norwegian coast. The machine rides on the pipeline, supported by wheels, and digs a trench under the pipe with a cutter 1.8 m in diameter and 2.1 m high, rotating about a vertical axis. The cutter and dredge pump are driven by a 1200 kW electro-hydraulic system, supplied with electric power through a 6.6 kV umbilical. The machine weighs 90 tons in air. It was designed so that it could work in the 350 m deep water off Norway and could be positioned without diver support. It has a sophisticated positioning and navigation system, and swims to the pipeline from a dedicated support vessel. When fitted with special cutting teeth, the machine can slowly chew through boulders.

This machine has performed well but has not been widely applied, probably for commercial reasons. The normal trench depth is about 2.5 m, but in tests a 4.5 m trench has been cut. The machine was successfully used to correct spans in the Statpipe system, to trench in the Gullfaks field, and to trench for the North Morecambe and Interconnector pipelines in the Irish Sea.

The Eager Beaver trenching machine developed by Heerema operates on a different principle. It has three inclined cutting chains, at 60° to the horizontal, two on one side of the pipe and one on the other. They cut a V-shaped trench as a dredge pump ejects the excavated material sideways. The machine is not supported by the pipeline and runs on tracks. In sand, the chains create a fluidized slurry within the trench. The slurry supports the trench sides long enough for the machine to move on and the pipeline to settle in the trench. The maximum trench depth is 2.5 m. The trenching speed is variable, but can reach 200 m/h. This machine was used in several North Sea projects, including the Unionoil Q1 pipeline in the Netherlands sector. The Allseas *Digging Donald* trenching machine is broadly similar to the *Eager Beaver*, but it has four cutting chains, rather than three.

12.5.2.4 Plows. Plows were used many years ago to trench pipelines, but they gained a bad reputation for poor depth control and a history of sinking. Their modern development began in 1975 with a program targeted on the trenching of a loading line in the Statfjord field.The first two plows built were pre-trenching plows, which cut a trench before the pipeline is set in place. Development of the post-trenching plow, which cut a trench under a pipeline already laid on the bottom, began in 1977. A post-trenching plow has a horizontal beam with a front end supported above the natural seabed by wheels or skids with a rear end that carries two hinged shares that rotate about axes parallel to the beam. The plows are lowered over the pipe in an open position with the shares hinged outward to clear the pipeline. When the plows are pulled forward by a cable fixed to the front end of the beam, the shares rotate and close under the pipe, resulting in a V-shaped trench. The pipe passes through a cut out in the shares.

Early post-trenching plow typically weighed about 150 tons and were 25 m long. The plows are either pulled along the bottom by an anchored barge or by a tug. The plowing force depends on the soil: for a 1.5-m trench, the force may be as little as 20 tons in clay or as much as 400 tons in sand.

In clay soils, the plowing force does not depend strongly on speed. Operating speeds have progressively increased as confidence in the ability to control the plows has grown. There is, however, an important speed effect in fine-grained dilatant sands and silts.[7] This soil type substantially increases the plowing force and has caused much trouble.

Recent post-trenching plows have an improved design and are lighter and smaller than their predecessors. These plows have a sophisticated control system and can be pulled safely as speeds averaging 1000 m/h. In the Northern Ocean Services' Advanced Pipeline Plow, a special control system (SMART) alters the trenching depth so as to smooth the trench bottom profile and eliminate overbend imperfections that might initiate upheaval buckling. Plows have become accepted as an alternative to jetting and have been used in many parts of the world.[8, 9]

12.5.2.5 Cutter-suction dredging. Cutter-suction dredging is the most widely used dredging technique. It can accept a wide range of soils, from mud and silt to soft rock. Cutter-suction dredgers are often used to cut pipeline shore approach trenches in shallow water and have the advantage that the depth of trench that can be cut is limited only by the maximum operating depth that the cutter can reach, generally about 20 m. A disadvantage is that the dredger needs a minimum depth of water to float safely and may need to cut itself a floatation ditch.

Cutter-suction dredgers vary in size and in capacity to cut hard materials. The largest have a production rate of 1500 m³/h in soft clay and sand and 400 m³/h in rock. A typical larger cutter-suction dredger *Beverwijk 31* has dimensions 66×15×4.35 m, installed power of 8.6 MW (11500 hp), and an 850 mm (33 in.) dredge pipe. This technique frequently is used to make deep trenches, needed because of possible future changes in seabed level. The largest dredgers can dredge to 50 m below the surface.

12.5.2.6 Bucket dredgers. Bucket-wheel and bucket-ladder dredgers cut a trench with moving buckets supported on a wheel. The spoil is either removed from the buckets on the wheel by gravity and a suction system, or the buckets are carried upwards on a ladder and their contents emptied into a hopper. An advantage of bucket dredging is that it can accept boulders and broken rock fractured by explosives. A typical bucket-wheel dredger *Scorpio* has the following parameters:

length(hull)	50.5 m
beam	13.6 m
draught	1.9 m
maximum dredging depth	18 m
power at bucket wheel	550 kW
diameter of suction pipe	800 mm

Production rates vary with equipment and material but are typically 1000 m³/h for mud, 500 m³/h for clay and 350 m³/h for rock. The bucket system is more sensitive to weather than the cutter-suction system, and this is thought to make bucket dredgers unsuitable for exposed locations.

12.5.2.7 Backhoe. A trench can be excavated by a backhoe on a barge. This method is relatively slow and requires considerable skill if the trench is to be completed efficiently. The production rate is low in the order of 100 m³/h. The method is sometimes used for pipeline trenches. An example is the 19 km Oresund crossing, where backhoes were used to excavate a 2–3-m trench in sand, chalk, and boulder clay over a period of three months.

12.5.2.8 Dragline. A dragline can be mounted on a barge, or a dragline bucket can be pulled between two barges. Production falls off rapidly in deeper water because of the increased cycle time needed to raise the bucket to the surface, slew it sideways, and dump the spoil and because the operator cannot see what he is doing. Expected production is less than 100 m³/h. This method is only suitable for short trenches in shallow water.

12.5.3 Burial

A buried pipe is far better protected than a pipe in an open trench. Burial secures complete protection against fishing gear, substantial protection against all but the largest dragging anchors and cables, and protection against most dropped objects. It much increases the thermal resistance between the pipeline and the sea so that the temperature of the fluid in the pipe remains high, which helps to reduce hydrate problems in gas pipelines and to minimize wax deposition and pumping costs in oil pipelines. If the cover is deep enough, burial can also eliminate upheaval buckling in hot flowlines operating at high internal pressures.

A drawback to burial is that the presence of the backfill makes it difficult to find a leak, and the backfill has to be removed before a repair can be carried out. For these reasons, it is unlikely to be desirable to have backfill follow immediately after trenching, although that has been done. It will generally be better to trench first, carry out the hydrostatic test, and only then to backfill.

Rock dumping is the most common method of burial and is now a developed technology offered by several contractors. The dumped material can be quarried rock or coarse gravel dredged from offshore banks. The rock is dropped through a steerable fall pipe, and acoustic profiling is used to locate the end of the fall pipe so that loss of rock is minimized. The method is frequently used to place rock for scour protection around platforms, to protect cable crossings, and to fill under pipeline spans. It has also been used to cover long lengths of pipeline, for example in the North Alwyn project, but doing so is expensive.

An alternative is to rock dumping is to backfill over the pipe with the excavated spoil from the trench. Plowing leaves the spoil heaped in a neat pile on either side of the trench. A backfiller can then pull the spoil back across the pipe.

The first application of a combined plowing and backfilling system was in 1986 for the 12 km long, 8-in. Auk pipeline. The line was trenched to a depth of 1 m by the Brown and Root trenching plows, and then backfilled by a new backfill plows. Trenching took 42 hours and required forces from 40 tons in loose sand to 110 tons in patches of stiff clay. The trench was backfilled in 36 hours. Most contractors now have backfill plows, and backfill is becoming routine.[10]

A potential problem with backfilling is that the pipeline can be partially lifted from the bottom of the trench during the backfill operation. This is partly a dynamic effect and can occur even though the pipe is heavier than the backfill. It is the subject of much current research.[11, 12] The mechanical properties of the backfill are important if it is relied on for uplift resistance, and the thermal properties are important if backfill is relied on as part of the insulation system.

References

1   London, C. J. (1991) "Forties Export Pipeline Project." *Proceedings of the Offshore Pipeline Technology Seminar.* Copenhagen.

2   *J-lay Step Ahead.* Video. Heeremac (1993).

3   McQuagge, C. H., S. Davey. (1991) "Bass Straits—an Australian Experience." *Proceedings of the Offshore Pipeline Technology Seminar.* Copenhagen.

4   Holm, E. (1998) "An Overview of Technical Aspects and Practical Operation of the Reel Pipelay System on the MSV Fennica." *Proceedings of the Offshore Pipeline Technology Conference.* Oslo.

5   Palmer, A. C., M. Carr, E. Lunny, K. Hulls, R. Hobbs. (1993) "Reeled Pipeline Bundles." *Proceedings of the Offshore Pipeline Technology Seminar.* Amsterdam.

6   Van Dongen, F. A. (1983) *"Het ingraven van onderzeese leidingen* (Trenching undersea pipelines)." *Civiele en bouwkundige techniek.* (7) 22–26.

7   Palmer, A. C. (1999) "Speed Effects in Cutting and Plowing." *Geotechnique.* 49 (3). 285–294.

8   Brown, R. J., A. C. Palmer. (1985) "Submarine Pipeline Trenching by Multipass Plows." *Proceedings of the 17th Annual Offshore Technology Conference.* Houston. 2. 283–291.

9   Palmer, A. C. (1998) "Innovation in Pipeline Engineering: Problems and Solutions in Search of Each Other." *Pipes and Pipelines International.* 43. 5–11.

10 Finch, M., R. Fisher. (1998) "The Right Tool for the Job: Selection of Appropriate Trenching and Backfilling Equipment." *Proceedings of the Conference on Subsea Geotechnics.* Aberdeen.

11 Cathie, D, J. B. Machin, R. F. Overy. (1996) "Engineering Appraisal of Pipeline Flotation During Backfilling." (OTC8136) *Proceedings of the 28th Annual Offshore Technology Conference.* Houston, TX.

12 Finch, M. (1999) "Upheaval Buckling and Floatation of Rigid Pipelines: The Influence of Recent Geotechnical Research on the State of the Art." (OTC10713) *Proceedings of the 31st Annual Offshore Technology Conference.* Houston, TX.

13 Shore Approaches

13.1 Introduction

A marine pipeline reaches a landfall by way of a shore approach. The shore approach is necessarily in shallower water than that of the rest of the pipeline, and is more exposed to wave action and to longshore currents. Experience shows that the engineering of pipeline shore approaches has to be carried out with special care, and that many of the most serious mishaps and cost overruns in marine pipeline construction have occurred in the difficult environment of very shallow water.

One of the difficulties is the extraordinary variability of the coastal environment This subject is considered in section 13.2. Another is that the available survey information is often inadequate. This shortcoming also is addressed in section 13.3. The last part of the chapter discusses trenched crossings of sandy beaches, horizontal drilling, rock shores, tidal flats, and tunnels.

13.2 The Coastal Environment

The varied forms of coastlines are described in standard texts on physical geography and coastal processes.[1, 2, 3] The topography of a coast is determined by a complex interaction between the marine environment and the geology and biology (and sometimes the effects of human intervention). All of these are dynamic and evolve with time. Over recent geological time, the sea level has changed with respect to the land, and the marine environment may become more aggressive as a result of changing climate or the erosion of offshore islands. The environment could become less aggressive as a result of sediment deposition or of deliberate or accidental human interference with coastal processes, such as the construction of coastal defense works, new breakwaters and piers, or damage to sand dunes or reefs. Some coasts alter very slowly, but others may be eroded by tens of meters in a single storm. In parallel with long-term secular changes in the shoreline, there are often seasonal changes between a summer profile and a winter profile. The summer profile is relatively steep and is determined by long-period swell components of the wave spectrum, often generated far away, whereas the winter profile has a smaller slope, and has alternating bars and troughs parallel to the shore.

The long-term effects of sediment transport may be significant during the design life of a pipeline. For instance, much of the eastern coast of England is receding, but the recession is sometimes prevented or retarded by coastal protection structures such as groynes and seawalls. There is a growing opinion that coastal stabilization may not be cost-effective, and that some coasts should be allowed to recede naturally, and that policy is beginning to be adopted, though obviously the economics have to be examined with care. Other parts of the coast aggrade, but their aggradations may be halted or reversed by stabilization further along the coast.

The physical factors that shape the coastline are waves, currents, and winds. The largest waves are generated offshore. As the waves move toward the shore, they become higher and steeper. Refraction alters the direction of propagation. This action is analogous to the refraction of light and happens because wave velocity decreases with decreasing depth.[2, 4, 5, 6, 7] Its effect is to rotate the wave propagation direction so that the direction becomes closer to the orthogonal to the depth contours. This action tends to concentrate wave energy on projecting capes and headlands and is one of the factors that make them undesirable as locations for landfalls.

A wave moving toward a shelving shore eventually becomes so steep that it breaks. Most waves break when the wave height is about 0.8 of the local water depth, though there is variation between waves and an interaction with the flow created by the previous wave. The form of the breaker varies and depends on the beach slope.[2, 5, 6] In a spilling breaker, on a nearly horizontal beach, the wave steepens until the crest becomes unstable and runs forward as a white, foaming mixture of water and air. In a plunging breaker on a steeper beach, the advancing face of the wave steepens until it is vertical, and the crest falls forward, creating a heavy and fast-moving jet that plunges into the trough ahead of the wave. On the steepest shores, a surging breaker rises and then collapses.

The region between the shore and the breaker line where waves begin to break is called the surf zone. The position of the breaker line depends on the height and direction of the incoming waves and on the state of the tide. Within the surf zone, broken waves reform and propagate toward shore and may break a second time. The breaker height within the surf zone is often roughly 0.4 times of the local depth, but highly variable. Breaking waves are particularly effective in entraining sediment, and longshore currents then transport the sediment both along the shoreline and further out to sea.

Wave action creates a number of hydrodynamic effects.[2, 4, 5, 6, 7] Waves that approach the shore obliquely induce a longshore current, which transports sediment stirred up by wave-induced movement of the water close to the bottom. The momentum flux towards the shore induces a rise in water level, called *set-up*, and may induce localized rip currents to seaward. These currents modify the seabed topography through sediment transport processes, and these changes in turn modify the pattern of wave breaking.

Occasional storm surges associated with tropical storms and deep barometric depressions have a major influence in shaping the coast, and the high currents and intense wave action that accompany them can determine the extreme conditions that govern a shore approach design.[5, 6] Three effects induce the currents and the sea-level rise: the direct effect of onshore wind stress, the barometric effect of low atmospheric pressure, and the bathystrophic tide that results from the interaction of strong longshore currents and the Coriolis acceleration associated with the earth's rotation. Altogether, they can raise sea level by several meters and induce longshore currents of the order of 1 m/s. The February 1, 1953, surge that killed 200 people in England and 1800 in the Netherlands raised the level of the North Sea 2.9 m above normal in about 15 hours. Topographically trapped shelf waves induced by barometric changes elsewhere can also induce longshore currents.

Biological processes play a part in the evolution of a coast. On some coasts, vegetation spreads into shallow water and encourages sediment deposition by slowing the flow of water, and the coastline gradually moves outward. This occurs in environments as diverse as tropical mangrove swamps and sheltered estuaries in the North Sea. On other coasts, the destruction of vegetation by pollution, overgrazing, or visitors on foot or in vehicles may destabilize sand dunes and cliffs and contribute to erosion of the coastline. The destruction of coral reefs by pollution and by alien species can reduce protection against waves and, in turn, create retreat of the coast.

Human activity has significant effects on beach processes, sometimes unintentionally. The dynamic equilibrium of a shore often depends on a steady process of active sediment transport along the shore. If the process in interrupted, for example, by the construction of a breakwater to protect a harbor entrance, sediment will build up on the upflow side of the breakwater; but on the other side, the shore will be starved by the absence of transported sediment. That situation results in the shore eroding back until a new equilibrium is reached. These effects can be significant during the construction phase of a pipeline.

The choice of landfall location is critical and demands a very careful engineering review of the coastal hydraulics, geotechnics, and environment in parallel with a study of the proposed construction method. The difference between a good choice of landing point and a bad one is measured in millions of dollars.

13.3 Site Investigation

Marine survey is carried out to determine the shore profile, the ocean and tidal currents, and the seabed bathymetry that determines local wave refraction. Geotechnical site investigation determines the geotechnical description and strength properties of the seabed material.

The planning of an efficient site investigation ought to be based on an initial reconnaissance to learn as much as possible about the topography, geology, and the development of the shore.[8, 9, 10] There are many useful sources of information, such as old maps (from which one can see whether the coast is retreating or advancing), old photographs (which may show that there is a difference between winter and summer profiles), navigation and fishing charts, previous surveys for other purposes, and local enquiry.

Most shore approaches are trenched. It is essential to determine geotechnical properties to at least the maximum depth of trenching because the geotechnics determine the choice of trenching method and the rate of progress. Experience shows that it is a false economy to skimp the geotechnical survey and that inadequate geotechnical information is a frequent source of disputes between the construction contractor and the owner. For example, a narrow rock ridge hidden just below the surface of an apparently uniform sandy shore cannot be ploughed or jetted. Dealing with the rock ridge can delay the trenching operation by months unless proper plans are made in advance. Similarly, the presence of large boulders impedes trenching.

Geotechnical information is gained by a judicious combination of core sampling, boreholes, in-situ tests by cone penetrometer or pressure meter, and acoustic sub-bottom profiling. Acoustic methods secure a continuous profile of the strata below the sea bottom and may make it possible to find a route that avoids the need to excavate rock. Core sampling leads to a positive identification of the seabed soils and recovers samples for mechanical testing. Sample disturbance is often a problem, and its effects can be avoided by in-situ measurements. Machin has prepared a guide to geotechnical site investigation for pipelines.[10] Contractors sometimes embark on the construction of shore crossings without adequate geotechnical investigation. Gambles of this kind often have disastrous consequences.

13.4 Beach Crossings

The most straightforward way of constructing a beach crossing is to excavate a straight trench from above high-water mark out to water deep enough to be safely reached by a lay barge or a reel ship. A winch is then installed on the beach at the head of the trench. A pull cable is taken out to the lay barge and shackled to a pull head on the end of the pipeline. The winch then pulls the pipe along the trench from the lay barge to the shore while the barge remains stationary. If tension is required to control the sagbend curvature, the winch pulls against the tension applied by the lay barge. When the pull head has reached the shore, the lay barge can start to move forward, away from the shore, laying pipe on the seabed behind it. *(laying away)*.

Almost all pipelines constructed across beaches and through surf zones are trenched into the seabed because a pipeline that is not trenched is vulnerable to changes of bed level induced by sediment transport and is exposed to large hydrodynamic forces during storms. The design of the trench is part of the design of the pipeline system. The engineer, therefore, has to integrate the installation scheme with the need to trench the pipeline. He can choose either to make the trench first and then pull or lay the pipeline into it or to pull or lay the pipeline first and then excavate the trench under it. It is practicable either to excavate the trench first or to set the pipe in position first and then excavate the trench under it. Both techniques are used in shallow water. In deep water, it is easier to place the pipe first and then make the trench.

The engineer can take advantage of the trench to make the design and installation simpler. In particular, the water moves more slowly in a trench than it does above the trench, and so the hydrodynamic forces on a pipeline in a trench are much smaller than the hydrodynamic forces on the same pipeline subject to the same waves and currents that is not trenched. Furthermore, the inclined sides of the trench increases the pipeline's lateral resistance: pushing the pipe up the sides of a trench requires more force that pushing the pipe horizontally across the seabed. As long as the pipeline is in a trench, it is much more stable than it would be without a trench.

A widely-applied construction technique for pipeline shore crossings is illustrated in the schematic diagram Figure 13–1. Note that the diagram is not to scale and has been compressed horizontally unlike a true scale diagram that would be much more extended horizontally. The diagram applies both to pipelines that are pulled into a pre-constructed trench and those that are pulled first and trenched afterwards.

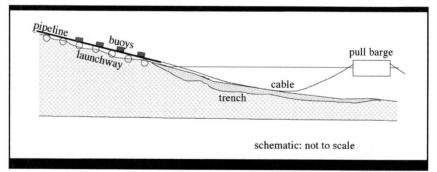

Fig. 13–1 *Shore Pull*

The pipeline is built in sections (*strings*) on shore on a make-up site roughly perpendicular to the shoreline. The seaward end of the pipeline is bolted or welded to a pull head (essentially a strengthened section of pipe with a large pad eye and swivel, usually with valves so that the pipeline can be flooded in a controlled way). One or more pull cables connect the pull head to a winch on an anchored barge. The winch tensions the pull cables, and the pipeline is dragged down a launch way into the sea. On the launch way, the pipe moves on rollers, which extend to the water's edge.

The section of pipe in the water often carries pontoons to reduce the submerged weight by buoyancy, but not to the extent that the pipe is lifted off the bottom. A pontoon is an air-filled tank often made from a section of steel pipe with welded ends. Each pontoon is connected to the pipeline by strong straps, often including a release mechanism so that they can readily be freed from the pipeline when the pull is complete. The pontoons may be attached on the launch way, or they may be attached at the shoreline or in the water. Sometimes the pontoons are replaced or supplemented by air bags, which are flexible plastic bags, usually taller than they are wide and open at the bottom so that they can be filled with air and attached to the pipeline by ropes and straps.

When the trailing end of each section comes opposite the leading end of the next section, the sections are aligned and welded together. The cycle continues until the leading end of the pipeline reaches the intended location. The barge may be relocated during the pull in order to reduce the required length of cable, and to reduce the force necessary to drag the cable itself across the bottom.

As the pipeline moves into the water, it passes through a shallow surf zone. Waves induce large velocities in the water and generate hydrodynamic lift and drag forces that can destabilize the pipe. These forces tend to lift the pipe and carry it sideways. Additional forces are induced by tidal- and wave-induced currents across the pipeline. If the waves break, the forces increase because plunging breakers generate intense jets that drive downward into the wave trough and generate intense forces. If the seabed consists of a mobile material such as mud, sand, shingle, or cobbles, waves and currents also destabilize the seabed itself and generate sediment transport. The sediment transport has both local effects, such as scour around boulders and rock outcrops, and large-scale effects in which sediment is transported both perpendicular to the shoreline and along the shoreline.

The force necessary to drag the underwater section of the pipeline is the product of the length of the pipeline, its submerged weight per unit length (allowing for the net buoyancy of any pontoons), and a longitudinal resistance (friction) coefficient. If the pipe is made heavier, it is harder to pull. If it is made lighter, it is easier to pull but also less stable against hydrodynamic forces. This compels the designer to find a middle way in which the pipe is heavy enough to be stable but light enough to be pulled into place without overtaxing the pulling system.

Prudent pull contractors design their pulling systems for a friction coefficient of at least 1 on the seabed, though they recognize that the actual pull force may correspond to a smaller coefficient. Once the pipe has started to move, the force required to keep it moving is usually smaller than the force required to start it.

The component of force necessary to drag the above-water section of the pipeline is the product of the length of the pipeline still above water, its weight per unit length in air (allowing for the weight of any pontoons), and a longitudinal resistance (friction) coefficient. This coefficient generally is much smaller than the coefficient on the seabed because the launch way rollers allow the pipe to move comparatively easily. The smaller friction coefficient depends on the design of the roller system, but is often taken as about 0.15.

Unless the barge is very close to the pull head, some of the pull cable is on the bottom and has to be pulled along the bottom. The force required to pull the cable is not negligible and has to be taken into account in calculations. To illustrate this, suppose the expected maximum pull force is 1.4 MN (140 tonnes) and the pull is through a single (one-part) wire rope. A prudent designer allows a factor of at least 2 on minimum breaking load so the wire rope has to have a minimum breaking load of at least 2.8 MN. The smallest wire rope that meets this condition is 60 mm nominal diameter and weighs 16.60 kg/m in air, which corresponds to 14.43 kg/m in seawater. If 1000 m of pull cable is being pulled across the seabed and if the friction coefficient is taken as 0.85 (the same as for the pipe and a reasonable value), the force required to pull the cable alone is 14.43◊1000◊0.85 = 12,300 kg (12.3 tonnes).

Pontoons or air bags temporarily reduce the submerged weight during the pull. However, the waves and currents induce hydrodynamic forces on the pontoons or bags themselves. The effect of adding the pontoons, therefore, is to reduce lateral stability because of the two effects of reducing submerged weight and increasing hydrodynamic forces. It follows that the designer must not add too much buoyancy because control of the pipe might be lost during the pull.

The engineer planning an installation by shore pull has to decide whether to construct the trench before the pull or after the pull and whether to protect the trench against wave and current action or to leave it unprotected. The options are the following:

Option I: pull first, trench second
Option II: trench first, pull second; do not protect trench
Option III: trench first, pull second; protect trench

In principle, there is a fourth option, which is to construct trench protection on either side of the pipeline, next to pull or lay the pipeline, and then to excavate the trench. To date, there is no record of a project having chosen this method.

The simplest imaginable possibility (option I) is to make up the pipeline onshore, either as a single section or in lengths, to pull it into the water and along the seabed, then to trench after the pull. The pipeline then makes a shallow groove in the bottom as it moves forward. The depth of the groove depends on the strength of the soil and the weight of the pipeline, but on most soils the groove will be quite shallow, much less than one pipe diameter deep. Even a small groove helps to increase the lateral resistance of the pipe.

Option I is risky if there is any significant probability of wave and current action. It is used in sheltered locations, but in exposed locations it is almost never chosen for several good reasons. The sea is never perfectly calm, and there is usually a longshore current. The pipe is subject to wave action. The magnitude of the hydrodynamic forces depends on the velocity of the current and on the height and period of the waves. Moreover, unless the seabed is rock, the waves and current can easily move the bottom sediment. The larger the waves are, the larger the forces are, and the more easily the bottom sediment can be moved. The movement of the seabed undermines the pipeline and makes it still easier for the waves to move it.

When the pipeline is being pulled along the route, the frictional force between the pipeline and the bottom is in line with the pipe (because the frictional force lines up to oppose motion). Unfortunately, a consequence is that the resistance to lateral movement is very small while the pipe is moving. Moreover, in shallow water the waves break, and breaking waves apply very high forces. The combined effect of all these factors is to make the pipe unstable, both when the pipe is being pulled and when the pipe is stopped.

An additional problem is that because the pipeline protrudes above the seabed, interaction between current and wave-induced, near-seabed velocities and the presence of the pipeline generates secondary currents which tend to erode any moveable material on which the pipe is resting. This generates scour under the pipe (tunnel erosion) and on either side (lee and luff erosion). The scour can be so effective that it creates a scour hole into which the pipeline sinks so that the pipeline becomes partially or completely buried.

An alternative (option II) is to excavate a trench, then pull the pipeline along the trench. The trench then shelters the pipe and reduces the hydrodynamic forces, giving the pipe additional resistance against lateral movement.

If the seabed is a hard and immobile rock, and there is no bedload or suspended sediment that can be carried into the trench, the trench will remain open.

If the bottom is mobile (or if the bottom is rock at the trench location, but the water carries sediment because there is mobile material nearby), any seabed sediment transport tends to destabilize and fill the trench. Waves and current can easily fill a trench overnight. If the pipeline pull has not been started when the trench fills in, the trench has to be re-excavated. If the pipeline pull is in progress when the trench fills in, the pipe rapidly becomes partially or completely buried in the sand, and the force required to continue the pull increases greatly. It becomes equally difficult to pull the pipeline back to shore, and the presence of the pipeline makes it hard to re-excavate the trench without damage the pipe.

For those reasons, option II is risky unless the location is sheltered and protected against waves and currents. A contractor might decide knowingly to accept the risk and to gamble on being able to complete the pull before the trench fills in.

Option III is the safest and most conservative. The trench is protected so that the sea cannot fill it in. A customary way of doing this is to construct a temporary steel trestle along the line of the trench, to drive two parallel lines of sheet piles from the trestle and to excavate the trench between the sheet piles. When the pipeline is in place, the sheet piles are cut off or pulled out, and the trestle is dismantled. The trench may be backfilled artificially with rock or it may be allowed to backfill by natural sediment transport.

An alternative to using sheet piles is first to construct rock berms (like rubble-mound breakwaters) along the line of the pipe and then to excavate the trench from equipment on the berms, pull the pipeline into position, and remove the berms. If the waves and sediment transport only come from one side, one berm on the weather side may be enough. This is a much less risky but much more costly alternative. A decision on how far to extend the trench requires a careful analysis of the sediment transport and the risk of storms during the construction period. Option III has been applied many times on the British and Dutch coasts of the North Sea where storms can occur at any time of year.

The method used to excavate the trench depends on the geotechnical conditions, the required trench cross-section and depth of cover, the depth profile, the extent of longshore sediment transport, and the exposure to storms. The deeper section is excavated by dredging techniques: by cutter-suction dredger, suction dredger, bucket-ladder dredger, or by dragline or clamshell excavator. The shallower section can be excavated by land excavation equipment such as a backhoe excavator, particularly if a high tidal range allows the equipment to move into the surf zone at low water. It is sometimes necessary to construct an embankment on one side of the trench and to operate the equipment from the embankment. If the trench is relatively shallow, it may be possible to excavate it using a plow.[11, 12]

Jetting or plowing after the pipe has been pulled into place is an alternative to excavation of a trench, provided that the soil is suitable and that the required depth is not too great. A jetting machine such as the Land & Marine TM4 can be pulled along the pipeline to fluidize the sand in front of the jets, entrain the sand/water mixture, and eject it to one side.

13.5 Horizontal Drilling

An important development over the past 20 years is the widespread application of horizontal directional drilling. The principle is to set up an inclined drilling rig, and to drill a pilot hole under the beach and the surf zone, controlling the direction by a steerable bit. A reamer is then pulled back to enlarge the hole, and the pipeline is pulled behind it. The hole is stabilized by a bentonite slurry. In some applications, a casing is pulled into the hole, and the pipeline is then pulled into the casing, which then becomes an extended J-tube.

The method was first developed to drill pipeline crossings under rivers in Texas and Louisiana. The first application to a shore crossing in Europe was by Amoco at Hoek van Holland in 1985. It has since become widely used, particularly in environmentally sensitive areas. For example, British Petroleum (BP) had to construct a pipeline bundle in the Wytch Farm field in southern England between a well site on Furzey Island and the Goathorn Peninsula. The area is of outstanding natural beauty and is important for birds and for recreational activities such as sailing. A plan to dredge a trench across the channel was considered to be disruptive and environmentally damaging. Therefore, BP decided to drill horizontally, i.e., to pull a 24-in. casing into the hole and to pull the bundle into the casing.

Among the advantages of horizontal drilling are the following:

- It is possible to avoid the beach and the surf zone, both notoriously difficult areas where the pipeline is exposed to wave action.
- The environmental impact is minimal.
- The whole operation is insensitive to weather and sea conditions.
- The impact on other uses of the shore is minimal. For example, in the Hoek van Holland shore crossing, one of the factors that weighed against other methods being used was that they would disturb the sand dunes that parallel the shore and form the first line of coastal defense against the sea flooding the low-lying land behind the dunes. The drill rig was positioned about 100 m back from the shore behind the first line of dunes so that the dunes remained undisturbed, and users of the beach would not be aware of the operation.
- The pipeline can be positioned well below the beach level, typically at a depth of 20 m so that changes in beach level produced by erosion have no effect.
- Most of the drilling operation is decoupled from most of the marine operation so that a project can be scheduled more efficiently.

Many contractors now offer this technique, and there is wide experience with it. The position of the breakout point can be accurately controlled within a few meters. The maximum practicable horizontal reach is about 1500 m, which is sometimes a limitation. The diameter is unlimited; and in a recent census of large directional drilling rigs, many contractors quote a maximum diameter of 1219 mm (48 in.).[13] Most completed jobs have been for pipeline diameters less than 24 in. (610 mm).

Horizontal drilling makes it possible to select shore-crossing sites that otherwise would be impracticable because of obstructions and environmental impact.

The first applications of drilled shore crossings were in soft ground, but it is now being applied in rock.[14, 15] The geotechnical conditions are important: Very permeable soils such as gravel allow the bentonite slurry to escape too easily, and boulders obstruct the drill.

13.6 Rocky Coasts

Rocky coasts are often exposed to severe wave action and pose difficult construction problems. Breaking waves impose very high forces on pipelines, and design of exposed pipelines in surf zones needs to be cautious.

Trenching is sometimes practicable by cutter-suction dredging or by blasting followed by clamshell excavation. If blasting is impractical, it may be possible to anchor the pipeline to drilled anchor piles. This technique has been used to anchor exposed outfalls in South Africa and an untrenched pipeline off the coast of Australia. An alternative is to hold the line down with very heavy saddle weights. A difficulty is that the wave forces are high on the weights themselves. The pipeline may also need to be protected against risks of boat impact and grounding.

An alternative is to construct a concrete protective culvert as a submerged-tube tunnel resting on prepared foundations fixed into the seabed, then to pull the pipeline into the culvert. This method was used for the Statpipe landfall at Karsto in Norway.

Increasing confidence in the practicability of horizontal drilling in rock is likely to make it an attractive method widely applied to rocky coasts. A 845 m pipeline crossing under the Penobscot River in the northeast United States was drilled through quartz and mica in 93 days in 1999.[15] In the different context of petroleum drilling, where the hole is much deeper, the record for extended-reached drilling currently stands at 10,700 m horizontally in the English Channel to reach part of the undersea portion of the Wytch Farm oil field in southern England. There is no reason to suppose that the limit has been reached.

13.7 Tunnels

Yet another alternative approach to shore crossings is to build a shaft/tunnel system from shore to a breakout point on the seabed. This method is relatively slow and expensive, but it can be used where the pipeline has to be given a high level of protection on an exposed shore and where geological conditions rule out horizontal drilling.

The pipeline can be installed through the tunnel in several ways. Kaustinen describes a scheme proposed for Arctic pipelines.[16] A vertical shaft is drilled at the end of the tunnel, and an elbow spool piece closed at both ends is lowered into the shaft and grouted in place. The tunnel is then excavated below the lower end of the shaft, and a chamber is excavated below the lower end of the spool piece. The pipeline is built in the tunnel and connected to the spool through a welded connection in the chamber. The other end of the spool is then connected to the marine pipeline by hyperbaric welding, by one-atmosphere welding, or by recovery and laying away. Koets describes a tunnel application completed in Norway.[17] An alternative method is to install a J-tube in the tunnel before breakout, then to pull the pipeline into the J-tube.

13.8 Tidal Flats

Pipeline construction across tidal flats has been a frequent source of problems. An example is described in chapter 15.

Tidal flats can be crossed by the trenching methods described in chapter 12. A difficulty is that the drainage patterns on tidal flats are irregular and unstable, and there is a possibility that the flow might be diverted into the dredged trench, leading to severe erosion and to a major realignment of channels. This possibility can be avoided by the *bathtub* method, which has been used to construct pipelines across mudflats in northern Germany and Holland.[18] A cutter-suction dredger excavates a channel wide and deep enough for a shallow-water lay barge. The lay barge follows the dredger and lays the pipelines. The pipeline is moored to anchors laid on the flats outside the channel, and the anchors are moved by work boats at high tide. Behind the lay barge follows a fill-in barge, which receives spoil through a floating pipe from the dredger and backfills the channel over the pipelines.

References

1   King, C. A. M. (1972) *Beaches and Coasts.* New York: St. Martin's Press.

2   Komar, P. D. (1976) *Beach Processes and Sedimentation.* Englewood Cliffs, NJ: Prentice-Hall.

3   Guilcher, A. (1963) *Coastal and Submarine Geomorphology.* London: Methuen.

4   Muir Wood, A., C. A. Fleming. (1981) *Coastal Hydraulics.* London: Macmillan.

5   U. S. Army Corps of Engineers (1984) *Shore Protection Manual.* Washington, DC.

6   Massie, W.W. , ed. (1976) *Coastal Engineering.* Coastal Engineering Group, Delft University of Technology. Delft, Netherlands.

7   Niedoroda, A. W., D. J. P. Swift. (1991) "Shore Face Processes." From *Handbook of Coastal and Ocean Engineering,* Vol. 2. Ed. J. Herbich, ed. Houston:Gulf Publishing Co. 736–770.

8   Palmer, A. C. (1979) "Application of Offshore Site Investigation Data to the Design and Construction of Submarine Pipelines." *Proceedings of the Society for Underwater Technology Conference on Offshore Site Investigation.* London. 257–265.

9   Niedoroda, A. W., A. C. Palmer, R. Pittman, J. Vandermeulen, J. Frisbee Campbell. (1985) "Oahu OTEC Preliminary Design: Sea Floor Survey, Advances in Underwater Technology and Offshore Engineering." *Proceedings of the Offshore Site Investigation 1985 Conference.* London: Graham & Trotman. 3. 15–27.

10 Machin, J. (1996) "Guidance Notes on Geotechnical Investigations for Marine Pipelines." *Proceedings of the Offshore Pipeline Technology Conference.* Amsterdam. IBC Technical Services. London.

11 Brown, R. J., A. C. Palmer. (1985) "Submarine Pipeline Trenching by Multipass Ploughs." *Proceedings of the 17th Annual Offshore Technology Conference.* Houston. 2. 293–291.

12 Palmer, A. C. (1985) "Trenching and Burial of Submarine Pipelines."
    Proceedings of the Subtech 85 Conference (Society for Underwater
    Technology). Aberdeen.

13 "Large Directional Drilling Rig Census." (2002) *Pipeline & Gas
    Journal*. 229. 72–76.

14 Bjorndal, T. A., A. Sharland. (1996) "The Construction of the Troll
    Oljerør Pipeline." *Proceedings of the Offshore Pipeline Technology
    Conference*. Amsterdam. IBC Technical Services, London.

15 McManus, G. (1999) "Gas Pipeline Buried under Penobscot."
    Bangor Daily News. July 27, 1999. Bangor, ME.

16 Kaustinen, O. M., R. J. Brown, A. C. Palmer. (1983) "Submarine
    Pipeline Crossing of M'Clure Strait." *Proceedings of the 7th
    International Conference on Port and Ocean Engineering under Arctic
    Conditions*. Helsinki. VTT Espoo. 1. 289–299.

17 Koets, O. J., J. Guijt. (1996) "Troll Phase 1 Pipeline: the Lessons
    Learnt." *Proceedings of the Offshore Pipeline Technology Conference*.
    Amsterdam, IBC Technical Services, London.

18 Park, C. A., A. C. Palmer, R. McGovern, J. P. Kenny. (1986) "The
    Proposed Pipeline Crossing to Vancouver Island." *Proceedings of the
    European Seminar on Offshore Oil and Gas Pipeline Technology*. Paris,
    IBC Technical Services, London.

14 Upheaval Buckling, Lateral Buckling and Spans

14.1 Introduction

This chapter considers two different but related phenomena in which a section of pipeline moves upward out of contact with the bottom, moves sideways, or forms a free span over a depression in the bottom.

A buried pipeline can sometimes arch upwards out of the seabed, forming a raised loop that may project several meters. That event is called *upheaval buckling*. A pipeline on the seabed can alternately snake sideways. That action is *lateral buckling* and can also happen if the pipeline is buried in very weak soil. These effects occur both to land pipelines, where they have been observed for many years, and to marine pipelines, for which this phenomenon has been recognized as an important one only in the last 10 years. The buckles sometimes occur with disconcerting suddenness. They have to be taken seriously because they can overstress the pipe wall, can occasionally lead to a rupture, and can lead to other difficulties such as excessive hydrodynamic loads if the line projects up into the sea.

A pipeline laid on an uneven seabed does not usually conform perfectly to the seabed profile. It bridges across low points in the profile and forms free spans in which the line is out of contact with the bottom. Some of the spans will persist after the pipeline is trenched. Pipeline engineers are concerned about spans because of the possibilities of fatigue damage induced by vortex-excited oscillations because a span might be overstressed and because it is vulnerable to hooking by fishing gear and anchors.

14.2 Upheaval Buckling

Most pipelines carry a longitudinal compressive force induced by the operating temperature and pressure. Upheaval buckling is caused by the interaction between that longitudinal compressive force and the local curvature of the pipeline axis. It is loosely analogous to the buckling of axially compressed columns, which is well known in structural analysis, and closely related to the buckling of axially constrained railroad track at high temperatures.[1, 2, 3]

If the pipeline is buried, it cannot easily move downwards or sideways, but there is much less resistance to upward movement. The pipeline, therefore, buckles upwards, almost invariably at overbends where the profile is convex upwards. Figure 14–1 illustrates the sequence schematically. Figure 14–2 depicts an upheaval buckle in a pipeline in Uzbekistan and illustrates the large scale of the problem. Figure 14–3 is a profile of the same buckle.

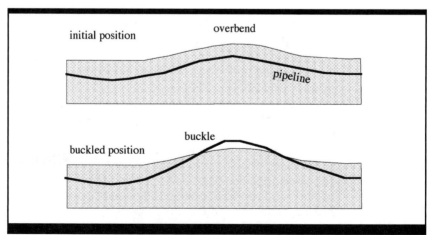

Fig. 14–1 *Upheaval Buckling (Schematic)*

Section 14.3 discusses the driving force for upheaval, lateral buckling, and section 14.4 addresses the analysis of upheaval movements. Section 14.5 examines the measures that can be taken to prevent buckling, and section 14.6 the corrective action that can be taken if buckling takes place.

Fig. 14–2 *Upheaval Buckle of 1020 mm Onshore Pipeline in Uzbekistan*

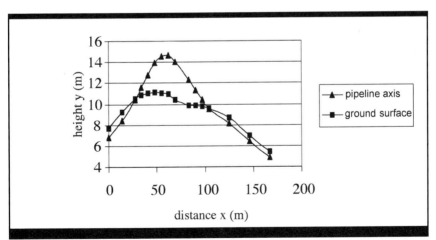

Fig. 14–3 *Buckle Shown in Figure 14-2, Profile of Pipeline Axis and Ground Surface*

14.3 Driving Force for Upheaval and Lateral Buckling

Upheaval and lateral buckling are driven by the longitudinal compressive force in the pipe wall and the fluid contents. Calculations of longitudinal stress easily go wrong if they are not tackled systematically. The calculations are best carried out using a thin-walled tube idealization; it is scarcely ever worthwhile to use thick-walled analysis, though it is not particularly difficult to do so if required.

In a case with no external pressure, the circumferential stress is statically determinate and is given by the following formula:

$$s_H = \frac{pR}{t} \tag{14.1}$$

where

p is the internal pressure, which is the Barlow formula Equation (10.3) in chapter 10), with D replaced by twice the mean radius R.

The longitudinal strain ε_L is given by the stress-strain relation for a linear elastic isotropic material:

$$\varepsilon_L = \frac{1}{E}\left(-\nu s_H + s_L\right) + \alpha\theta \tag{14.2}$$

where

ε_L is longitudinal strain

s_L is longitudinal stress

s_H is circumferential stress

E is Young's modulus

ν is Poisson's ratio

α is linear thermal expansion coefficient

θ is change of temperature (increase positive)

The effect of the radial stress in the pipe wall has been neglected (which is consistent with the thin-walled shell idealization).

The longitudinal stress s_L is not statically determinate but depends on the extent to which longitudinal movement is constrained. If there is complete axial constraint

$$\varepsilon_L = 0 \qquad (14.3)$$

and Equations (14.1), (14.2), and (14.3) together give the longitudinal stress:

$$s_L = \frac{vpR}{t} - E\,\alpha\theta \qquad (14.4)$$

Longitudinal stress, therefore, has two components: the first related to pressure and the second to temperature. The pressure component is positive (tensile), and the temperature component is usually negative (compressive).

The cross-section area of the pipe wall is $2\pi Rt$. The longitudinal force in the pipe wall is as follows:

$$2\pi\,Rts_H = 2v\pi R^2 p - 2\pi\,RtE\alpha\theta \qquad (14.5)$$

There is an additional component of longitudinal force in the pipe contents. The cross-sectional area of the contents is πR^2, and the longitudinal stress in the contents is $-p$ (consistently counting tension positive). Therefore, the longitudinal force in the contents is the following:

$$-\pi R^2 p \qquad (14.6)$$

Adding Equations (14.5) and (14.6), the total longitudinal force is represented by the following:

$$-(1-2v)\pi R^2 p - 2\pi\,RtE\alpha\theta \qquad (14.7)$$

which has both a pressure term and a temperature term. In most cases, θ and p are both positive, and $(1-2v)$ is always positive. Then both terms in Equation (14.7) are negative and, therefore, compressive.

The presence of a compressive pressure term suggests that pressure alone can cause upheaval buckling, and this is confirmed by theory, by laboratory-scale experiment, and by field experience. Palmer and Baldry discuss this in more detail and demonstrate this kind of buckling by a simple experiment.[4]

If external pressure is present, p in Equation (14.7) is replaced by the difference between the internal pressure and the external pressure.

Residual lay tension may be significant. In order to find the axial force in operation, it is necessary to trace through the whole history of tension in the pipeline, including laying and pressure testing, and carefully to avoid double counting.

Equation (14.7) gives the resultant force in a fully constrained pipeline in which all axial movement is prevented. At the ends of pipelines, expansion loops and expansion doglegs allow some longitudinal expansion movement to occur; and the longitudinal force is then less compressive. At an expansion loop, the resultant longitudinal force is small because the longitudinal tension in the pipe wall almost balances the compression in the constrained fluid.

14.4 Analysis of Upheaval Movements

There has been much research on upheaval buckling, and at least three strategies have evolved.

The first strategy is the most closely related to classical buckling theory in structural mechanics. In the case of a perfectly straight pipeline that rests on a perfectly straight and horizontal foundation, Hobbs drew on earlier research on railway track buckling by Martinet and Kerr. Hobbs considered the conditions under which the initially straight pipeline could remain in equilibrium as a raised loop, taking into account longitudinal movement toward the loop from the pipeline on either side.[1, 2, 3] That resolves the problem of how the buckled pipe can remain in equilibrium but does not explain how the pipeline jumps from its initial configuration into a buckled configuration.

In mechanics terms, the perfectly straight pipeline has an infinite buckling force coupled with an infinite degree of imperfection sensitivity. The imperfection of the seabed profile that the pipe rests on is, therefore, a central feature of the problem. The next step was to consider different kinds of profile imperfection, each characterized by a height, a length, and a mathematically defined shape. Thus, for example,

$$y = (1/2)H(1-\cos(2\pi x/L)) \quad 0 < x < L \tag{14.8}$$

where

y is height

x is the horizontal distance

is a sinusoidal profile imperfection H high and L long

Analysis then determines the conditions under which the pipeline can become unstable and lift away from the profile.[5] Palmer et al. developed this idea in detail and derived the universal design curve shown in Figure (14–4), in terms of two parameters, a dimensionless download parameter is calculated as follows:

$$\Phi_q = \frac{qF}{HP^2} \tag{14.9}$$

A dimensionless length is calculated as follows:

$$\Phi_L = L \sqrt{\frac{P}{F}} \tag{14.10}$$

where

q is the total download (the pipe weight plus the uplift resistance if the pipe is buried;

F is the pipe flexural rigidity

H is the profile imperfection height

P is the longitudinal compressive force derived earlier in this chapter

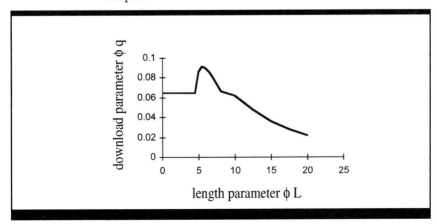

Fig. 14–4 *Design Chart*

This approach has several limitations. It assumes the pipe to be elastic, and it relies on rather simple idealized imperfection shapes, whereas actual profiles turn out to be complicated and not easily idealized.

The second strategy was developed at the same time.[6] It uses a dedicated finite-element program UPBUCK to solve the problem numerically. That makes it possible to incorporate more complicated imperfection profiles, to take account of inelastic response of the pipe, and to follow the behavior of the pipeline in detail as the operating temperature and pressure are increased. UPBUCK has been commercialized as PCUPBUCK.[7]

The third strategy is to ask a simpler question: If one can measure or calculate the initial profile of the pipe, what external force is required to hold the pipe in position when it goes into service, and the longitudinal force increases? For a section of pipe in an arbitrary profile defined by a height y (measured positive upward from a datum height) which is a function of horizontal distance x, Figure (14–5) shows an element dx of the pipe.

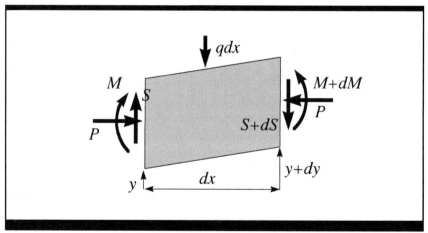

Fig. 14–5 *Element of Pipeline*

In this diagram, P is longitudinal force (compressive positive), S is shear force, q is external vertical force per unit length, and M is bending moment. S and M vary along the length. From vertical and moment equilibrium of the element

$$q = -\frac{dS}{dx} \tag{14.11}$$

$$P\frac{dy}{dx} + \frac{dM}{dx} - S = 0 \tag{14.12}$$

and, therefore

$$q = -P\frac{d^2y}{dx^2} - \frac{d^2M}{dx^2} \qquad (14.13)$$

If the pipe remains elastic so that the bending moment M is proportional to the curvature

$$M = F\frac{d^2y}{dx^2} \qquad (14.14)$$

where F is the flexural rigidity derived in chapter 10, and therefore

$$q = -P\frac{d^2y}{dx^2} - F\frac{d^4y}{dx^4} \qquad (14.15)$$

In Equations (14.13) and (14.15), the first term on the right is a curvature term, the product of the longitudinal force and the curvature $\frac{d^2y}{dx^2}$, which is negative in overbends (where the pipe tends to push upwards and, therefore, requires a positive value of q to hold it down) and positive in sagbends. The second term is less obvious and is proportional to changes in shear force.

Therefore, Equations (14.13) and (14.15) make it possible to determine from the pipeline profile the force needed to hold it in that position without the more complex analysis required to determine how it moves and ultimately becomes unstable and jumps into a new position. This is the best approach to most problems that arise in practice and lends itself directly to the design of countermeasures.

14.5 Measures That Can Be Taken to Prevent Buckling

The essential factor that determines whether upheaval buckling will occur is the smoothness of the seabed profile. Nothing can replace factual quantitative data. It is sometimes suggested for a design to be based on a standard imperfection height such as 0.3 m, but this has no rational basis.[8] An assumed height of 0.3 m would be a pessimistic design choice for a level bottom on uniform medium clay but a hopelessly optimistic choice for an uneven seabed of sandstone.

It follows that good profile information is essential. This information can be secured from ROV survey with precision bathymetry or, sometimes, by divers. If the pipeline is already in place, its profile can be accurately measured by an inertial navigation pig such as Geopig.

Many alternative strategies have been investigated, and several have been applied in construction. The simplest strategy is to reduce the driving force, either by reducing the operating temperature and pressure or by reducing the pipeline wall thickness to the minimum possible value. It can be seen in Equation (14.7) that the temperature term in the driving force is proportional to t, and it turns out that the improvement that comes from reducing that component more that outweighs the adverse effect of also reducing F (which is proportional to t). This was one of the major motivations for the development of the allowable strain design methods described in chapter 10.[9] Alternatively, the driving force can be reduced by laying the line in a zigzag, by introducing cooling loops that allow the fluid to cool by heat transfer to the sea, or by incorporating expansion loops at intervals along the pipeline.[10]

A second strategy is to make the pipeline profile smoother. This can be done by selecting the route so as to avoid rough areas. That tactic also helps to reduce the length and number of spans. The profile can be further smoothed by careful trenching, particularly with a *smart* plow that trenches more deeply on high points of the profile.

If these measures are not enough by themselves, the pipe must be buried, either under rock or, if the pipeline is trenched by plowing, under natural seabed soil previously excavated when the trench was cut. The uplift resistance of a pipeline buried in cohesionless soil or rock is usually calculated as follows

$$q' = \gamma HD \left(1 + f\frac{H}{D}\right) \qquad (14.16)$$

where

γ is the submerged unit weight of the soil

H is the cover from the top of the pipe to the surface of the soil above the pipe centerline

D is the pipe outside diameter

f is an uplift coefficient determined experimentally, generally about 0.7 for rock and 0.5 for sand, but occasionally much smaller in loose sand.

Uplift resistance has been investigated in several research programs.[11, 12, 13, 14]

One practicable but costly strategy is to place rock cover along the whole length of the pipeline. Another is to place rock at regular intervals. A more economical strategy is to identify the critical overbends where upheaval might initiate and to place rock on those overbends alone.[5, 15] That technique uses less rock but obviously depends on the ability to locate the overbends and to return to the same spot to place the rock. The rock must be chosen so that it is stable against seabed velocities and so that it does not sink into the seabed. Geotextiles under the rock enhance uplift resistance.[8]

Rock can be placed in shallow water by dropping it from split-hopper barges or by pushing rock over the side of a specialized dump vessel. These operations demand great care because if too much rock is dropped at once, it falls as a heavy cloud, which can damage a pipeline severely. In deep water, rock dropped freely from the surface scatters wastefully; and it is better to drop the rock down a fall-pipe. Specialized vessels are available to do this.

14.6 Corrective Action if Buckling Takes Place

Not all buckles need to be corrected, but sometimes the pipeline is buckled upwards into the water to an extent that it is vulnerable to increased hydrodynamic forces or to hooking by fishing trawls and anchors. If it is not overstressed, the most economical solution may be to stabilize the pipeline in its new position by placing rock or mattresses around the pipeline of sufficient weight to prevent further movement. The rock and/or mattresses must be stable themselves.

In other cases, it may be judged necessary to cut and remove the buckled section of pipe and to replace it with a new spool piece connected by hyperbaric welding, surface tie-in, or mechanical connection. It will obviously be necessary to make sure that buckling will not repeat itself, by adding more cover or by incorporating an expansion spool.

14.7 Lateral Buckling

Buried pipelines almost invariably buckle upwards because there is less resistance to upwards movement than there is to sideways movement. If a pipeline is not buried, on the other hand, it is usually easier for it to

move sideways. The resistance to sideways movement is the submerged weight multiplied by some lateral friction coefficient, usually less than 1. Moreover, in marine pipelines (in contrast to land pipelines), lateral departures from a straight line in plan (as seen on a plan view as in a map) are often larger than profile departures from a straight line in elevation.

The driving force for lateral buckling is the same as the driving force for upheaval. Many pipelines buckle laterally to some extent, but lateral movements frequently go undetected. If lateral buckling occurs, it initiates at lateral snaking imperfections in the line of the pipe, as seen in plan. The amplitude and wavelength of these imperfections depend on the method of construction. Few data are available, but some degree of snaking is inevitable, particularly in lay barge pipelay in shallow water under small tension.

Lateral movements are often harmless, because the lateral movement occurs over a substantial distance, the bending stresses are small, and the buckle does not localize into a sharp kink. In one instance, a large-diameter gas pipeline was laid in a 3000 m radius curve away from a platform and was effectively anchored by rock dumping over a length of 500 m. After the line had been in operation for some time, it was found to have moved laterally in six locations, roughly equally spaced, always moving towards the outside of the curve. However, the movements were small (between 1 and 2 m) and occurred over a substantial distance (from 100 to 200 m). Straightforward calculations showed that the associated bending stresses were small and that the total bending stress was well below yield. No action needed to be taken. Indeed, it could be argued that lateral buckling had a beneficial effect because it relieved the longitudinal compressive force that might otherwise have caused upheaval.

In some instances, however, lateral movements may be larger, and longitudinal movement of the pipeline towards the buckle (*feed-in*) may lead to a localization in which all the movement is concentrated in one buckle. At the point where the lateral movement is largest, the pipe may form a localized kink in which the strain is large enough for the wall to rupture.

Lateral buckling has been less studied than upheaval. Palmer et al. review the state of knowledge of lateral buckling.[16] Miles and Calladine investigated lateral movements through physical and finite-element modeling.[17]

A lateral buckling incident in Brazil in 2000 has generated further concern. A hot pipeline buried in soft mud in a shore approach buckled sideways and kinked. The thin wall folded, and the pipe ruptured, leading to a damaging oil spill. Much attention is now being given to hot deep-water lines in the Gulf of Mexico and to the possibility of deliberately initiating lateral buckling at regular intervals so that the bending is distributed rather than concentrated at one location.

14.8 Span Formation

A span is formed when a pipeline bridges across a depression in the seabed. It is important not to exaggerate the importance of spans in comparison to other problems such as instability and upheaval buckling, which in practice have caused far more trouble. No North Sea pipeline has failed as a result of spans, and in some parts of the world spans are given little attention. Very long spans are known to have developed in several North Sea pipelines, among them the following three cases:

1. A pipeline that after laying was found to have an unsupported span some nine joints long (about 110 m)

2. A pipeline that had been in operation for several years, with spans some 70 m long and up to 2 m off the bottom

3. A pipeline that was found to be in operation with a span 550 m long, and turned out to have been in that condition for some time, perhaps several months

These examples are adduced to set the problem in perspective. In cases 2 and 3, the pipeline concerned is in operation today, many years after the spans were discovered, with the originally spanned section still in place. There is no reason to suppose that each of the three lines is not completely safe. In the first instance, the long span disappeared as soon as the pipe was hydrotested. In the second instance, the spans were stabilized by rock dumping to protect them against the risk of hooking, but no other action was taken. Only in the third instance were major intervention and stabilization measures necessary.

A few spans do need corrective action. Operators annually assign significant resources to correction operations that stabilize spans by rock dumping and the installation of mattresses and grout bags. Similarly, large amounts of money are spent on presweeping to smooth the seabed profile before a pipeline is laid. The amount of work that is to be done depends on the number of spans that require correction and on the rule that decides whether correction is necessary. It is unnecessary and wasteful to require extensive correction and documentation of large numbers of small spans below any length at which technical difficulties could occur.

There have been several research programs. It is now possible to arrive at rational span criteria that an engineer can apply confidently to assess spans in existing pipelines and in new pipelines. The potential problems of fatigue, overstress, and hooking are considered in turn.

14.9 Vortex Shedding

Water flow across a pipeline span induces the formation and shedding of vortices, alternately at the top and bottom of the pipe at a rate determined by the flow velocity. A small hydrodynamic force accompanies the formation of each vortex on the pipeline; and, therefore, the vortices together induce an oscillating force. The vertical component of the force induced by a vortex shed near the top of the pipe is in the opposite direction to the vertical component of the force induced by a vortex shed near the bottom of the pipe, whereas the smaller horizontal components of these forces have the same direction. It follows that the overall excitation force has a vertical (cross-flow) component that oscillates at the vortex-shedding frequency; and a smaller horizontal (in-line) component with twice that frequency, superimposed on a steady in-line component.

The pipeline span is free to oscillate in flexure, and indeed, in other modes; but they have much higher natural frequencies. Because the pipe is axially symmetric and the end conditions are usually approximately the same for vertical and for horizontal deflection, the natural frequencies for horizontal and vertical motions are almost equal (though this is not so for very long spans). If one of the frequency components of the external hydrodynamic force is close to a natural frequency of span oscillation, it is possible for oscillations to occur. Those motions might induce fatigue damage in the pipe itself, the girth welds, or the coating.

This is a simple outline of a very complicated phenomenon.[18] In particular, the motion itself modulates the flow so that oscillations occur over a wide range of flow velocities and do not require exact coincidence of the vortex-shedding frequency and a natural frequency. A *lock-on* phenomenon introduces hysteresis so that the response depends on the increase or decrease in flow velocity, not just on its present value. The proximity of the bottom under a span alters the flow. The amplitude is influenced by structural damping, hydrodynamic damping, and damping at the ends, all of which tend to reduce or eliminate the oscillations.

Vortex-excited oscillations have been investigated in several research programs. The most directly applicable results are those secured in two full-scale test series carried out at a site in the Severn Estuary by Hydraulics Research (HR). The first was for Polar Gas Project, and second was for the UK Department of Energy. The site lends itself to tests because it has peak tidal currents of 1.4 m/s and dries at low tide but has 6 m of water at high tide. The tests are described in detail in reports and papers.[19, 20, 21, 22, 23] In the Polar Gas tests, spans were simulated by supporting lengths of 20-in. pipe on steel rods. In the Department of Energy tests, the pipe sections were longer and were mounted on universal joints, representing simple supports at each end. A limitation of these arrangements was that they gave a very low level of structural damping, either within the pipe or at the ends. A true pipeline span resting on a sandy seabed has much higher damping at the ends and higher internal damping if it is concrete-coated.

Typical results are shown in Figure 14–6, which plots motion amplitude divided by pipe diameter against reduced velocity V_R, a dimensionless parameter defined by the following:

$$V_R = \frac{U}{ND} \qquad (14.17)$$

where

U is the flow velocity

N is the (lowest) natural frequency of oscillation

D is the pipe outside diameter

This is the conventional way to report measurements of vortex-excited oscillations.[18]

The responses show that in-line horizontal motion begins at a reduced velocity of about 1.4, but its amplitude remains small. Cross-flow vertical motion begins at a reduced velocity between 2 and 3 and reaches much larger amplitudes. The reports examine the differences between smooth pipes and rough pipes with a surface simulating concrete and the effect of the gap between the pipe and the bottom.[19, 20] Measurements show differences between the vortex-shedding frequencies for rough and smooth pipes. This happens because roughness trips the boundary layer and induces a transition to fully turbulent conditions upstream of the vortex separation point. This underlines the significance of Reynolds number and indicates that results from model tests at much lower Reynolds numbers should be approached skeptically. However, the response thresholds for large movements are not very different for rough and for smooth pipes.

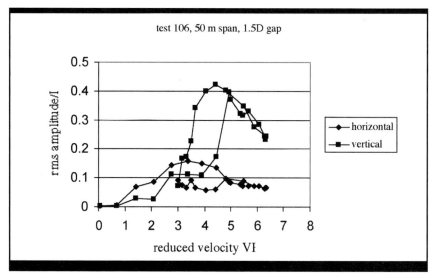

Fig. 14–6 *Observed Relationship Between Reduced Velocity and Oscillation Amplitude (Redrawn from Reference 20)*

Since the amplitude of vortex-excited oscillations depends on the reduced velocity U/ND, an analysis requires an assessment of span natural frequency N. The natural frequency of a span idealized as a uniform linear-elastic beam with no axial force is calculated as follows:

$$N = \frac{C}{L^2} \sqrt{\frac{F}{m}} \qquad (14.18)$$

where

L is the span length, and

C is a constant that depends on the end conditions.

F is the flexural rigidity, defined by Equation (10.31).

m is the mass per unit length

The mass per unit length m has to allow for the surrounding fluid. This is usually done by adding a mass to the pipe mass that is equal to the mass of water displaced. A single span does not have complete fixity against rotation at the ends of the observed span (the length over which the pipeline is visibly out of contact with the bottom); and it is customary to allow for this by taking C as 2.45, the *exact* value for a span that is fixed at one end and pinned at the other (fixed/pinned), rather than as 3.56, the value for a span that is fixed at both ends (fixed/fixed). This is rather arbitrary. It is preferable to treat the span as fixed-ended, to take C as 3.56, but to allow for lack of end fixity by taking L somewhat larger than the observed span length.

Measurements give some support to this approach. Acting on behalf of British Petroleum, DEI measured the natural frequency of pipeline spans by applying electromagnetic force excitation and analyzing the dynamic response. The results have not been published in full but are available in a limited form. They show that the measured natural frequencies are invariably higher than those calculated from the fixed/free assumption about end conditions. These measurements are an extremely valuable contribution to the understanding of the actual behavior of spans.

Axial force has a significant effect. The effective axial force is almost always compressive and reduces the natural frequency. If the axial force were to reach the critical Euler axial force for the span, the natural frequency would fall to zero. It is sometimes argued that this will happen when the span length reaches the level at which the critical axial force equals the effective axial force corresponding to full axial constraint; but this argument is incorrect, at least for spans that are progressively increasing in length, because deflection in the span alters the axial force. The pipeline sags, the sag induces a tensile axial strain component, and the axial force becomes less compressive, though this effect is partly balanced by longitudinal movement towards the span necessary to equilibrate axial forces.

Any movement induces oscillating stress, but small movements generate stresses so small that they fall below the fatigue limit. Returning to Figure 14–6, for instance, a reduced velocity of 2 induces an root mean square (rms) response amplitude of 50 mm. In a 40-m span in a 20-in. pipe, simply-supported and bending in a mode corresponding to a half-cycle of a sine wave, this displacement amplitude corresponds to a rms bending stress amplitude of 16 MPa (2320 psi), which is far too low to cause fatigue damage. Cross-flow motions at larger reduced velocities do, however, induce much larger movements; and they might indeed lead to fatigue.

Since fatigue is the concern, the most direct approach is to calculate the fatigue life explicitly and to compare it with the design life. The DnV 2000 rules, clause 5 D 709, for instance, require the limit damage ratio not to exceed 0.1 for safety class high so that in that instance the fatigue life must be at least 10 times the design life.[24]

Therefore, a simple procedure to determine the fatigue life of a span is as follows:

1. Determine natural frequency (by calculation or measurement).

2. Estimate a frequency distribution of transverse velocity.

3. For each level of velocity, determine V_R, and determine the response amplitude, using the responses measured in the HR tests.

4. Estimate a mode shape, and combine it with the response to determine the level of fluctuating stress.

5. Determine the level of damage per cycle, using S-N curves for the relation between stress range and fatigue life in the usual way and taking account of fatigue limits.

6. Repeat steps 3 through 5 for each level of velocity in the frequency distribution and combine by Miner's rule to determine cumulative damage in the conventional way.

Galbraith and Kaye describe an explicit span assessment carried out in this way.[25] The analysis was divided into two phases. The first preliminary analysis was carried out on the survey vessel that found the spans. A more detailed final analysis followed on shore. Figure 14–7 shows typical results: 178 spans were found, and 20 of them failed the preliminary analysis. The final analysis eliminated nine more, so that only 11 out of the original 178 required correction.

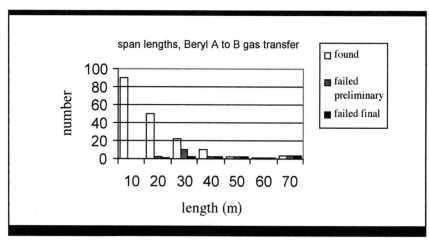

Fig. 14–7 *Span Lengths and Span Assessment (Redrawn from Reference 25)*

There are, of course, more sophisticated ways of carrying out the fatigue calculation. A fracture mechanics approach starts with an actual or assumed crack or a welding defect and calculates the progressive growth of the crack by the Paris law, which relates crack growth per cycle to the stress intensity range.[26] This method allows the growth of a crack to be followed explicitly and for the end of fatigue life to be identified with growth of a crack to a maximum acceptable size. Celant has published data on Paris law parameters for pipeline steels.[27]

There has been continued controversy about how to account for the interaction of wave-induced bottom velocities with spans. Wave effects were not explicitly investigated in the HR program, but the site was exposed to wave action in southwesterly gales. Waves about 1 m high with periods between 2 and 3 s were present during a test on February 2, 1984, and it was observed that they advanced the cross-flow response, in the sense that the threshold reduced velocity calculated from the steady velocity component was about 30% lower than when there were no waves.

A simple approach is to say that the motion amplitude during a wave cycle is determined by the peak transverse velocity during the cycle because in critical cases, the pipe natural frequency is substantially higher than the peak of the wave spectrum. Wave-induced velocities then have a significant influence. However, it is overly conservative to determine a design seabed velocity by adding a maximum wave-induced velocity to a maximum tidal current. By doing so, one ignores the fact that waves and tides are statistically uncorrelated so that the largest waves will not occur at the same time as the largest tidal currents.

This uncorrelated relationship between large waves and large tidal currents can readily be accounted for by an extension of the explicit calculation of fatigue life described earlier. The frequency distribution of bottom velocity can then be estimated from a joint probability distribution of wave height and tidal current, assuming they are uncorrelated. Numerical experiments then show that in deep water the inclusion of wave-induced currents has little effect on allowable spans. The reason is that only the largest waves induce significant seabed velocities, and they tend not to occur at the same time as the largest tidal currents. The relationship between wave-induced velocities and vortex shedding is the subject of continuing research, but it appears to be much less important for seabed pipelines than it is for risers and surface-piercing structures.

14.10 Overstress

The deformation of a pipeline span reflects its complete history, including manufacture, mill hydrotest, laying, trenching, filling, hydrotest, dewatering, commissioning, and final operation. It would be a demanding task to unravel the whole history, particularly because plastic deformations can occur at several stages; and the effects of earlier deformations are overwritten by later deformations. For instance, local plastic deformations invariably occur close to girth welds, but the residual stress distributions that they leave behind are partially erased by further local plastic deformations that occur during the subsequent hydrostatic test.

Fortunately, it is unnecessary to follow through the deformation history. It is necessary only to be certain that the stresses imposed in spans do not lead to gross deformations such as the occurrence of buckling and the development of large strains that might induce rupture. Adopting a limit state approach, which was done long ago in conventional structural design and is now adopted in pipeline codes, it turns out to be extremely unlikely that static deformation of a span could lead to a limit state.[24, 28] It might well lead to local plastic deformation—indeed, plastic deformations often occur during hydrotest—but plastic deformation does not in itself threaten the safety of a pipeline. Buckling only occurs some way into the plastic range.

This maxim is illustrated by an order-of-magnitude estimate of the strain that might occur in an overstress situation. Imagine, for example, a span in a 762 mm (30-in.) pipeline with a 19.1 mm wall, and that sediment transport leads to a growth of the span to a length of 60 m, at which point the span collapses plastically by the formation of three plastic regions (at the ends and the center). Before the span collapsed, the gap under it was 1 m; it collapses until it touches down in the middle. The plastic regions are not highly localized plastic hinges because the geometry and mechanics of plastic bending require the deformation to be distributed (over a distance that depends on the details of loading and the strain-hardening properties of the steel). The length of each plastic region can be estimated at 5 diameters. The curvature in each plastic region is 0.0175 /m, and the longitudinal bending strain about 0.0067, more than three times the yield strain but much less than the buckling strain, which is 0.0150 as seen in Figure 10–6, taking the maximum yield to tensile ratio as 0.85 and the girth weld factor as 1.

A conclusion is that for the kinds of spans encountered in the southern part of the North Sea and in other areas where the seabed is sand made uneven by sediment transport, overstress is unlikely to lead to a limit state. That conclusion does not necessarily apply in other parts of the world such as northern Norway, the Gulf of Mexico, Canada, and the Arabian Gulf, where the seabed is sometimes much rougher and rockier and has deeper and sharper features. In those locations long spans are much more frequent, and a pipeline in a span might be bent to an unacceptable extent. Detailed elastic-plastic analysis is necessary. That analysis is carried by a finite-element program such as ABAQUS.

14.11 Hooking

In many areas of the world, there is intensive fishing for fish that live on and in the seabed. The principal method is trawling, where a sock-shaped net is dragged across the bottom by a trawler. The mouth of the net is held open by trawl boards that act as hydrofoils or by a beam. This equipment can be large and heavy. If it encounters a pipeline, the boards or the beam might catch under it.

Hooking can threaten the safety of the fishing vessel and the loss of its gear, as well as damage the pipeline. A 2000 horse-power (1.5 MW) trawler can exert a steady bollard pull of about 200 kN, and a significantly larger force as it decelerates after a trawl is hooked. The breaking load of a typical 30 mm trawl warp is about 600 kN, and maximum pull-over forces measured in trawl/pipe interaction tests are about 200 kN. Applied near the center of a long pipeline span, these forces are large enough to lead to substantial plastic deformation in bending. They are also large enough to have a significant effect on the stability and motion of the vessel towing the trawl.

There are different kinds of trawl boards. A research program has studied the mechanics of the interaction between trawl boards and pipelines in model tests in a flume. The tests have shown that some kinds of boards behave better than others.[29] Similarly, beam trawls can be modified so as to make them more likely to ride over cables and pipelines.

In practice, span hooking of large-diameter pipelines has been not been a significant problem, but opinion is changing because of an incident described below. The justification for this statement is that trawls cross large pipeline spans very frequently, and if many of these crossing were to lead to hooking, then hooking incidents would be very frequent.

The span crossing frequency can be quantified by a simple order-of-magnitude argument. If the trawling intensity is I (measured in trawling time/unit seabed area per unit elapsed time), and the trawling speed is V,

all the trawling is in the same direction, and the trawling direction makes an angle of incidence ϕ to the normal to the pipeline, then the expected number of crossings per unit time per unit pipeline length is

$$c = IV \cos \phi \qquad (14.19)$$

where I and V are defined in consistent units, such as trawling years/km^2 year for I and km/year for V, which gives c in crossings per year per km of pipeline. If trawling directions are random, $\cos\phi$ can be replaced by its average value $2/\pi$; if trawling directions are not random, $\cos\phi$ can be replaced by a weighted average that allows for the frequency distribution of ϕ.

This does not make much difference unless the trawl direction is aligned with the pipeline, but that often happens, because fishermen know that fish are attracted to pipelines (by food, shelter, and, perhaps, warmth), and therefore choose to trawl along pipelines.

If now the proportion of the total pipeline length that is part of a span is denoted p, the expected number of crossings of spans per unit time per unit pipeline length is represented by the following equation:

$$sc = (2/\pi)pIV \qquad (14.20)$$

Taking typical values for a heavily-fished UK sector block in the North Sea:

I	is	1.14×10^{-3} /km^2 (10 hours/km^2 year)
V	is	7.88×10^4 km/year (2.5 m/s, 5 knots)
p	is	0.05

then the expected number of span crossings is 2.9 /km year. A pipeline 300 km long will, therefore, have nearly 1000 span crossings a year. Over the North Sea pipeline system as a whole, span crossings by trawls must occur several times an hour, and no reasonable variation in the data alters this conclusion.

The conclusion is based on existing practice, which is to trench small-diameter pipelines and to leave only large-diameter pipelines untrenched. Since trawls can easily get into pipeline trenches, however, there must also be a significant number of contacts between trawls and small lines spanned in trenches. Those contacts appear not to lead to hooking incidents. This is probably because it is rare for small-diameter trenched lines to span above the trench bottom to a significant height because of the combined effect of the relative flexibility of small lines and the profile smoothing that accompanies most trenching processes.

A recent incident suggests that hooking may occasionally occur with very serious consequences. The trawler *Westhaven* capsized with the loss of several lives. The capsize has been attributed to its hooking a pipeline. An increased risk of span hooking does appear to be a primary argument against the relaxation of existing rules requiring the trenching of small lines. Burial is naturally much better than trenching.

14.12 Span Correction

On mobile sand and silt seabeds, many spans correct themselves naturally. Enhanced scour under the touchdown points at the ends lengthens the span, and the span deflects downward under its own weight until it touches down near the center, forming two short spans instead of one long one. Active sediment transport often fills in the gap under a pipeline span and perhaps forms new spans somewhere else. The net effect is often a progressive lowering of the pipeline so that the number of spans decreases during its operating life.

Long spans may need to be corrected, both on mobile seabeds and on harder rock seabeds. The available techniques include sand bags, grout bags, mattresses, trenching, rock dump, and mechanical supports. These techniques deal with vortex-excited oscillations and the possibility of overstress. Sometimes combinations of methods are used. Vortex shedding can be suppressed by spoilers, shrouds, or strakes.

Sand bags are readily available almost everywhere and can be placed by divers. They can be filled with a cement/sand mixture so that they harden in place. The sandbags themselves have to be stable against wave action, scour, and settlement. Small bags are often not heavy enough. Though elaborate piles of sand bags held together by steel stakes are sometimes installed, they often shift, making this method an unsatisfactory solution in the long term.

Grout bags are larger and more stable. A bag can be pulled under the pipeline by divers or ROVs, and then filled with cement grout pumped from the surface. The bag rises under the pipeline like a pillow and prevents oscillation or settlement. Inflated bags may be unstable and roll easily until the grout sets.

Mattresses consist of rectangular or hexagonal concrete units linked together by rope. They can be dragged under a pipeline span or laid across the line, providing support and damping to prevent oscillations. Stability of the mattress itself has to be addressed. Wave action can lift the upstream edge of a mattress and flip it over, and there have been instances in which a mattress was carried completely away.

An alternative strategy is to reduce the length of a span by lowering the pipeline at the ends, rather than supporting the middle. This obviously can be done only if the seabed is soft enough to be excavated. The trenching techniques described in chapter 13 can be applied, or sand can be removed by jets held by divers or ROVs.

Another idea is to stop a span moving by dumping rock around it. The pieces of rock have to be large enough to be stable. Rock can be placed through a fall-pipe system. Alternatively, rock can be dropped from the surface, but the pipe can be damaged by the rock unless this is done very cautiously. Yet another possibility is to install adjustable steel supports under the pipeline. This method is uncommon but has been used in steep and rocky shore approaches in Sicily and Canada.

If vortex-excited oscillation is the only concern, it is possible to reduce the excitation by adding various spoiler strake and shroud devices. They work either by breaking up the span-wise correlation of hydrodynamic excitation forces (so that the forces act upwards on one section of pipe but downwards on another section a little further along), or by increasing hydrodynamic damping. These devices are often used to suppress wind-excited oscillations and are sometimes applied to risers. A disadvantage of this method is that the devices increase hydrodynamic drag. They can be installed by ROVs or divers. If they are to continue to be effective, they have to be kept free of marine growth.

References

1 Hobbs R. E. (1984) "In-Service Buckling of Heated Pipelines." *ASCE Journal of Transportation Engineering* 110. 175–189.

2 Martinet A, (1936) "Flambement des Voies Sans Joints sur Ballast et Rails de Grands Longueur" (Buckling of Jointless Track on Ballast and Long Rails)." *Revue Generale des Chemins de Fer.* 55/2. 212–230.

3 Kerr, A. D. (1979) "On the Stability of Railroad Track in the Vertical Plane." *Rail International* 9. September 1979. 759–768.

4 Palmer, A. C., J. A. S. Baldry. (1974) "Lateral Buckling of Axially-Compressed Pipelines." *Journal of Petroleum Technology* 26. 1283–1284.

5 Palmer, A. C., C. P. Ellinas, D. M. Richards, J. Guijt. (1990) "Design of Submarine Pipelines Against Upheaval Buckling." *Proceedings of the 22nd Offshore Technology Conference.* Paper OTC6335 presented at the 22nd Annual Offshore Technology Conference. Houston, TX. 2. 551–560.

6 Klever, F. J., L. C. van Helvoirt, A. C. Sluyterman. (1990) "Dedicated Finite-Element Model for Analysing Upheaval Buckling Response of Submarine Pipelines." *Proceedings of the 22nd Offshore Technology Conference.* Houston, TX. 2.

7 Andrew Palmer and Associates. (1996) PCUPBUCK, PLUSONE documentation.

8 Palmer, A. C., M. Carr, T. Maltby, B. McShane, J. Ingram. (1994) "Upheaval Buckling: What Do We Know, and What Don't We Know?" *Proceedings of the Offshore Pipeline Technology Seminar.* Oslo. IBC Technical Services. London.

9 Klever, F. J., A. C. Palmer, S. Kyriakides. (1994) "Limit State Design of High-Temperature Pipelines." *Proceedings of the Offshore Mechanics and Arctic Engineering Conference.* Houston, TX.

10 Rich, S. K., A. G. Alleyne. (1998) "System Design for Buried High Temperature and Pressure Pipelines." *Proceedings of the Offshore Technology Conference.* Paper OTC8672 presented at the Offshore Technology Conference. Houston, TX. 341–347.

[11] Faranski, A. S. (1997) *Uplift Resistance of Trench Backfill in a Soft Atlantic Silty Clay*. (Unpublished MPhil dissertation) University of Cambridge.

[12] Baumgard, A. J. (2000) *Monotonic and Cyclic Soil Responses to Upheaval Buckling in Offshore Buried Pipelines*. (Unpublished PhD dissertation) University of Cambridge.

[13] Fisher, R., T. Powell, A. C. Palmer, A. J. Baumgard. (2002) "Full-Scale Modelling of Subsea Pipeline Uplift." *Proceedings of the Physical Modelling in Geotechnics Conference*. St. John's, Newfoundland.

[14] Palmer, A. C., D. J. White, Baumgard, A. J., M. D. Bolton, A. Barefoot, M. Finch, T. Powell, A. S. Faranski, J. A. S. Baldry. (2001) *Uplift Resistance of Buried Submarine Pipelines: Comparison Between Centrifuge Modelling and Full-Scale Tests*. TR. Cambridge University.

[15] Locke, R. B., R. Sheen. "The Tern and Eider Pipelines." *Proceedings of the Offshore Pipeline Technology Seminar*. Amsterdam. IBC Technical Services. London. (1989).

[16] Palmer, A.C., C. R. Calladine, D. Miles, D. Kaye. (1997) "Lateral Buckling of Submarine Pipelines." *Proceedings of the Offshore Oil and Gas Pipeline Technology Conference*. Amsterdam. IBC Technical Services. London.

[17] Miles, D. J., C. R. Calladine. (1999) "Lateral Thermal Buckling of Pipelines on the Seabed." *Journal of Applied Mechanics 66*. 891-897.

[18] Blevins, R. D. (2001) *Flow-induced Vibration*. Malabar, Florida: Krieger Publishing.

[19] Author unknown. (1977) *Vibration of a Pipeline Span in a Tidal Current*. (Report EX 777) Hydraulics Research. Wallingford, England.

[20] Author unknown. (1984) *Vibration of Pipeline Spans*. (Report EX 1268) Hydraulics Research. Wallingford, England.

[21] Raven , P. W. J. (1986) "The Development of Guidelines for the Assessment of Submarine Pipeline Spans—Overall Summary Report." *Offshore Technology Report OTH 86 231*. J. P. Kenny and Partners for the Department of Energy, HMSO. England: Her Majesty's Stationery Office.

22   Grass, A. J., P. W. J. Raven, R. J. Stuart, J. A. Bray. (1983) "The
     Influence of Boundary Layer Velocity Gradients and Bed Proximity on
     Vortex Shedding from Free Spanning Pipelines." *Proceedings of the
     15th Annual Offshore Technology Conference*. Houston, TX. 1. 103–112.

23   Raven, P. W. J., R. J. Stuart, J. A. Bray, P. S. G. Littlejohns. (1985)
     "Full-Scale Dynamic Testing of Submarine Pipeline Spans."
     Proceedings of the 17th Annual Offshore Technology Conference. Houston,
     TX. 3. 395–404.

24   Offshore Standard OS-F101:"Submarine Pipeline Systems."Hovik,
     Norway. Det Norske Veritas2000

25   Kaye, D., D. Galbraith, J. Ingram, R. Davies. (1993) "Pipeline
     Freespan Evaluation: A New Methodology." (Paper SPE 26774)
     Proceedings of the Offshore European Conference. Society of Petroleum
     Engineers. Aberdeen.

26   Hellan, K. (1984) *Introduction to Fracture Mechanics*. New York:
     McGraw-Hill.

27   Celant, M., G. Re, S. Venzi. (1982) "Fatigue Analysis for Submarine
     Pipelines." *Proceedings of the 14th Annual Offshore Technology
     Conference*. Houston, TX. 2. 40–49.

28   Ellinas, C. P., A. C. Walker, A. C. Palmer, C. R. Howard. (1989)
     "Subsea Pipeline Cost Reductions Achieved Through the Use of Limit
     State and Reliability Methods." *Proceedings of the Offshore Pipeline
     Technology Seminar*. Amsterdam. IBC Technical Services. London.

29   *Otterboard Pipeline Interaction*. Video. Trevor Jee Associates, Tunbridge
     Wells, England (1997).

15 Internal Inspections and Corrosion Monitoring

15.1 Introduction

Inspection may be destructive or non-destructive. Destructive inspection entails physical degradation of samples so that they are unsuitable for service. This is often called testing, rather than inspection. Tensile and Charpy tests are examples of destructive techniques. On the other hand, non-destructive inspection applies techniques that do not detract from the continued functioning of the system that is being inspected. Radiography and ultrasonic wall thickness measurement are examples.

Non-destructive inspection and testing is routinely carried out throughout the life of a pipeline. Inspection is necessary before and during construction to ensure that the pipeline is constructed as intended. Inspection during operation checks that the pipeline remains fit for its intended purpose. Though cumulative inspection data may be used to evaluate historical corrosion rates, inspection is generally regarded as providing *snapshots* in time of the material condition, and the hard factual data from which the condition of the pipeline can be deduced and the remnant life estimated.

Corrosion monitoring serves a different function.[1] Monitoring is often done on a more frequent basis than inspection, and it provides more global information. Traditional monitoring techniques are intrusive, requiring contact or probes or samples of the pipeline fluids. Monitoring provides an estimate of the corrosion rate from which a corrosion control strategy can be decided. Monitoring may also be used to determine changes in the corrosiveness of the fluids.[2] Corrosion rates obtained from monitoring data are regarded as indicative of trends and are not generally used on their own to evaluate the condition of a pipeline, though the corrosion information will be a vital adjunct to inspection. Developments in both inspection and monitoring techniques, particularly the advent of real-time techniques allowing semi-continuous, non-intrusive assessment, have blurred the distinction between inspection and monitoring.

Inspection data and corrosion rate measurements are specific to the location of the measuring probe, and extrapolation to the whole pipeline must be with circumspection. There are statistical techniques available that may be used to give reasonable estimates of worst case metal loss within a pipeline based on limited local data.[3] Increasingly, computer models are being used to quantify the various domains within a pipeline system, and this information can be used in conjunction with monitoring and inspection data to determine the corrosion profile along the pipeline.[4] However it is generally necessary to inspect the complete pipeline periodically.

In the future, corrosion monitoring by itself is likely to be phased out, and it will be replaced by semi-continuous inspection techniques. Monitoring will only be used as an adjunct to inspection and for specific functions, for example, to evaluation corrosion inhibitors that are beyond the scope of inspection techniques. The philosophical difference in approach between inspection and monitoring is likely to remain. As an example, inspection engineers will use the information on residual wall thickness to calculate the allowable operating pressure of the pipeline, while corrosion engineers will use the change in wall thickness to estimate the rate of corrosion.

An important current trend is to link the inspection program to risk assessment. This approach is intended to reduce the cost of inspection by reducing the level of general inspection, and at the same time allowing the operator to target the more critical pipelines or parts of pipelines. Though the risk-based approach is easy to discuss in theoretical terms, soundly based application requires a very deep knowledge of the corrosion processes occurring in the pipeline or pipelines.

Risk is defined as the product of the probability of an occurrence multiplied by the consequence of the occurrence. In a corrosion risk assessment, it is necessary to quantify the probability of a corrosion process occurring. Consequences are generally easier to define than probabilities because the consequences are based on costs. Evaluating the probability of an occurrence is complicated and uncertain. This issue is examined further in chapter 16. Statistical modeling may be helpful, but it does not eliminate uncertainty. However, it is more important to make the right decisions than to generate a faultless analysis, and often the decision is not strongly sensitive to the details of modeling.

Operational conditions change with time, and the level of the corrosion hazard also changes. Inspection data are often less useful here than would be expected. Trends taken from inspection data are historical and cannot take into account the effects of changes in operational parameters, the accumulated effects of past history, or the impact of future changes in future operational parameters.

It becomes necessary, therefore, to develop a system that incorporates a pro-active model that can simulate the corrosion processes underway in the pipeline. This allows the effects of changes in operational parameters to be assessed and gives advanced warning of the emergence of critical corrosion conditions.[5] Schematics of a typical feedback model and the relationship of inspection to pipeline condition assessment are given in Figure 15–1. The need for computer-based simulation modeling increases as the complexity of a network of pipelines increases. The sheer size of a pipeline network can overwhelm a conventional approach for prioritizing inspection and monitoring programs.[6]

15.2 Access

Conventional corrosion monitoring techniques require access to the fluids, and, therefore, a point of entry to take samples or insert probes into the pipeline must be created. Access into the pipeline may be through a valve system or by specialized access fittings. Most access systems are designed to allow the insertion and retrieval of the measurement probes without a shutdown of the pipeline. For example, intelligent pigs are introduced into the pipeline through pig launchers and are recovered in pig receivers. The valve systems required to operate the launchers and receivers are complex. Tethered or umbilical inspection devices usually require a shutdown and depressurization of the pipeline.

Valve systems and access fittings for corrosion monitoring are more modest than those required for pigging. Access fittings are installed with the pipeline but may be retrofitted to in-service pipelines by hot tapping. In this procedure, the fitting is welded onto the pipeline and a cutter inserted through the fitting to trepan an insertion hole through the pipe wall. The disc cut in the wall is held and removed by magnets, and the hole is then reamed to size.

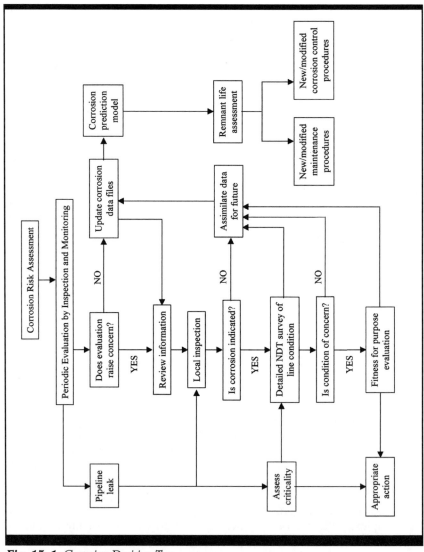

Fig. 15–1 *Corrosion Decision Tree*

The trend has been to develop inspection and monitoring techniques that do not require access into the pipeline and that do not interfere with pipeline operation. Recent developments include thin layer activation (TLA), hydrogen patch probes, vacuum cells, the field signature technique, and continuous ultrasonic monitoring. The classification of some of these techniques as inspection or monitoring is somewhat arbitrary and depends on the purpose for which the technique is being applied.[7] Table 15–1 is a summary of techniques.

Table 15–1 *Inspection and Monitoring Techniques*

Access	Frequency	
	Continuous or frequent	Intermittent or infrequent
Intrusive	Electrochemical monitoring techniques Pressure surge monitoring	Weight-loss monitoring Fluid sampling Intelligent pigging Umbilical inspection tools
Non-intrusive	Thin layer activation Field Signature Method Fixed ultrasonic probes Hydrogen patch probes	Radiography Conventional ultrasonics Automated and time-of-flight ultrasonic systems

Land pipelines can be accessed wherever it is safe and secure. This obviates the need for many of the more sophisticated (and expensive) techniques. Submarine pipelines can be accessed directly only on the platform or at the landfall though TLA, fixed ultrasonic surveys, vacuum cells, and fingerprint techniques may change this restriction. Even though diver access for maintenance will still be required, the feedback to the data processing point is straightforward.

New developments in signal transmission allow information to be sent along the pipeline using low-frequency radio waves. These signals can be used to trigger the activity of probes mounted on the pipeline surface, resulting in a transmission of the readings through the pipeline to a detector at the platform or surface. Probes can measure CP potentials, current densities, wall thickness, hydrogen damage, heat transfer, pressure, and temperature. On land pipelines, the probes can be powered from the CP system, and a similar power source is under investigation for subsea sacrificial systems.

15.3 Inspection Techniques

15.3.1 Radiography

Radiography uses high-energy electromagnetic radiation generated by either an electrical machine, X-rays, or from radioactive sources, gamma rays. Whenever feasible, X-ray radiography is preferred because the inspector has control over both sensitivity and reproducibility. An additional bonus is that the source can be tuned to identify the minimum acceptable defect and not show the non-significant defects.

During installation of the pipeline, the radiation source may be situated on one side of the pipe and the film located on the opposite side. The X-ray photograph is taken through the double wall. This procedure is only used for small diameter, thin-wall pipe. Alternatively, the source may be installed inside the pipeline, and the photographic film wrapped around the outside surface of the pipe. This is the most common procedure and gives vastly improved identification of weld and material defects. The photographs are read as the negatives, and positive images generally are not made. Material defects—cracks, porosity, and lack of fusion—are essentially areas of missing metal where there is less absorption of the radiation. They show up as white areas against the black of the un-developed negative film if the source is correctly matched to the nominal pipeline thickness.

When the pipeline is in service, it is only feasible to locate the radiation source on one side of the pipeline and place a suitable photographic film on the other side. Areas of metal loss absorb less radiation than the uncorroded pipe wall, and the density of conversion of the film reveals this metal loss. Radiography is cumbersome and only used to investigate specific areas. Usually, it is used for bends and manifolds. Modern techniques can use cameras, rather than film. Cameras permit inspection in real time. Very high voltage systems have been developed that allow detection of other integrity factors, such as the degree of closure of a jammed valve.

The advantages of radiographic techniques for evaluation of metal loss in service are speed of execution and provision of a permanent record. Also mastic at field joints, insulation, and pipeline coatings are transparent and may not need to be removed. A disadvantage is that only local areas can be inspected, so it takes a long time to cover a long length of pipe. However, X-ray cameras may overcome this limitation to some extent, for example real time camera radiography is used to investigate corrosion under thermal insulation). The technique only provides three-dimensional (3-D) information on the planar projection of the defect. The location of the defect within the metal is not always clear. Free access is also necessary for the radiation source and film to be placed.

15.3.2 Ultrasound

Ultrasonic (U/S) systems provide 3-D information on the location of defects and are used alone or as an adjunct to radiography. Ultrasound is high-frequency sound in the kHz to MHz frequency range generated by a vibrating piezo-electric crystal located in a probe that is introduced into the metal through a couplant. The sound waves are reflected from areas of change in density, e.g., from the back wall of the pipe or at any flaws inside the metal. Some of the reflected signal is detected, and the time between the injection of the signal and the receipt of the echo is used to determine the thickness of the metal and/or the location of the flaw. The higher the frequency of the sound, the shorter the wavelength and, consequently, the better the definition; however, the wavelength is limited to the average size of the grains of metal, and at shorter wavelengths the noise to signal ratio becomes excessive.

Simple wall thickness measurements are made using D-meters that operate with compression waves and give a wall thickness or depth of metal to flaw. The sophisticated manual ultrasonic systems employ cathode ray tubes that portray the signal and reflected echoes. Localized U/S surveys are rapid when access is available, and most techniques are very flexible. The 3-D information allows critical engineering assessments of defects to be made. However for large areas, the manual techniques are slow and labor-intensive, and automated techniques are increasingly used.

Automated D-meters can rapidly scan the pipeline surface, and the data can be logged on a computer and represented as a color or color density plot of the metal thickness. The technique can cover 3 m^2/hr with a high level of discrimination, representing relatively good value for the money as data points per dollar. The graphics allow identification of patterns of corrosion that might be missed if manual inspection were used, e.g., corrosion associated with flow patterns.

External inspection of operational pipelines may be carried out by U/S examination. This is usually done at a field joint because mastic is relatively easier to remove than reinforced concrete weight coating, and there is also a general concern about weldment areas. Once exposed and cleaned, the U/S probe, mounted on a frame clamped to the pipe, is automatically moved around the pipe surface. Subsea devices are available that can fit into

a 50-mm (2-in.) gap between the pipe steel surface and the seabed at the areas where the field joint has been removed. However, the automated compression wave systems cannot detect corrosion at the root of welds; angled probes are needed to direct the signal around the weldment.

The time of flight method uses angled U/S emitter and receiver probes that are moved in parallel across the pipe surface. The sound beam is refracted around the internal flaws, and the echoes are analyzed by computer to generate a tomographic pattern of the internal surface or volume of the scanned section. The technique has been used for the inspection of hyperbaric repair welds on pipelines and risers at rates up to 1 m/minute. This technique has the potential to replace radiography as a method of evaluation of production welds and, because the digital information can be machine analyzed, the impact of a defect on the fitness for purpose of a weld could be decided in real time. This technique would allow an engineering critical assessment approach, which would reduce the number of production welds that must be repaired.

U/S systems have also been developed to identify hydrogen damage described in chapter 7. Shear and compression waves are introduced into the steel, and the change in either sound velocity or attenuation between the two waves is used to identify the extent of micro-cracking and, hence, the extent of hydrogen damage. These techniques have been extensively applied from the outside of the pipes. Recently, they have been included in intelligence pigs for identification of internal stress corrosion cracking. Unfortunately, the pipelines at most risk of cracking are gas pipelines. Because the technique requires a couplant between the sensor and pipeline wall, the pig must be sent through the pipeline within a slug of liquid, or the pipeline must be flooded with a liquid.

15.3.3 Magnetic particle inspection (MPI)

MPI is used to detect surface-emergent cracks. It can be applied to steel underwater and is regularly used for inspection for fatigue cracks on the structural nodes on jackets. The pipe surface is magnetized using electrical or permanent magnets. The magnetic field is distorted at the cracks and other surface-emergent defects that are not parallel to the magnetic flux. A fluid containing ferromagnetic (usually iron) particles is painted onto the metal surface or blown gently over the surface. The ferromagnetic particles concentrate at the areas of magnetic flux distortion, thereby identifying the cracks. The ferromagnetic particles may be encapsulated in a plastic coating that glows under UV light. This technique is used to identify lamination in the weld bevels and to detect fatigue cracks and other external cracking of pipelines in service.

The main advantage of the MPI technique is the detection of fine cracks not visible to the eye. However, it requires a clean surface and is relatively slow and messy. If it is used to detect laminations, the pipeline may need to be de-magnetized afterwards to prevent the residual field from deflecting the welding arc when the girth weld is made.

15.4 Pigs

15.4.1 Introduction

The earliest pigs were crude devices such as bales of straw tied up with wire, driven along pipelines to clean them. There is now a well-established and sophisticated technology of pigging. Many kinds of pig are in use to clean pipelines of construction debris, wax, asphaltenes and corrosion products, and to separate batches of different fluids. Figure 15–2 below illustrates some of the types available.

This section of the chapter is concerned with pigs used to monitor the condition of pipelines. These pigs are termed *intelligent* or *intelligence pigs* or *tools*. It is often much more convenient to make measurements from the inside of a pipeline than to try to make the same measurements from the outside. External access is needed only at the ends of a pipeline section, and the same inspection systems can be used offshore and onshore. Inspection pigging is a rapidly developing technology and has attracted large investments. It seems likely that more and more conditioning monitoring will be performed by intelligent pigs.

15.4.2 Magnetic flux leakage (MFL) inspection pigs

MFL was the first technique integrated into inspection pigs for complete pipeline inspection. MFL pigs house electric or permanent magnets, a ring of detectors, a recording unit, and a power supply. The pig travels through the pipeline at a speed of 1–5 m/s, driven by the fluid pressure. The magnets must be sufficiently powerful to fully magnetize, or saturate, the pipe wall. At areas of metal loss, there is a distortion of the magnetic field. The detectors pick up the strength and rate of change of distortion of the field and pass the data to the recording section. When the pig is recovered, the data are downloaded for analysis. The pig can distinguish between internal and external defects by measurement of residual magnetism, either by a repeat run with the magnets off or removed or by use of a second array of sensors placed after the magnets.

There are several suppliers of MFL pigs. Pigs are classified according to sensitivity and resolution and, from time to time, a beauty competition is run to classify the pigs. All pigs will identify with confidence metal loss above 30% of wall thickness. Low-resolution pigs cost the least and will identify 100 mm^2 defects. High-resolution pigs identify smaller areas of metal loss (~ 5 mm^2), but the improvement in sensitivity to loss of wall thickness is less marked, typically starting at 10–20% of wall thickness.

The geometry of the defect can affect the sensitivity of the detectors. Sharp-edged defects give a pronounced magnetic distortion that can be read by the pig as a deeper metal loss than actually exists. Soft-edged defects

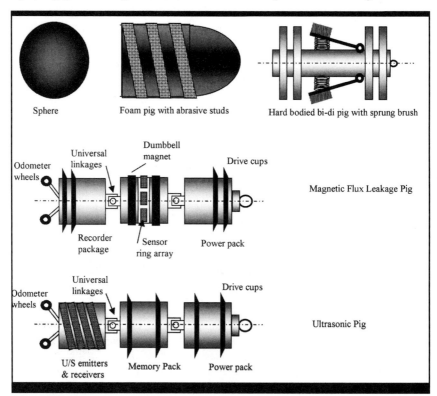

Fig. 15–2 Pigs

cause less distortion of the magnetic field, and the sensors can underestimate the depth of those defects. It is usually necessary to quantify the defects using external U/S examination of representative defects.

MFL pigs can be used in all types of pipelines because a couplant between sensors and pipe wall is not required. Though the pigs can operate in dirty pipelines with only small loss of sensitivity, internal scale must be removed before pigging. MFL pigs cannot detect cracking, hydrogen blistering, or external fatigue cracking and are limited in the detection of longitudinal flaws.

15.4.3 U/S inspection pigs

Intelligent pigs incorporating U/S inspection systems were developed in the 1980s. The pig carries an array of U/S probes that emit high frequency sound pulses and use the time delay of the echoes from the inside and outside of the pipe wall to evaluate the wall thickness. U/S pigs are more expensive to operate than high resolution MFL pigs, but some people claim that they are more sensitive, able to detect 10% loss of wall thickness and upward in increments of 1–2%; metal loss resolution is about 5 mm^2.

As with MFL pigs, the U/S pig gives information on the complete length of the pipeline with minimum interference with pipeline operations. A couplant is required between the sensors and the pipe wall; consequently, the pigs cannot be used in gas pipelines without either a liquid slug around the pig or the pipeline flooded with a liquid. U/S pigs require a higher level of cleanliness compared to MFL pigs and may not detect pits that are filled with impacted debris. The conventional U/S pig will not detect stress corrosion cracks but can detect severe cases of hydrogen blistering. New pigs are available that use angled sensors, and these are able to detect circumferential cracking but cannot as yet identify longitudinal cracks.

15.4.4 Eddy current (E/C) inspection pigs

E/Cs are induced to flow in the pipe wall by oscillating electrical currents generated in a probe head. Surface-emergent defects alter the induced currents, and a detector measures the changes. E/C probes are widely used for inspection of very small diameter tubing in heat exchangers and process pipe work, and the system has been adapted for use in small-diameter pipelines.

The pigs are sensitive to travel speed and are less sensitive than U/S pigs; sensitivity is 20–30% of wall thickness. The technique does detect transverse cracks. Power consumption is high and the pigs are only suitable for short length pipelines or in tethered/umbilical units.

15.4.5 Tethered pigs

Tethered pigs are an alternative to free running pigs that are driven through the pipeline by the fluid pressure. The measurement head is attached to an umbilical cable, which provides power and signal transmission and pig recovery. The pig is launched into the pipeline and crawls along the pipeline using a hydraulic motor. Once at the maximum distance, the pig is withdrawn by retracting the umbilical at a controlled rate. Measurements may be made during either the entry or the withdrawal of the pig, though ultrasonic measurements are usually made during the withdrawal. Typical measurement heads include video cameras, ultrasonic scanners, and E/C detectors.

A typical tethered pig can inspect 1–2 km of pipeline. A tethered pig is used when it is not practicable to use an intelligent, driven pig; for example in small diameter pipelines with tight riser bends, in heavy-walled pipelines that cannot be fully saturated by MFL pig magnets, and in short pipelines. In all cases there must be some method of safely launching the pig. This generally means that production pipelines must be de-pressurized and de-gassed. For video, the pipeline needs to be empty, though water injection pipelines can be investigated when full if they are reasonably clean internally. For ultrasonic inspection, there must be a couplant present; and the pipeline must be filled with liquid or the pig introduced within a liquid slug.

15.4.6 Caliper and geometry pigs

A pipeline engineer needs to know about any deformations present and the final pipeline configuration. He needs to know if the line has been dented, if it has become ovalized, if it has moved unexpectedly (as in upheaval buckling), and if it is in a span. He will often want to check that there are no dents and blockages before he runs one of the intelligent pigs described previously so that he can be sure that it will not get stuck at a bend.

The simplest kind of measurement uses a gauging plate, a soft aluminum plate a little smaller than the internal diameter of the pipeline, is mounted on the front end of a hard pig. (See Figure 15–2.) If the pig runs past a dent, the edge of the gauging plate bends back when it comes out at the downstream end, and visual inspection then gives the depth of the largest dent.

More refined measurements are made by a caliper pig, which carries a number of radial feelers. A dent deflects the feelers, and their deflections at different points around the circumference are recorded as a measure of the cross section. The feelers pick up internal weld beads and, counting welds, fix the position of the pig when it detects diameter anomalies.

Another possibility is to record the position of the pig by using inertial navigation. An inertial navigation system measures the acceleration of the pig by accelerometers and gyroscopes, integrates the acceleration with respect to time to get the velocity, and integrates the velocity to get the position. Military requirements for aircraft and ballistic missiles have brought inertial navigation to an extraordinary level of refinement. Pigs carrying inertial navigation systems can produce a precise three-dimensional configuration of a complete pipeline and can detect movements of the order of 100 mm.

15.5 Corrosion Monitoring: Intrusive Techniques

15.5.1 Pressure surge monitoring

A pipeline acts in a similar manner as an organ pipe. Pressure surges that arise during operation run back and forth along the pipeline and eventually attenuate. Sensitive pressure-monitoring devices are installed at both ends of the pipeline, and the pattern of the pressure surges is recorded on a local personal computer (PC). Over a period of time, a continuous record of the pressure patterns arising during normal operations is established. Should a leak occur, the pressure patterns change, and the PC program identifies the change.

The technique is low cost, and claims are that the technique can detect integrity losses of ~0.1%. By analyzing the changes in the pressure surge pattern, the location of the leak can be calculated to within 5–10 m.

15.5.2 Weight loss coupons

Weight loss coupons are used for evaluating corrosiveness and inhibitor performance over the long term because the coupons are relatively insensitive. Increasingly, weight loss is being augmented and/or replaced with electrical resistance and electrochemical monitoring methods.

A small sample of cleaned and pre-weighed steel is exposed for a defined time period in the pipeline. It is then removed, and the change in weight noted.[8, 9] Corrosion rates are generally calculated based on the whole sample area, which may give optimistic values. Calculation of the corrosion rate over the corroded area of the coupon will give a more realistic rate. Weight loss coupons are either small rectangular plates that project into the fluid streams in the pipeline or are small discs (penny washers) that are fitted flush with the pipe. (See Figure 15–3.) Flush coupons ensure that the conditions are as representative as possible of the conditions at the pipe wall and also avoid risk of damage during pigging. When retrieved, the coupons can be scraped to remove corrosion product for chemical and microbiological analyses and, after cleaning, examined for pitting.[10]

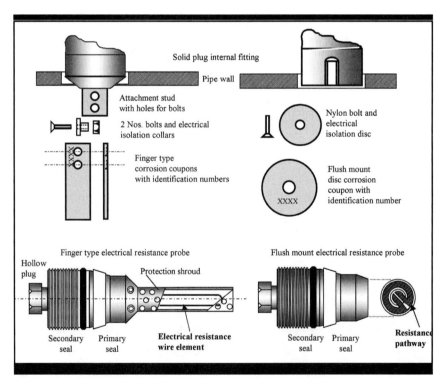

Fig. 15–3 Weight Loss and Electrical Resistance Probes

Slightly radioactive samples are now available, and the loss of weight is determined by loss of radioactivity while the samples are in-situ or after their removal. Very low corrosion rates can be measured using this technique.

Weight loss coupons are simple and provide a *hard copy*, giving visual evidence of pitting and other corrosion morphology. The coupons can be used in any system, though corrosion would only be expected when free or active water is present. The coupons must, therefore, be placed at the bottom of pipelines and in areas where water is likely to collect. Coupons are relatively insensitive and do not provide kinetic information, though a sequence of weight loss data can be used to give some indication of kinetics. Almost all pipeline operators provide corrosion monitoring access fittings for insertion of weight loss coupons.

15.5.3 Electrical resistance (ER) probes

An electrically isolated wire, tube, or plate is exposed to the fluid (Fig. 15–3). A fixed electrical voltage is imposed across the metal, and the current flow in the wire or tube is measured. As the tube or plate corrodes, it thins. The electrical resistance increases and, consequently, the current flow reduces. If the wire in the probe is thin then the sensitivity will be high, but the probe will have a short life. Thick wires, tubes, and plates will last longer but are less sensitive.

ER probes are the second most popular monitoring technique and can be used in a wide range of environments, including high resistance systems, e.g., in gas pipelines. A new type of probe includes a cooling element so that water condensation from the gas phase can be induced. This probe allows measurement to be made on the platform where the gas is hot while simulating the subsea pipeline conditions where water is condensing. ER probes are simple and reliable and reasonably sensitive in all environments. The probes are sufficiently sensitive to provide kinetic information. Pitting is generally not determined.

15.5.4 Linear polarization resistance measurement (LPRM) probes

With the LPRM technique, a probe comprising two small, electrically wired steel fingers is exposed to the environment, and a low frequency AC voltage of a few millivolts (⁄ 20 mV) is imposed on the fingers. The current response is measured, and the instrument converts the current into a nominal corrosion rate, using a resistance factor input by the user (Ohm's Law is assumed).[11] The technique is sensitive to the resistivity of the environment and is good in water systems and wet crude systems but less useful in gas systems. LPRM is the third most common technique for use on pipelines, though it is generally used as a special technique when evaluating the effectiveness of corrosion inhibitors.

The LPRM technique is relatively simple and sensitive and gives kinetic information. Visual and weight loss information is also available by removal and examination of the fingers. Short-term and periodic operation of the technique does not affect the electrode surface so the probes can be used for long-term monitoring.

15.5.5 Hydrogen probes

Hydrogen probes are used to estimate corrosion in sour systems. Corrosion reactions on the steel probe surface generate atomic hydrogen; some atomic hydrogen migrates through the steel into a void within the probe where it combines to form hydrogen gas that cannot escape. The void is formed by drilling a blind hole into a steel rod or as an annular space between a close fitting outer tube over an inner rod. Over time, the pressure in the void builds up and is measured and related to the corrosion rate. Periodically the pressure must be released.

15.5.6 Potentiodynamic polarization

Using the potentiodynamic polarization method, a probe containing several small samples of pipeline material and a reference electrode is introduced into the pipeline. The electrochemical potential of one of the metal samples is altered continuously or in a series of steps using a potentiostat.[12] The potentiostat is an electronic device that maintains the potential of the metal sample at a prescribed value compared to a reference electrode. It does this by increasing or reducing the electrical current input into the electrolyte by a third, auxiliary electrode. As the potential is altered, the current response is measured and the potential (E) and current (i) plotted as an E-log i plot.[13]

The potentiodynamic polarization technique is widely used in laboratory studies of material corrosion mechanisms and corrosion rates and provides mechanistic information. It detects the propensity to pitting of the material, and gives information on corrosion mechanisms and the mode of operation of corrosion inhibitors. It is a complex technique requiring considerable skill to use and interpret results.[14] It is relatively slow if quality data are required. However, new computer-directed systems have simplified the technique, and it is likely that automated systems will become available for regular use on difficult and critical systems.

15.5.7 Advanced electrochemical techniques

15.5.7.1 Impedance spectroscopy. With this technique, a probe containing steel electrodes, similar to those used with the LPRM technique, is used. A millivolt (< 50 mV) AC voltage difference is imposed between the electrodes. The frequency of the AC voltage is altered in steps from a very high frequency (kHz) to a very low frequency (mHz). A corrosion process on the electrodes generates a double layer of charged ions that behaves rather like a leaky capacitor, showing both impedance and resistance. The current response and the phase angle shift between voltage and current are recorded.

Computer analysis is used to convert the data sets into Bode plots, so called *snail curves*, which record the change in relationship between impedance and resistance as the frequency varies. The intercepts of the curve with the resistance axis give the solution resistance and the corrosion rate. The overall shape of the plot is characteristic of the electrochemical process occurring on the metal surface. For example, a diffusion-controlled process is identified by a straight-line relationship between impedance and resistance at an angle of 45° at low frequencies.

The technique provides kinetic and mechanistic information. The low voltages do not affect the electrode surface, and so the probe can be used to give semi-continuous information. It can be used in high-resistance conditions, for instance to measure corrosion of steel in reinforced concrete or in an oil-water emulsion. It is a complex technique using expensive equipment, and the interpretation of the results is sometimes questionable. It is slow if good data are required, and, consequently, is rarely used except for specialist studies. Limited cross-correlation data is available to compare the technique with the conventional techniques.

15.5.7.2 Electrochemical potential and current noise. To employ this technique, LPRM-type probes are used connected to sensitive voltmeters that measure the electrochemical potential of the steel fingers exposed to the corrosive environment. The corrosion processes induce small voltage oscillations (micro volts) between the fingers, and these are amplified and digitized. Over a period of time, typically from 1 to 5 minutes, the data are analyzed; and the voltage-frequency spectrum derived. The spectrum of the signals is similar to pink noise and has a characteristic relationship to the dominant corrosion process. For current noise monitoring, two samples are used wired through a zero resistance ammeter, and the current signals are converted into voltages that receive similar analytical treatment to potential noise. Measuring both the

potential and current noise simultaneously gives more information on the chemical processes and a measure of the corrosion rate.

This technique is rapid and gives information on kinetics and pitting propensity of a material. At present, the technique is expensive and subject to many errors from external signals because of the small voltages. To date the technique has a limited track record, but it shows promise for identification of active-passive transformations, e.g., risk of chloride induced pitting of duplex and austenitic stainless steels and pitting of steel in sulphide environments.

15.5.7.3   Zero-resistance ammetry (ZRA).   Initially developed for studying galvanic corrosion, the ZRA method uses an LPRM probe. The metal fingers are connected together through a device that compensates electronically for the resistance of the current measuring circuit, so that the samples behave as if short-circuited together. Modern circuitry has allowed the technique to be used to study single metal corrosion processes. The technique is often used for measuring corrosion in water injection systems.

ZRA is continuous and simple. It is a well-established technique that is presently enjoying a revival with modern electronic circuitry. The technique uses simple standard probes. The main disadvantage is that the technique is limited to measurement in conductive fluids only, as is LPRM.

15.5.7.4   Potentiostatic polarization probe.   The probe includes two steel electrodes and a reference electrode. The steel electrodes are connected to an electronic device that maintains one of these electrodes in a slightly more corrosive state than would normally occur in the pipeline. The current response of the sample is continuously measured. The more active potential makes the sample more likely to corrode if adverse conditions occur so that the probe gives advance warning of changes in the environment. The technique is useful for studying pitting propensity on materials that suffer active/passive pitting, for example duplex and austenitic stainless steels. The sample will detect the onset of pitting before it is measurable on the pipeline surface. It is continuous and provides indications of corrosion kinetics and pitting.

15.5.7.5   Galvanostatic polarization.   The probe used in the galvanostatic polarization technique contains two metal electrodes connected to a device that passes a fixed current between the electrodes. The changes in voltage between the electrodes are recorded. The technique is widely used for studying galvanic effects, the effect of inhibitors, and the behavior of sacrificial anodes. Galvanostatic polarization can be used for continuous monitoring and is simple and useful for testing oxygen contamination in water injection systems. It is, however, limited to use in conductive fluids and has been largely superseded by simpler galvanic probes.

15.5.8 Sand monitoring

One form of sand-monitoring probe is an ER probe fabricated from a corrosion-resistant alloy. As erosion wears away the CRA, the resistance alters and is monitored periodically. The thickness of the resistance element, usually formed as a spiral, determines the sensitivity and the life expectancy. The technique is simple and reliable, but the probe must be located in an area where erosion is expected to occur. The erosion of the corrosion resistant material has to be correlated to the erosion of the carbon steel pipeline.

A second type of probe employed by this technique uses loss of gas pressure as an indicator. A hollow tube fabricated from corrosion-resistant alloy is gas-filled and sealed. The probe is inserted into the fluid stream at a location where erosion is expected. When erosion wears through the wall thickness of the tube, the gas pressure is lost and an electrical signal transmitted. The technique is simple and reliable but gives a single value point only. The wall thickness determines the sensitivity and the life, and there is a need to balance them.

The presence of sand in a pipeline can be detected using a conventional U/S probe in the passive mode. Sand striking the probe generates a signal with a characteristic frequency-amplitude distribution; the signal is evaluated using filter circuits in the electronic package. The main disadvantage of this system is the extensive back-wiring required, and there is also a limit on the lower detection level that can be identified.

15.5.9 Installation of intrusive monitoring probes

Weight loss coupons and probes are installed in pipelines through corrosion monitoring access fittings as illustrated in Figure 15–4. The on-line use of these fittings is described later in this chapter. An access fitting is a heavy-schedule machined metal tube that is attached to the pipeline by welding. (Flanged fittings are also available.) A 45 mm hole is trepanned through the base of the fitting to allow access into the pipe. The fitting has internal and external threads and is sealed by an internal threaded plug that holds the weight loss coupons or corrosion probes. A steel outer cap is fitted to provide a secondary seal and to prevent damage to the external threads.

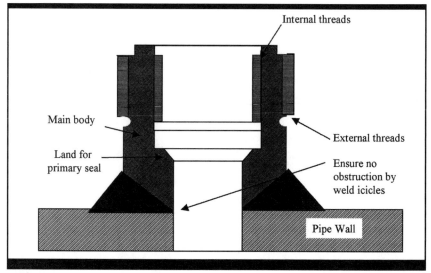

Fig. 15–4 *Corrosion Monitoring Access Fitting*

For installation, the fittings are usually fabricated into spool pieces that are installed in suitable locations in the topside section of the pipeline or associated pipe work. Care should be taken that the spools are installed so that the access fittings do not face out to sea or toward other pipes or structural members, as in these cases either scaffolding will be required for access to the fittings or access will not be possible. Access fittings may also be installed into existing pipe by welding and then trepanning though the pipe wall, though it is usually necessary to shut in the pipeline to permit welding.

It is generally good practice to install the corrosion monitoring fittings at the bottom of the pipeline where the highest rate of corrosion is expected. Corrosion injection (CI) fittings are similar to the access fittings but include a side tee fabricated in a corrosion resistant alloy. The inhibitor is introduced through an isolation and non-return valve into the access fitting and from there through an injection quill or atomizer into the process stream. CI fittings are usually installed on the top of the line.

So that a comparison can be made between uninhibited and inhibited corrosion rates, corrosion monitoring access fittings should be installed both upstream and downstream of the CI fitting. The downstream fittings need to be from 5 to 10 pipe diameters downstream of the CI fitting to give enough time for the inhibitor to disperse into the process fluids. The cost of installation of access fittings during pipe installation is low, and it is prudent to install them in pairs in order to provide an additional monitoring point to apply a second technique or a fallback in case a fitting is damaged.

Only competent, accredited contractors should be allowed to install the corrosion monitoring fittings. After installation the fitting is sealed using a solid plug. The seal between the solid plug and the corrosion-monitoring fitting is achieved using a soft, large face gasket and a secondary O-ring seal. The fine threads of the plug are not the sealing element, and fluorocarbon tape is not required to establish a seal. It is most important that the solid plugs in the fittings are not over tightened. Experience indicates that about 50% of access fittings are irretrievably damaged by over tightening of the sealing plugs into the access fittings before they enter service. With new installations, only solid internal plugs, as used for corrosion weight loss coupons, should be fitted because they can be changed out for ER and LPRM probes by the corrosion monitoring teams when the system is ready to commission. External plastic caps should be fitted to protect the external threads on the access fittings. A plastic cap will fail during the hydrotest if the installation procedure of the solid plug or access fitting is faulty. The plastic caps will be replaced with steel or alloy caps when the fittings are commissioned on the operational pipeline.

On-line retrieval and replacement of weight loss coupons and probes is illustrated in Figure 15–5. A combination of isolation valve and retriever is used in the process. The retriever is a dual pressure vessel: an outer tube/shell contains the process pressure while an inner tube can be manipulated to unscrew the solid plug or probe from the access fitting. Three types of retriever are available: simple double chamber, telescopic mechanical, and hydraulic. The most widely used (and less problematical) is the double chamber type. Retrievers can be used up to 240 Bars (3,500 psia). With gas plant, however, it is usual to arrange retrievals when pressures are below 140 Bars (2,000 psia) or when the system is shut in and de-pressurized. The length of the retriever depends on the overall length of isolation valve, internal corrosion fitting, and, depending on the location of the corrosion-monitoring device, on the diameter of the pipe. If bottom-of-line access fittings are installed then the length of the retriever is independent of the pipe diameter. This is a major advantage when both several diameters of pipe and large diameter pipelines must be surveyed.

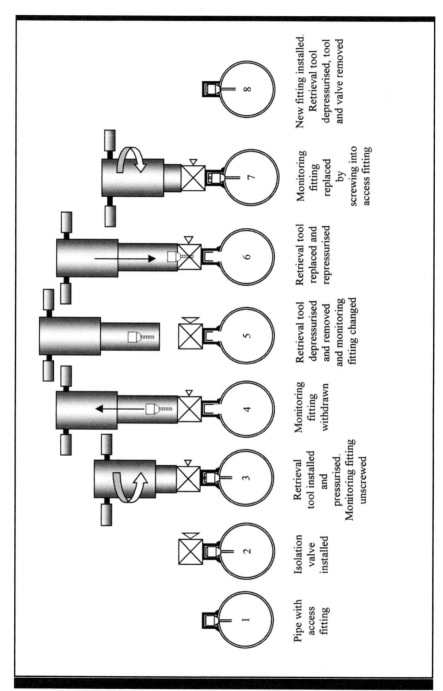

Fig. 15–5 *Installation and Retrieval of Monitoring Devices*

15.6 Corrosion Monitoring: Non-Intrusive Techniques

15.6.1 Hydrogen patch probes

Hydrogen patch probes are used to measure the corrosion in sour systems. A cylindrical cell is clamped onto the outside of the pipeline and presses a palladium foil against the pipeline surface. The cylinder is filled with an alkaline solution. Atomic hydrogen generated by the internal corrosion processes passes through the palladium foil and is reduced at the inner face of the foil in the alkaline solution. The current required to reduce the hydrogen is measured and can be related to the corrosion process within the pipe.

The hydrogen reduction technique is semi-continuous and reasonably sensitive. It can be used only for sour systems and is an indirect technique and requires regular maintenance of the alkaline solution.

15.6.2 TLA—irradiated spools

Radiation is extensively used for evaluation of wear in rotating equipment. Thin layer activation (TLA) is an extension of this technique. The internal surface of a spool of pipe is irradiated with a very small level of radioactivity. The spool is welded into the pipeline. Residual radiation of the spool is measured periodically, and the loss equated to metal loss, making allowance for the natural decay of radiation. At least two North Sea submarine pipelines have been fitted with diver accessed TLA spools, and the technique has been proven to work well in practice. Sensitivity is very good, though the system has a relatively short service life because the depth of irradiation is very limited. The system will also detect erosion from solids or sand.

The TLA-irradiated spools technique is simple and extremely sensitive. It does not require access into the pipeline. However, adherent corrosion products and radioactive scale deposition would interfere. It is expensive and may require certification. The technique is essentially a large, working, weight loss coupon; and no mechanistic information is obtained.

15.6.3 Fixed U/S probes

The nuclear industry has developed long life U/S probes that can be welded onto vessels or pipelines. The resolution of these probes is far better than hand-held probes, and they typically have a resolution to 0.1 mm. The probes are wired to a control center for continuous monitoring or can be interrogated periodically with diver or ROV operated meters. These probes have been installed on one North Sea submarine pipeline.

Fixed U/S probes are simple and have good sensitivity; however, the measurements made are sensitive to temperature. The method does not require access into the pipeline and correlates with conventional inspection. The probes are expensive and perhaps vulnerable to damage as they protrude 3–4 in. from the pipe. When installed within a protective package, the overall assembly is very bulky.

15.6.4 Vacuum shields

Vacuum shields are a type of hydrogen permeation probe. A plate is glued or welded to the bottom of the pipe and fitted with a vacuum gauge. The annulus between the plate and the working pipe is evacuated. Atomic hydrogen generated by internal corrosion migrates through the pipe wall and into the annulus, thus reducing the vacuum. The instrument gauge is calibrated to give an approximate corrosion rate.

The technique is very simple, non-intrusive, and essentially safe. Claims are that it is sensitive and has a potential use for measuring corrosion in sour service. The large calibration factor needed to relate the hydrogen permeation rate to the corrosion rate would prevent any accurate relationship between being established without independent inspection and monitoring data. There is risk of false readings if HIC is present, and experience shows that the vacuum needs frequent refreshing for adequate sensitivity.

A recent development uses a semiconductor that is very sensitive to hydrogen contamination. Air is sucked off the surface of the pipe at a controlled rate, and the concentration of hydrogen measured.

15.6.5 Field signature monitoring (FSM) technique

This is a variation of the electrical resistance method. Studs are brazed onto a spool piece at regular intervals. The studs are wired back to a multiplexed voltmeter. When a measurement is to be made, a large current is injected into the pipe upstream and downstream of the stud array. The voltage between each set of studs in the array is recorded and converted to a metal thickness measurement.

Once installed, the technique purports to give a rapid and accurate evaluation of the condition of a working section of the pipeline. At present the pipe sections are expensive. The technique needs a high current input for accuracy. The multiple leads and current input requirement limit the location of the device. The location needs to be very carefully considered to gain valuable results.

15.6.6 U/S non-intrusive sand monitor

The presence of sand in the flow stream in a pipeline can be detected using ultrasound arrays. These may be conventional U/S probes used in the passive mode. The sound generated by sand impingement has a characteristic frequency-amplitude distribution that can be detected and evaluated using filter circuits in the electronic detection equipment. The technique is claimed to be reliable although there is also a limit on lower level of detection. The probes can also be used in the active mode to detect metal loss, including corrosion. The probes form a protruding array, and the extensive back-wiring required may limit the locations available.

15.7 Fluid Sampling

Considerable information about the condition of a pipeline can be gleaned from analysis of the transported fluids. Indeed, it is prudent to review the fluids on a regular basis to determine if there have been any changes in chemical or operational parameters that may change the corrosion rate or invalidate the selection of corrosion inhibitor. It is always necessary to review the changes in fluid composition in conjunction with the operating conditions of the pipeline. A pipeline engineer should review the parameters in listed in Table 15–2.

Table 15–2 *Operating Parameter Linked to Corrosion*

Chemical Parameters	Physical Parameters
Oil gravity	Inlet pressure
Gas to Oil Ratio (GOR)	Outlet pressure
Basic salt & water (BS&W)	Inlet temperature
Carbon dioxide	Outlet temperature
Hydrogen sulphide	Flowrate of oil
Microbiological count	Flowrate of gas
Iron and manganese counts	

The major parameters can be used in the corrosion rate prediction equations to determine if the corrosion rate is likely to have increased. For example, an increase in carbon dioxide or operating pressure would be expected to increase the corrosion rate; an increase in water cut combined with a reduction in the oil flow rate could result in the initiation of corrosion in an otherwise non-corroding pipeline. If the changes are major, it may be necessary to run a hydraulic analysis of the pipeline to identify the nature of the flow regime.

Microbiological counts for SRB are only required if sulphate is present in the water. In water injection pipelines, microbiological counts are usually done every week. In production pipelines, the iron and manganese concentration can be used in conjunction with the water cut to evaluate the overall corrosion of the pipeline. It is also usual to test water removed during pigging operations.

References

1   ASTM G15: "Definitions of Terms Relating to Corrosion and
    Corrosion Testing." American Society for Testing and Materials.

2   NACE Recommended Practice 0175: "Control of Internal Corrosion
    in Steel Pipelines and Piping Systems." National Association of
    Corrosion Engineers.

3   ASTM G16: "Applying Statistics to Analysis of Corrosion Data."
    American Society for Testing and Materials.

4   Horner, R. A. (1966) "The Technical Integrity Management of Sour
    Service Ageing Facilities." *UK Corrosion.* London.

4   Commercial Programmes including Electrical Corrosion Engineer,
    Intetech, Waverton, UK; Corrosion Watch, Corrosion Watch Inc.,
    Calgary, Canada; Predict, InterCorr, Houston.

5   King, R. A. (1988) "Production Flowline Analysis as an Aid to
    Corrosion Management of Networks." *Advances in Pipeline Technology.*
    IBC Gulf. Dubai.

6   BP Exploration, Costs of Corrosion 1990 to 1992, Marine Offshore
    Management Ltd., Aberdeen, 1993.

7   ASTM G1: "Preparing, Cleaning and Evaluating Corrosion Test
    Specimens." American Society for Testing and Materials.

8   NACE Recommended Practice 0775: "Preparation and Installation of
    Corrosion Coupons and Interpretation of Test Data in Oil Production
    Practice." National Association of Corrosion Engineers.

9   ASTM G46: "Examination and Evaluation of Pitting Corrosion."
    National Association of Corrosion Engineers.

10 ASTM G59: "Potentiodynamic Polarisation Resistance
    Measurements." National Association of Corrosion Engineers.

11 ASTM G5: "Standard Reference Method for Making Potentiostatic
    and Potentiodynamic Anodic Polarisation Measurements." National
    Association of Corrosion Engineers.

12 ASTM G3: "Conventions Applicable to Electrochemical Measurements in Corrosion Testing." National Association of Corrosion Engineers.

13 ASTM G102: "Calculation of Corrosion Rates and Related Information from Electrochemical Measurements." National Association of Corrosion Engineers.

16 Risk, Accidents, and Repair

16.1 Introduction

Pipelines have an excellent record for safety and reliability, but failures do occur from time to time. Sometimes they lead to leaks, and sometimes to serviceability failures such as blockages that restrict or close off flow. Occasionally, accidents are more serious. In the United States and Russia, there have been land pipeline accidents that have led to loss of life, and offshore accidents in the UK and United States that have resulted in fatalities, pollution and the associated clean-up and repair costs, plus the loss of production and operator embarrassment.

The most instructive approach to minimizing risk is to examine specific incidents and to try to learn lessons from them. Incidents and lessons are described in in section 16.2. They can be supplemented by reliability theory methods, which are being widely applied to rationalize codes and design practice, to gain a deeper understanding of the factors that govern reliability, and to provide a basis for the incorporation of new research. Section 16.3 introduces reliability theory concepts. The ultimate task is to reduce the risk of failure, and this is explored in section 16.4. Section 16.5 discusses repair.

16.2 Failure Incidents

16.2.1 Introduction

The incidence of failure varies through the life of a system and follows the pattern illustrated in Figure 16–1, sometimes called the *bathtub* curve. It plots the incidence of failure, the number of failures per unit time, against the age of the system.

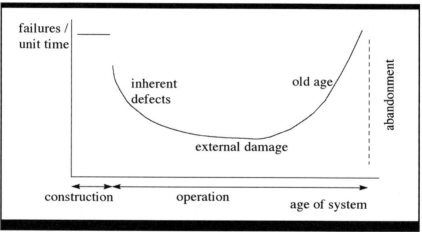

Fig. 16–1 *Age Dependence of Incidence of Failure*

In this form, the diagram does not distinguish between different kinds of incidents nor represent their severity; but it is obviously possible to redraw the diagram, perhaps with a third axis that represents the severity of the incident.

There is a particularly high risk of failure during construction. Once the system has been commissioned, the first year or two of operation reveal inherent defects in the design or in construction; and the number of failures per unit time is relatively high. Later, the incidence of failure declines, and failures are generally associated with external damage coming from such things as anchors, dropped objects, unusually severe storms, earthquakes, and sinking ships. Ultimately old age sets in, and the number of failures increases due to time-related factors such as corrosion and fatigue. Repairs and modifications are themselves sources of damage incidents.

This pattern is common to mechanical systems (cars, dishwashers, space stations) and to biological systems (ourselves, cats, houseplants). As in those examples, the number of failures linked to inherent defects can be reduced by inspection and treatment during operation and by better design and better control of manufacture and installation. The number of failures associated with external factors can be reduced by better protection and by care in operation. The right-hand old-age part of the curve can be shifted to the right by inspection, care in operation, and preventive maintenance.

Some examples are listed below. Several of the incidents are well known and have been described in published reports and papers. It is a good practice to record what happened so that the lessons learned are not forgotten and so that the industry and the community benefit lessons learned. In other instances, the pipelines and the operators are deliberately left unidentified. There is no intention to imply any kind of criticism or to impute fault in any of the cases quoted.

Unfortunately, some operators report as little as possible, often out of a justifiable fear of ill-informed public criticism and legal challenge. The knowledge is then lost to the broader pipeline community; and often it is expensively rediscovered, occasionally by the same people that lost it the first time.

A few of the summaries are composites that represent more than one incident.

16.2.2 Incidents during construction

16.2.2.1 During the construction of an oil pipeline in the North Sea, girth weld X-rays revealed white lines parallel to the pipe axis on either side of some of the welds. At first this was attributed to a fault in the radiography system, but later it was realized that the lines were longitudinal crack defects, which had been present in the pipe as supplied and were related to rolling defects in the plates from which the pipes were made. As a result, it was decided to discard 42 km of pipeline. This incident emphasizes the need for inspection and for thorough investigation when unexpected findings occur. It often happens that unwelcome findings are dismissed as a fault in the detection system.

16.2.2.2 The dynamic positioning system of a lay barge malfunctioned. The barge moved sternward toward the pipe, tension was reduced, and the sagbend increased in curvature until it buckled.

Several incidents of this kind have occurred. They show that quite small uncontrolled movements of a lay vessel can cause the pipeline to buckle, and confirm the value of redundancy in the control system.

16.2.2.3 A pipeline was being pulled along a trench across a mud tidal flat in Iraq. The pipe was made up on shore in 1130 m (3700 ft) strings, and pulled seaward along a flotation ditch by an anchored pull barge, each string being welded to the trailing end of the preceding string. The first failure occurred as the seven strings were being pulled. The advancing tide banked up against the southern side of the line, which acted as a dam over a length of some 3000 m (10,000 ft). The lateral force of the water was 503 kg/m (338 lb/ft). The tension in the pull cable rose until it broke at 1.39 MN (312,000 lb). 4300 m (14,000 ft) of pipe were carried sideways and fractured, at first partially and then completely when the falling tide induced further movement.

The pipe was salvaged, not without difficulty. It was decided to pull the pipeline along a ploughed ditch. The pull resumed three months after the first failure. A storm occurred after the seventh string had been pulled, with a southeast wind of 18 knots gusting to 23 knots generating an offshore wave height of 2.7 m (9 ft). Wave action mixed the seawater with the mud and formed a dense liquid slurry suspension in the trench. A sample taken from the trench at 0808 in the morning gave a density of 1041 kg/m³ (65 lb/cu ft), and another sample taken at 0812 gave 1084 kg/m³ (67.7 lb/cu ft). At 0812, the pipe started to rise out of the ditch because the mean density of the pipe was 1062 kg/m³ (66.3 lb/cu ft), which was less dense than the slurry. Within 10 min., by 0816, 3000 m (10,000 ft) of pipe were out of the trench and were being carried sideways across the mud flat by the water.

In this incident the pipeline was constructed to be extremely light, just denser than seawater. A very low submerged weight reduces the pull force required to move the pipe but makes it vulnerable to wave action and to increases in the density of the surrounding fluid. The incident also illustrates how quickly things can go wrong. Cutler and Beale describe this project in an exemplary paper.[1]

16.2.2.4 A jet sled was incautiously lowered onto a pipeline instead of across it and dented the line severely. This incident shows the need for care in positioning any heavy equipment near a pipeline.

16.2.2.5 A pipeline laid through an area of strong tidal currents lost concrete weight coating and floated to the surface. This incident occurred many years ago, and the weight coating reinforcement did not meet modern standards. The coating probably fatigued as a result of oscillations of spans, formed by a combination of rough seabed and high currents. The route was a contributory factor. Another operator had come first, and had selected an easier route through the same area, but had not left room for a second line. The second line then had to go through an area of strong currents.

16.2.2.6 A laybarge dropped a pipeline under construction in very deep water. Gilchrist describes this incident in a paper.[2]

16.2.2.7 A pipeline was to be pulled from shore into a pre-excavated trench through a surf zone. The seabed in the surf zone consisted of sandstone with an irregular upper surface; the hollows were filled with sand. The trench was excavated by blasting, but this was found to be very difficult. During the pull, the pipe became buried in the sand and the pull could not continue.

16.2.3 Inherent defects

16.2.3.1 A routine survey of a North Sea pipeline was carried out some years after the line went into service. The survey was carried out by a survey vessel with an echosounder zigzagging back and forth across the pipe. Anomalous reflections from within the water column instead of from the bottom were at first attributed to equipment malfunction, but investigation by divers showed that some of the line had floated partway to the surface forming a very long span. Diver video surveys showed that some of the concrete weight coating had been lost.

Research found that the relevant section of the line had been laid in severe weather. Inspectors on the lay barge had reported that the line was banging against the stinger rollers and that some of the concrete had fallen away. It was concluded that part of the pipe had been light and had formed a short span, only just above the bottom and too low to be detected in post-construction surveys. A severe storm (approximately a 10-year storm) had occurred four months before the span was discovered. The wave-induced oscillatory velocities just above the bottom pushed the span backwards and forwards, and the lightly reinforced concrete weight coating progressively fatigued and fell off so that the length of the span increased.

This incident demonstrates the value of tough and fatigue-resistant concrete coating. It illustrates the value of regular surveys, confirms the need for active investigation when inspection reports reveal construction damage, and again shows that unexpected inspection findings have to be investigated.

16.2.3.2 A routine survey of a riser by an inspection pig found a short spool-piece section in which the internal diameter was larger than in the rest of the riser. At first this anomaly was attributed to a defect in the pig, but it was decided to have divers carry out an external wall-thickness measurement. They found that the wall was anomalously thin.

An investigation concluded that a short length of different and thinner pipe had been built into the riser, though it was not mentioned in the as-built records. The increased internal diameter had occurred when the pressure test before commissioning had expanded the pipe (though not to the extent that it ruptured).

Like the incidents in 16.2.2.1 and 16.2.3.1, this shows that anomalous inspection results ought not be taken lightly and attributed to equipment failure. This problem of actively pursuing the identity of anomalous results is likely to increase with the wider use of automated or semi-automated inspection systems. The incident also demonstrates the need for rigorous construction supervision and for genuine as-built records of what was actually built, rather than what was supposed to be built.

16.2.3.3 A pipeline in a shallow trench buckled upwards, forming a raised bend above the level of the seabed.[3]

The buckle was formed by upheaval, induced by the axial compressive force in the pipeline generated by operating temperature and pressure. Upheaval buckling had been known for many years as a problem that occurred in land pipelines, but on no clear grounds had it been generally believed not to occur in underwater pipelines. Since this first incident in the North Sea, upheaval buckles and lateral buckles have been observed in many marine pipelines, and much effort has been given to understanding the phenomenon and taking measures to eliminate it. Chapter 14 describes this phenomenon.

16.2.4 External factors

16.2.4.1 A fishing boat hit and ruptured a gas pipeline in shallow water in the Gulf of Mexico. The gas ignited, and several members of the crew were killed.

This incident confirms that pipelines under internal pressure are vulnerable to impact damage. If a pipe projects above the bottom in shallow water, there is always a possibility that a boat might strike it. The kinetic energy of even a small fishing boat moving at several knots is enough to rupture a pipeline, and it is never safe to place a pipeline where a boat might strike it. The solution is to trench the pipeline, to protect it by a rock berm, or to reroute it. In part as a result of this incident, the United States authorities decided to require pipelines in shallow water to be buried.

16.2.4.2 A ship dragged an anchor across a pipeline in the approaches to a port. The pipeline was severely damaged.

It is normally impracticably expensive to make pipelines completely safe against damage by ships' anchors, which can exert forces of the order of 1 MN. There is no risk in deep water because ships there do not have anchor cables long enough to anchor effectively. In moderate depths (> 35 m) away from ports, the risk is very small because ships rarely attempt to anchor in the open sea. The risk is much higher in the approaches to ports because ships anchor while waiting for the tide or for a berth. Though pipelines are marked on navigation charts, ships may overlook them, be using outdated charts, be mistaken about the ship's true position, or take the risk of anchoring in order to avoid some other hazard such as stranding or collision.

16.2.4.3 A ship sank across a pipeline in the Netherlands sector, and the pipeline began to leak. A similar incident occurred in the approach to the port of Singapore.

As in other systems, there is an irreducible risk of rare events that against which a pipeline cannot reasonably be protected. It is not practicable to design pipelines so that ships can sink across them without causing damage.

16.2.4.4 In the fire that followed the explosion on the Piper Alpha platform in the North Sea in 1988, three gas risers were heated externally without being significantly depressurized. The yield stress of steels falls as the temperature increases. Forty minutes after the explosion, the yield stress of each riser fell to the hoop stress induced by the internal pressure, and the riser expanded and ruptured. The gas then flowed uncontrolled into the fire, and the fire became worse.[4]

The initial explosion was not related to the risers or pipelines. However, the incident showed that flow from pipelines can contribute to accidents on platforms. The public inquiry recommended that operators install subsea isolation valves in pipelines on the seabed at a short distance from platforms. An isolation valve closes automatically if a fire or an explosion occurs on the platform, and only the contents between the valve and the platform can flow to the platform.

The Piper Alpha risers had no protection against fire except for the platform deluge system, which did not operate. The inquiry was critical of the lack of study of passive fire protection systems, which insulate risers externally so that the increase of riser wall temperature is slowed. The riser may then never reach the rupture temperature because the fire dies down or firefighting efforts are successful or because there has been time for the line to be depressurized.

16.2.5 Old age

16.2.5.1 A pipeline became partially blocked by corrosion products. Two spheres were put into the line to clear it, and half of one sphere came out. Somewhere in the line were one and a half spheres and a pile of debris.

The line may not have been adequately designed against corrosion. It is imprudent to delay sphering or pigging because over time the volume of material pushed in front of the sphere becomes so great that it blocks the pipe and prevents further movement.

16.2.5.2 An insulated pipeline carrying heated heavy oil in the intertidal zone in the Mersey estuary in northwest England began to leak. The first report of the leak came from a ship in the estuary. After some delay, the pipe was depressurized and clamped, but not until the local fire brigade had made a video showing a fountain of oil spurting into the air. The video was shown on television. An inquiry concluded that the line leaked as a result of external corrosion damage at a field joint, which followed damage to the field joint coating after cyclic longitudinal movements induced by changes of temperature. The operator was criticized and fined heavily, and its public reputation suffered.

A study by the UK Department of Energy showed that the anti-corrosion coating over the girth welds had been damaged by repeated longitudinal movements induced by changes in the temperature of the line.[5] The insulation over the girth welds was also damaged and allowed water to reach the unprotected outer surface of the pipe causing external corrosion. This situation demonstrates that external coatings need to be mechanically tough. Old age and inherent defects interact, and some inherent defects are only revealed by old age. The study also showed that both the design and the operating practices were open to criticism. There was no sensitive system for detecting leaks.

16.2.5.3 Loss of a riser clamp (tentatively attributed to fatigue) accentuated the bending moments in the riser induced by wave loading, and the riser fatigued and broke off.

This event again confirms the value of regular and thorough inspection.

16.2.6 Repair and modifications

16.2.6.1 A platform riser was cut above water to carry out a modification. It was thought that the pipeline and the riser were free of gas and liquids, but they were not. Condensate flowed out of the cut and ignited, leading to a fire that killed several people.

Hydrocarbon systems are always a potential fire risk, and the risk is often greatest when the system is nominally empty. This fact is completely familiar to engineers who carry out maintenance operations on refineries or tankers but occasionally is overlooked in the context of pipelines.

16.3 Reliability Theory

Another possible technique for gaining understanding of failures is to apply structural reliability theory.[6, 7] The extent to which it can be applied usefully to pipelines is highly controversial. For statements of contrasting positions, see Sotberg and Palmer.[8, 9]

The theory examines the risk of failure as a function of the statistics of the different features of the system that might contribute to failure. Consider first a simple and completely deterministic system. The strength of the system is characterized by a single quantity R. The load on the system is characterized by a single quantity S. The system fails if the load S is greater than or equal to the strength R and does not fail if S is less than R.

Now suppose that the strength of the system is not a single value but is described by a probability distribution. At least in principle, this distribution can be determined by constructing a large number of nominally identical systems and testing each one by loading it until it fails. The variability derives from variation in the strength of the component materials, the dimensions, and manufacturing processes such as welding and bolt tightening. The strength is described by a probability density function (pdf) $f_R(x)$, which means that the probability that the strength lies between x and $x+dx$ is $f_R(x)dx$. The probability that the strength is less than a value R is the following:

$$P(\text{strength} < R) = \int_0^R f_R(x)dx \qquad (16.1)$$

The probability of failure under a load S is the probability that the strength is less than S so the following equation applies:

$$P(\text{failure under load } S) = \int_0^S f_R(x)dx \qquad (16.2)$$

Suppose instead that the strength is a single value R, but that the load is described by a pdf $f_S(x)$ so that the probability that the load lies between x and $x+dx$ is $f_S(x)dx$. Again, this function could be measured by recording the load at regular intervals on a large number of occasions. In the case of the road bridge, for example, the load would sometimes be zero (no traffic in the early hours of the morning) and sometimes heavy (a stationary line of trucks during a traffic jam). The probability that the load is greater than the strength is then

$$P(\text{failure if strength is R}) = \int_R^\infty f_S(x)dx \qquad (16.3)$$

Finally, suppose that both the strength and the load are described by pdfs. Failure under a load between y and $y+dy$ requires that both of the following are true:

1. The load lies between y and $y+dy$, which has probability $f_S(y)dy$
2. The strength is less than y, which has the probability given by Equation (16.2) with y in place of S

Accordingly:

$$P(\text{failure under load between } y \text{ and } y+dy) = f_s(y)\left(\int_0^y f_R(x)dx\right)dy \quad (16.4)$$

The overall total probability of failure is determined by integrating Equation (16.4) for all values of y and is, therefore:

$$P(\text{failure}) = \int_0^\infty f_S(y)\left(\int_0^y f_R(x)dx\right)dy \qquad (16.5)$$

This scenario assumes one strength parameter and one load parameter, but it can readily be generalized for multiple strengths and multiple loads.

In principle, this formula offers a way of quantifying failure probabilities. Consider, for example, the problem of determining the probability that a pipeline will rupture because of inadequate circumferential strength. A simple model of rupture might assume that

1. The circumferential stress is determined by the Barlow formula (section 10.2).
2. Rupture occurs when the circumferential stress reaches the yield stress Y.

The rupture pressure is therefore:

$$P_{rupture} = \frac{2tY}{D} \qquad (16.6)$$

where

t is the wall thickness

D is the diameter

Suppose that the diameter and the wall thickness have fixed invariable values and that the only source of variability is the yield stress Y, whose pdf is $f_Y(x)$. From Equation (16.2)

$$P(\text{failure under pressure p}) = \int_0^{pD/2t} f_R(x)dx \qquad (16.7)$$

and if the pdf for pressure is fp(x), the overall probability of failure is

$$P(\text{failure}) = \int_0^{\infty} f_p \left(\int_0^{pD/2t} f_Y(x)dx \right) dp \qquad (16.8)$$

This formula can be generalized to incorporate pdfs for D and t and to allow for other kinds of failure, such as bending buckling (which depends on the same parameters) or internal corrosion (which depends on different parameters, such as contents composition, operating temperature, and steel composition, all of which have their own pdfs).

The difficulty with structural reliability theory is not with the theory itself but with its application to real cases. A pipeline has to be designed so that the risk of failure is very low. The DnV 2000 rules [reference 10, clause 2 C 603] give a target probability of failure of 10^{-5} per pipeline per year for safety class high and ultimate, fatigue, and accidental limit states. This means that 1000 pipelines could operate for 100 years, and there would be only one failure. It follows inevitably that a real application of results like Equation (16.8) is entirely governed by the distant tails of the statistical distributions, which correspond to unusually high operating pressures or unusually low yield stresses. The central parts of the distributions are irrelevant. By their nature, these tails are extraordinarily difficult to determine, and are out of reach of any conceivable measurement program.

Reliability analysts have attempted to get round this difficulty by conjecturing that the distributions follow some standard form, such as the Gaussian (normal) or log-normal distribution. They then argue that one can extrapolate to the tails from a small number of measurements, most of which necessarily fall close to the center. This is spurious: there is no reason to suppose that the tails follow the same distributions as the center or that any definite form applies. It has been said that mathematicians believe the Gaussian distribution to be a law of physics and that physicists believe it to be a law of mathematics. An alternative and equally unconvincing argument is that the probabilities are notional and not meant to be given a literal frequentist interpretation.

On the contrary, there is a more convincing argument that routine manufacturing and construction practices are designed to eliminate the distribution tails, or at least to modify them. Pipes manufactured for offshore pipelines are almost invariably all subjected to a mill hydrostatic test to a pressure at least 1.25 times the maximum operating pressure.[10] That test identifies and eliminates all sections of pipes that have a rupture pressure less than the hydrostatic test pressure and reduces to zero the failure probability in that range (except for a small probability that the test equipment or procedures are defective). When the pipeline is completed, the entire line is subjected to a pressure test, again to a pressure at least 1.25 times the operating pressure, which tests the girth welds, identifies damage during construction, and retests each length of pipe. Immediately after the test, the failure probability in the range up to the test pressure is again zero. It may, of course, later be degraded by corrosion or external damage.

Palmer discusses the role of proof testing, which needs further research.[11]

It would be wrong to suggest that reliability theory has no value at all. It confirms the intuitive notion that variability is important. It may be better to choose a pipe with a lower mean yield stress but with a tightly-controlled manufacturing process that leads to a uniform product, reflected in small standard deviations of yield stress and dimensions, rather than to choose another pipe that has a higher mean yield stress[10] but a much larger variability.

This approach is reflected in the 2000 DnV rules, which allow a designer to take advantage of demonstrated small variability in yield stress.[10] The defining strength parameters for steel—characteristic yield and tensile strengths—are related to the specified minimum yield and tensile strengths by the following:

$$f_y = (SMYS - f_{y,temp})\alpha_U \qquad (16.9)$$

$$f_u = (SMTS - f_{y,temp})\alpha_U\alpha_A \qquad (16.10)$$

$f_{y,temp}$ allows for temperature derating of both yield and tensile strength and is given in a guidance note in clause B604. For C-Mn steel, the guidance note value in the temperature range above 50 °C is 0.6 MPa/deg C for temperatures between 50 and 100 °C, but alternative information can be used if it is available.

α_A is an anisotropy factor, taken as 0.95 for the axial direction but 1.00 for other cases and, therefore, is of no consequence for pressure containment.

α_U is a material strength factor, given in Table 5–1 within clause B602. Its objective is to penalize materials with a high variability of yield stress. Materials with a normal degree of variability have α_U equal to 0.96. Materials with a low level of variability meet supplementary requirement U, which ensures increased confidence in yield strength, and α_U can then be increased to 1.00. The conditions under which supplementary requirement U is satisfied are set out in section 6 clause D500 and E800.

Reliability methods come into their own in rational analysis of problems, such as inspection and preventive maintenance,[12] which are inherently statistical.

16.4 Minimizing Risk: Integrity Management

An important engineering management task is to minimize the risk of failure incidents like those described in section 16.2.

Most failures result from mistakes (see, for example Perrow, Reason, Matousek, Bignell).[13, 14, 15, 16] They are sometimes the consequence of ignorance, sometimes the result of conscious risk-taking to save money or to hasten completion of a project, and sometimes the result of equipment or material failure. Often, there are precursor incidents that should have been warnings but were overlooked or discounted.[16] Only a very small fraction of failures are unavoidable, out-of-the-blue incidents that could not have been reasonably anticipated or protective measures taken. The incident in case 16.2.4.3 is an exception, but such cases are rare.

Guarding against mistakes is a central part of the management of design and construction. It has various components, among them training, calculation checking, software validation, peer review, inspection, testing, and quality assurance procedures that ensure the documentation and traceability of design decisions and material procurement.

Once a pipeline system has gone into service, it is a valuable asset whose soundness must be safeguarded by an integrity management system. The principles of integrity management have much in common with good management generally.[17]

16.5 Repair

16.5.1 General remarks

Several general principles apply to repair. "Do no harm" is a maxim propounded in antiquity and learned by generations of doctors. It is only too easy to attempt a repair incautiously and, in doing so, to make matters worse. A repair requires careful engineering at least as much as new construction does. A repair is almost invariably unexpected, and what is to be done needs to be examined very carefully. For both of these reasons, it is generally best not to act in haste.

Occasionally, there is no alternative. If a broken riser is on fire and threatens the safety of a platform, somebody needs to act decisively and at once. But often people feel themselves in the spotlight and under pressure to act immediately when, in fact, there is no need to do so; and it would be better to wait and consider the options thoughtfully and only then to proceed.

A repair is not often a good moment to try something new. There is less experience with a new procedure than with tried and tested state-of-the-art procedures; and there may be surprises, particularly in the uncertain and incompletely planned context of a repair.

Finally, preparedness pays off. Many repairs require additional lengths of pipe. If the operator has to go to a pipe mill and try to secure new pipe with the required diameter, wall thickness, grade, and compatible weldability and have that pipe coated and shipped, he or she will incur a significant expense due to the great amount of time involved to take all that action. A prudent operator buys slightly more pipe than the original project requires, typically 1–2% of the length, and stores the extra pipe accessibly. Accountants may look on this with a baleful eye and argue that the extra length is wasted; but if a repair is needed, the pipe will pay for itself many times over.

Similarly, it is valuable to have repair welding procedures pre-qualified and to have available repair equipment such as clamps and alignment frames. This is particularly true in remote locations.

16.5.2 Repair techniques

The engineer can bring to bear the whole technology of marine pipelines, keeping in mind the general points of 16.5.1. Small pinhole leaks can sometimes be mended by clamps, though in most cases this is considered only a temporary measure. Clamps can be welded to the pipeline and become a permanent part of the system. Alternatively, if the pipeline is thought to be dangerously weakened by corrosion or fatigue but is not actually leaking, a strong sleeve can be connected around the line by bolting or welding, and the narrow annular gap filled with epoxy or cement grout. The grout then transfers some of the internal pressure load to the sleeve, and the sleeve supplies the missing hoop strength.

Many repairs are made by the hyperbaric welding technique described in chapter five. The line is flooded, the damaged portion is cut out by a diamond saw or rotary milling cutter operated by divers or an ROV, the alignment is adjusted and the ends prepared, and a new spool piece is welded into the pipeline. The damaged section may be temporarily or permanently bypassed by a section incorporated into the pipeline by underwater hot tapping.

Alternatively, there are many proprietary mechanical connection systems that can be deployed. The scheme consists of an articulated spool piece with mechanical connectors at each end. The articulation allows the connectors to be aligned with the undamaged pipe ends. The connectors grip the pipe through collet connectors or make a swaged connection.

Yet another alternative is to flood the pipe, cut out the damaged section, attach buoyancy to the end sections of the undamaged pipe, and lift the ends to the surface by side davits on the barge. The lifting scheme is arranged so the pipe ends are just above the surface and bend so that the pipe axis is horizontal. A spool piece is welded between the ends by conventional welding since the ends are above water. Then the raised loop is laid sideways so that it does not go into compression, by moving the barge sideways as the davits lower the pipe. This surface tie-in technique is often used in shallow water and was applied in 1974 by Saipem in 110 m of water to connect the two sections of the 32-in. Forties I pipeline. Studies have shown that it can be applied in deep water.[18]

There is little experience of repair in very deep water or of repairs to bundles. Building a new pipeline may be less expensive than repairing the old one.

References

1   Cutler, R. C., P. A. H. Beale. (1972) "Twin Submarine Pipelines to Khor-al-Amaya." (Paper 7445S) *Proceedings of the Institution of Civil Engineers*, London. 17–43.

2   Gilchrist, R. (2000) *Proceedings of ASME J-Lay Workshop.* Houston. February 2000.

3   Nielsen, R., B. Lyngberg, P. T. Pedersen. (1990) "Upheaval Buckling Failures of Insulated Buried Pipelines: A Case Story." (OTC6488) *Proceedings of the 22nd Annual Offshore Technology Conference.* Richardson, TX. 4. 581–592.

4   "The Public Inquiry into the Piper Alpha Disaster" (Report) Her Majesty's Stationery Office (1987).

5   "Investigation Report of the Hot Oil Pipeline Failure at Bromborough on Saturday, August 19, 1989"(Report) UK Department of Energy, Her Majesty's Stationery Office (1990).

6   Thoft-Christensen, P., M. J. Baker. (1982) *Structural Reliability* Theory. Berlin, Germany: Springer-Verlag.

7   Melchers, R. E. (1987) *Structural Reliability: Analysis and Prediction.* Chichester, England: Ellis Horwood.

8   Sotberg, T., B. J. Leira. (1994) "Reliability-based Pipeline Design and Code Calibration." *Proceedings of the Offshore Pipeline Technology Conference.* Oslo. IBC Technical Services, London.

9   Palmer, A. C. (1996) "The Limits of Reliability Theory and the Reliability of Limit State Theory Applied to Pipelines." (OTC8218) *Proceedings of the 28th Annual Offshore Technology Conference.* Richardson, TX. 4. 619–626.

10 Offshore Standard OS-F101: "Submarine Pipeline Systems." Hovik, Norway: Det Norske Veritas, 2000.

11 Palmer, A. C., C. Middleton, V. Hogg. (2000) "The Tail Sensitivity Problem, Proof Testing, and Structural Reliability Theory. Structural Integrity in the 21st Century." *Proceedings of the 5th International Conference on Engineering Structural Integrity Assessment.* Cambridge. 435–442.

12 Lamb, M., A. C. Palmer. (1996) "Reliability Analysis: Methodologies, Limitations and Applications." *Proceedings of the Conference on Risk, Reliability and Limit States in Pipeline Design and Operations.* Aberdeen.

13 Perrow, C. (1984) *Normal Accidents.* New York: Harper-Collins.

14 Reason, J. (1996) *Human Error.* Cambridge, England: Cambridge University Press.

15 Matousek, M. (1977) "Outcomings of a Survey of 800 Construction Failures." *IABSE Colloquium on Inspection and Quality Control.* Zurich: Institute of Structural Engineering, ETH.

16 Bignell, V., G. Peters, C. Pym. (1977) *Catastrophic Failures.* Milton Keynes, England: Open University Press.

17 Peters, T. J., R. H. Waterman. (1982) *In Search Of Excellence.* New York: Harper and Row.

18 Kaustinen, O. M., R. J. Brown, A. C. Palmer. (1983) "Submarine Pipeline Crossing of M'Clure Strait." *Proceedings of the 7th International Conference on Port and Ocean Engineering under Arctic Conditions,* Helsinki. VTT Espoo, 1, 289–299.

17 Decommissioning

17.1 Introduction

Underwater pipelines ultimately come to the end of their useful operating lives. Often they are not needed any more because the fields from which they carry petroleum are depleted or because transportation patterns have changed. Sometimes they are severely corroded or blocked with wax and corrosion products. The required period of service is hard to predict at the time a line is designed, because it is influenced by the oil price and by the potential of tying in new fields, possibly still to be discovered. A gas pipeline from a small field may only need to operate for six or seven years, whereas a long-distance transmission line might operate for 100 years.

It is irresponsible for an operator simply to shut down a pipeline and walk away, though in the distant past that was sometimes done. The line will slowly decay, can release hydrocarbons to contaminate the environment, and ultimately will break up to leave untidy steel and concrete litter. The litter might interfere with other users of the seabed such as fishermen and recreational divers, though it might also form sheltered habitats for fish. In most countries, regulation and legislation mandate that the operator has the continuing responsibility for maintaining a safe and non-contaminating environment. Decisions about underwater pipelines belong within the wider context of decommissioning offshore installations generally. The deplorable Brent Spar episode showed how decommissioning decisions could become hugely controversial and political and that once upset has occurred, it requires much patient effort to arrive at conditions in which a calm and rational dialogue can take place.

Deciding on the right course of action engages linked political, legal, environmental, financial, tax, and technical factors. The technical factors are perhaps the easiest to deal with. The legal and political background varies from country to country. The engineer who has to select a policy is well advised to seek advice from specialists at the earliest possible stage. It is inevitable that a decision will involve competing values and that the decision-makers will need to arrive at a balance between values attached to the environment, to sustainable development (not the same thing), to immediate costs, to operator company reputation, and to the long term against the short term.

Section 2 of this chapter discusses the legal background. Section 3 examines the various technical alternatives, starting with the consequences of simple abandonment without doing anything.

17.2 Legal and Political Background

The legal framework within which decommissioning decisions are made is extremely complicated, and there is little established case law. An engineer is well advised not to attempt to make his own decisions about the law but to seek specialized professional advice at the earliest opportunity.

In most countries, pipeline abandonment is governed by legislation. In the United Kingdom, for instance, the Submarine Pipelines Act includes a definition of abandonment, requires the Secretary of State to be notified when a pipeline is abandoned, and empowers the Secretary of State to impose conditions on the state of the abandoned pipeline. International law such as the United Nations Law of the Sea and the convention on dumping may be relevant, though it is not clear whether the abandonment of a pipeline can be construed as dumping. The OSPAR Convention for the Protection of the Marine Environment in the Northeast Atlantic came into force in 1992 and has been acceded to by all the western European countries that border the North Sea and the Atlantic. OSPAR Decision 98/3 stipulates that "the dumping or leaving wholly or partly in place of disused maritime installations within the maritime area is prohibited." However, the decision allows derogation from this principle for large steel and concrete structures, and it is not clear whether it is intended to cover pipelines, just as it is not clear whether it is intended to cover shipwrecks. In countries governed by the common law, there may be the possibility of legal action by individuals whose interests have been damaged by an abandoned pipeline. Lissaman and Palmer discuss the UK legal background in greater detail. [1]

17.3 Alternatives

17.3.1 Doing nothing

Doing nothing is not an option that a responsible person would normally choose, but it focuses attention on decay mechanisms. Imagine a large-diameter oil pipeline with an anti-corrosion coating and a concrete weight coating, cathodically protected by anodes and initially half-buried in a sandy seabed. Then also imagine the oil flow stopping and no action taken.

When the flow stops, water in the oil will drop out and collect at the bottom of the pipe. Internal corrosion will begin there and later progress to other segments of the circumference. External corrosion will initially be prevented by the external coating and by the CP system. CP is usually designed rather conservatively for a significant percentage of coating breakdown and for a design life of the order of 40 years. Eventually, the anti-corrosion coating will break down, and the anodes will be consumed. The time in which this will happen is hard to estimate because polymeric anti-corrosion coatings are new materials that have only been known for 40 years or so, but the time might be of the order of 100 years. The steel will then corrode, and in the end, the pipeline will begin to leak oil. Oil will flow out and contaminate the environment and will slowly be displaced by seawater (because of the density difference). Concrete lasts longer than steel (plenty of Roman concrete has survived 2000 years, but little Roman steel remains), but ultimately it will break up when the steel inside has corroded away.

The ultimate effect is to transfer the oil to the sea where it becomes subject to dilution and to biological breakdown mechanisms.[2] This mechanism leaves rust on the seabed, broken concrete, and broken anti-corrosion coating. After a long period, the coating remnants will break as a result of chemical and biological processes.

17.3.2 Stabilization and trenching

A second option is to remove the oil as much as possible by repeated pigging. The pipeline is then filled with water, to which are added a corrosion inhibitor, a biocide, and an oxygen scavenger to retard internal corrosion but not eliminate it. In addition, it is possible to supplement the anodes by new groups of anodes (connected to the old anodes by Cadwelding or a similar process as described in chapter 8).

This option releases to the environment only the residual small amount of oil that cannot be removed by pigging. If that corresponds to a 1mm layer on the inside of a 32-in. line, the residual volume is 15 bbl per km of pipe, released over say 200 years. That rate might be tolerable and within the capacity of the natural environment to break it down. The ultimate fate of the pipe itself is the same as in the do-nothing option, but it takes much longer.

Some North Sea operators have chosen this option for temporary stabilization of lines for which operation has been suspended but have not been permanently abandoned. Whether this option can be accepted as a long-term solution still has to be decided.

Additional trenching and burial is a substantial improvement on this option for the long term. The trenching depth is chosen so that the top of the pipe is below the natural seabed. The trench will fill back naturally in a time much shorter than the time for the pipe to break down. When the pipe wall collapses, the sand will fall into the collapsed pipe, and back fill will then pile up in the new trench and cover the fragments. Therefore, this option will leave the fragments of the original pipeline buried under the sand, and they will not appear on the seabed unless there is a large secular change in seabed level. The steel will corrode, and ultimately the concrete will degrade.

17.3.3 Re-use

In keeping with the philosophy of sustainable development, re-use is the ideal option. Unfortunately, existing lines are generally unsuitable for re-use because they are composed of the wrong material, are the wrong diameters, have too low an operating pressure, or are damaged by corrosion. A semi-rational prejudice against second-hand systems is often a factor.

Small-diameter lines have occasionally been re-used by towing them to a new location. They could also be reeled up from the seabed onto a reel ship and then re-laid after repair to damaged coatings and replacement of anodes occurs. Underwater pipelines could also be recovered, perhaps recoated, and then re-used in less demanding service onshore. Land pipelines generally have lower operating pressures than do marine pipelines.

17.3.4 Recovery for scrap

Recovery for scrap recycles the materials content of the pipeline, leaves no debris on the seabed and, in that sense, is environmentally friendly and consistent with sustainable development. Recovery consumes more energy than re-use or stabilization, and has a greater immediate environmental impact. Political and legal factors might make this option an attractive one.

Recovery can be accomplished by tow, by reversed reeling onto a reel ship, or by a lay barge using its tensioners to pull the pipe up the stinger, and then cutting the pipe into sections with a shear. The pipe is then taken to shore and divided into its components. Steel can be recycled as scrap, aluminum and zinc anodes can be separated and melted, polymer coating can be recycled as plastic scrap, and concrete broken up and used as fill or as aggregate. The value of the recovered materials is not enough to pay for the cost of recovery. Carbon steel scrap is only worth about $100/tonne, and corrosion-resistant alloy (CRA) scrap about $650/tonne at the time of writing.

In the context of another offshore installation, the Ekofisk tank, an unusual concrete structure in the central North Sea, detailed studies of recycling compared to leaving in place or dumping show that the risks associated with recycling are substantially higher, the energy requirement is greater, and the associated CO_2 emissions are greater.[3] Recycling is not by any means necessarily the best option from the environmental point of view.

References

1 Lissaman, J., A. C. Palmer. (2000) "Decommissioning Marine
   Pipelines." *Pipes and Pipelines International* 44. 6. 35–43.

2 Van Bernem, C., T. Lubbe. (1997) *Öl im Meer* (Oil in the sea).
   Darmstadt, Germany: Wissenschaftliche Buchgesellschaft.

3 *Information on the Ekofisk I Cessation Project* (Brochure). Phillips
   Petroleum, EPKE 10 (2

18 Future Development

18.1 Introduction

Things would be very dull if technology stood still. Happily it does not, and pipelines are no exception. The technology has made extraordinary progress, and there is still much to be done. The customers in the petroleum industry rightly ask for pipelines to be built more rapidly and more cheaply without sacrificing security or risking damage to the environment. There are developments in many areas, and this considers in turn design, materials, joining, and construction, as they appear in July 2003. Inevitably, this chapter will rapidly become outdated. That is as it should be.

18.2 Design

Rational design methods nowadays focus on the idea of a limit state. That is an old idea in geotechnics and in structural engineering, but a relative newcomer to pipelines. A limit state is a condition that limits the continued safe operation of a pipeline. The emphasis is on directness, rather than on abstract conditions. Thus, bursting is a limit state: a pipeline that has burst has plainly become unusable. On the other hand, reaching a certain circumferential stress is not in itself a limit state, even though it might be linked to a limit state such as bursting.

Limit state concepts have proven useful in deriving rational design methods and in achieving economies. They are not just the reformulation of old methods in new language. The best-known pipeline example is the replacement of allowable-stress methods for considering the combined effect of longitudinal and hoop stresses once it was realized that old limits on allowable stress did not actually correspond to a limit state. The old limits could be dropped and replaced by a much less restrictive requirement on plastic strain (which can be linked to limit states such as rupture under alternating plasticity). Chapter 10 and Appendix B discuss this in detail.

A limit state philosophy informs modern codes such as DnV 2000, API RP 1111, and the 1993 Germanischer Lloyd rules. It has been most influential in the area of structural design and span assessment, and less so in internal hydraulics and in external hydrodynamics and stability.[1, 2] A systematic approach has much to offer in those areas too.

The design process also needs to be made more efficient, less costly, and less time-consuming. Almost all pipeline design calculations can now be carried out very rapidly and economically on PCs. It is not fanciful to think of the calculation portion of the design of a major system being carried out in a few minutes or, indeed, a few seconds. Likewise, it is not beyond reason to think of design being made essentially automatic through the application of intelligent systems and of the design being documented automatically.

However, calculation itself is only a small part of the design process. The design has to be documented efficiently, there have to be specifications of all the materials and components, there need to be tender documents so that contractors and manufacturers have an accurate and complete picture of what they are asked to supply, there has to be tender evaluation, and so on. That part of the process is still remarkably primitive and inefficient. There is much to be done through the application of information technology (IT) and intelligent documents.

18.3 Materials

A perceptive outside observer of the pipeline scene remarked that the yield stress of steels used in pipeline construction increases by about 2 MPa/year, less than 0.5%. Why, he wanted to know, do pipelines not apply the high-strength steels used in other areas of steel construction, such as the automotive industry? Those steels correspond to X120 and up, and high strength is achieved without unacceptable sacrifice in ductility or impact resistance. At the time he was writing, the maximum grade of line pipe commercially available was API X80, with a minimum yield strength of 551 MPa, minimum ultimate tensile strength of 620 MPa, and Charpy toughness of 68 J (average on three specimens). He pointed out that the same operator had already carried out a successful trial on grade 165 drill pipe, 1140 MPa minimum yield.

At long last, that issue is being considered seriously through wider applications of X70 and X80 steels and through the X100 project supported by several majors. In parallel, much is being done to apply a wider range of corrosion resistant alloys.

A more ambitious approach to reducing corrosion in pipelines while retaining or increasing material strength is to replace steel with composites. Composites have many advantages, among them light weight, the absence of corrosion, and lower thermal conductivity. Glass-fiber epoxy composites have been used to construct serious long-distance pipelines, among them a 28-in. oil pipeline in Algeria. Glass-reinforced polymer (GRP) is widely used for outfalls and intakes, and high-density polyethylene (HDPE) has captured much of the market for gas distribution and sewage outfalls. Composites technology has moved much further than many pipeliners realize.

There remain several major difficulties in applying composites to pipelines. The primary structural function of a pipeline is to carry internal pressure by circumferential tension as described in section 10.2. The hoop tension per unit longitudinal length is $pD/2$, where p is the internal pressure and D the diameter. They are usually determined by operational requirements, and the designer can usually only reduce p if she increases D. If the allowable circumferential stress is fY, the volume of the pipe wall per unit length is $(\pi/2)pD^2/fY$. If the cost per unit volume of the material is C, the cost per unit length of the pipeline structure is calculated as follows:

$$\frac{\frac{(\pi/2)pD^2}{fY}}{C}$$

Therefore fY/C is a figure of merit for the material: the larger fY/C is, the lower the cost of the pipeline structure. This calculation, incidentally, confirms the attractiveness of higher-strength steels. Higher-strength steels have a higher figure of merit provided that the cost per unit volume increases less than proportionately to the yield stress (though other factors such as buckling may come to govern the design).

Composites find it hard to compete on this basis. For steel with design factor f 0.8, yield stress Y 448.2 MPa (X65), density 7850 kg/m^3, and cost C 1200 \$/tonne, fY/C is 0.038 MN/m \$. Competing composites appear to be about 60% more expensive, but an increasing market and production volume will reduce this difference.

This simplified argument is unfair to composites. Corrosion of steel pipelines has frequently been underestimated, and there are depressingly many examples of lines that have had to be taken out of service after only a few years of operation. It is obvious that if a low-cost, corrosion-resistant alternative can make it possible to (say) double the operating life of the pipeline, that option can become economically more attractive. For that reason, composites are beginning to take a significant proportion of the market for downhole tubulars. Development of a reliable girth joint remains a problem, and this an active research area.

Another possibility for developing improved materials for pipelines is to combine the good properties of steel (ease of making connections by well-understood techniques) with the good properties of composites (resistance to corrosion, low weight, low thermal conductivity). A recent article described experience with composite reinforced line pipe (CRLP™) applied to 48-in. (1219.2 mm) outside diameter land pipelines.[4] The pipe consists of an 11.71 mm (0.461 inch) X70 steel pipe externally wrapped with an isopolyester resin-glass fiber composite. A 150 mm cutback in the composite at each end of each joint allows the lengths to be joined by welding, and the field joints are then protected with a pre-impregnated glass fiber roving cured by dielectric heating. One of the advantages of this scheme is that it makes it possible to use large diameter X70 pipe without exceeding the capacity of North American pipe mills. Recalling from chapter 10 that the mean longitudinal stress in a uniform pipeline is less than half the circumferential stress, and extending the argument that leads to that result, the longitudinal force per unit length in the circumferential direction is much smaller than the circumferential force per unit length in the axial direction. The longitudinal force at a field joint can be carried by the steel alone.

Yet another possibility is to apply a composite internally. That method has the advantage of providing some internal thermal insulation so that the pipe temperature is reduced, which reduces the problem of lateral and upheaval buckling.

18.4 Connections

The speed of many pipelining operations is controlled by welding. There are peculiar inconsistencies: one sometimes sees, for example, a highly sophisticated lay barge with all kinds of electronics and automated positioning and move-up systems, incongruously accompanied by old-style stick welding and a stone-age procedure for infilling field joints with hot mastic. That irony must generate the creative dissatisfaction that drives progress.

One way to increase construction progress is to select alternative welding systems. Many options have been investigated, among them electron-beam welding, friction welding, flash-butt welding and homopolar welding.

Flash-butt welding (FBW) is the dominant technique used to construct the large-diameter gas pipeline system in the FSU. FBW is completely different from conventional welding. An internal welding machine grasps the two ends of the pipe and generates a pulsed arc around the whole circumference as the pipe faces move slowly together. When the welding temperature has been reached, the machine pushes the ends together, expelling a surface molten layer containing impurities and forming a continuous weld. The machine then mills away the upset and flash formed on the inside surface. The system is highly automated and depends on control of welding parameters such as current and upsetting force, rather than on the skill of the operators. FSU figures indicate that productivity per man shift is much higher than that with automatic welding and that the need for repairs is reduced.

FBW was developed in the Ukraine for land pipelines. In North America, FBW is widely used for railroad rails but not at all for pipelines. Bitter opposition from welders' unions is said to be a factor. McDermott carried through a major program to develop it for use offshore and was enthusiastic.[5] At least one western major has picked up the FBW idea again and is considering it further, but progress is slow.

A joint industry project (JIP) based in the United States has been examining homopolar welding and has achieved promising results on pipe up to 12 in. in diameter.

A more radical welding approach is to abandon the idea that a pipeline has to be welded and to join the lengths by mechanical connection systems. Engineers who move into pipeline engineering from drilling are astonished to find that pipelines are almost invariably made from lengths of pipe joined by welds.

Some of the many reasons for the conservatism of the pipeline industry are not rational, but some are sensible. Pipeline engineers have been preoccupied with the risk of corrosion and the fear of leaks and are understandably sensitive to the long-term integrity of a line that may contains thousands of joints. Only one joint has to fail for there to be environmental damage, the possibility of fire and injuries, and a costly interruption of service. That said, drilling engineers are used to threaded connections, trust them as sound and reliable, and recognize that the connections now available are the product of a huge amount of research and development and of high-quality manufacturing systems. Most other people who deal with pipes use non-welded connections extensively, even in critical applications such as nuclear submarines.

A joint-industry project was completed in 1998. It responded to various concerns, among them the compulsion to minimize costs and speed construction, the new spirit of readiness to question received ideas, as well as to the courageous projects that have moved forward and applied new kinds of connectors. The industry also wishes to use materials such as 13Cr and X120 steels, which have better corrosion resistance and higher strength than conventional pipeline steels but are difficult or impossible to weld.

The project has looked at all the implications of mechanical connection systems and has explored most of the problem areas. Most of all, it has sought to determine the strength of non-welded connections. In a conventionally welded pipeline, the girth welds are normally at least as strong as the pipe joints themselves. If it can be proved that the same is true of a non-welded connection, a major hurdle will have been jumped. The 1998 project showed that some non-welded connections meet that test.

This is an idea whose time has come, provided that the system cost can be made competitive. It is not the end of the story, but it may be the end of the beginning, so the industry can move forward, cautiously at first, to more applications. A small number of projects have already used systems of this kind, among them the BP Harding project.

18.5 Construction

The three principal methods of construction (lay barge, reel, and tow) are well established and compete vigorously. The needs of customers are moving into deeper and deeper water, and all these methods have been applied in that environment. There is much interest in the J-lay variant of the lay barge method, whose advantages and disadvantages are examined in chapter 12. In deep water, the advantages of a dynamically-position J-lay vessel are overwhelming, particularly if the inherent difficulty of being able to work at only one or two stations can be overcome by rapid, one-shot connection systems with multiple joints.

Some of the accepted conventions of pipelaying have to be questioned. Engineers moving into marine pipelining from other areas invariably ask why pipelines are almost always laid full of air at atmospheric pressure. They are always told that the pipe would be unacceptably heavy if it were laid water-filled. In shallow water this is true; but in deep water, an air-filled pipe must be heavy anyway because its wall thickness has to be high enough to sustain the external hydrostatic pressure against buckling. The cost of the additional wall thickness is very high. It complicates other problems such as upheaval and welding, and it limits the number of pipe mills that can compete to produce the required pipe. Moreover, that extra wall thickness, in fact, is required for only a short portion of the pipeline's life—the installation period—because once the line has been hydrostatically tested and commissioned, the internal operating pressure is almost invariably higher than the external hydrostatic pressure so the risk of buckling disappears.

Studies have shown that in deep water there are great advantages offered by alternative strategies to pipelaying, e.g., design the pipe with a lower wall thickness and to balance the external pressure by laying the pipe wholly or partly filled with water (or with a lighter liquid such as pentane, the lightest hydrocarbon that remains liquid at atmospheric pressure and sea-water temperature).[6, 7] With this strategy, the wall thickness and steel cost are hugely reduced. In intermediate depths, the submerged weight during laying is greater than it is if the pipe is laid air-filled but remains within acceptable limits. In great depths, the submerged weight for the liquid-filled pipe with the lighter wall turns out to be *less* than for the air-filled pipe with the heavier wall.

The air-filled strategy seems a promising way of achieving savings. It does, of course, have operational and repair implications. An oil line never needs to be depressurized too much below the external hydrostatic

pressure, even at great depth, because of the relatively small density difference between water and oil. A gas line is more problematic because it might be accidentally depressurized, and the external hydrostatic pressure then would be much greater than the internal pressure. Isolation valves and check valves could give protection against that kind of failure.

Shore approaches and shallow water remain frequent sources of delays and huge cost overruns, and more and more questions are being raised about the damaging environmental impacts of some techniques. Horizontal drilling has taken a large share of the market. However it has much room to grow, particularly if shallow horizontal drilling, currently thought to be limited to about 1500 m horizontally, can take advantage of developments in extended-reach drilling directly to the reservoir. That technique has already reached 11000 m horizontally in the Wytch Farm field in southern England.

References

1   Palmer, A. C. (1996) "The Limits of Reliability Theory and the
    Reliability of Limit-State Theory Applied to Pipelines." (OTC8218)
    Proceedings of the Offshore Technology Conference. Houston, TX. 4.
    619–626.

2   Palmer, A. C., D. Kaye. (1991) "Rational Assessment Criteria for
    Pipeline Spans." *Proceedings of the Offshore Pipeline Technology Seminar.*
    Copenhagen.

3   Palmer, A. C., K. Asi, M. Bilderbeck. (1992) "Intelligent Documents
    and Standardisation." *Proceedings of the Conference on Offshore
    Standardisation.* Aberdeen.

4   "TransCanada Installs Demonstration Section of Composite
    Reinforced Line Pipe." *Pipeline & Gas Journal 230*, 50–54, June 2003.

5   "Flash-butt Welding." Video. McDermot International, 1986.

6   Palmer, A. C. (1997) "Pipelines in Deep Water: Interaction Between
    Design and Construction." *Proceedings of the Workshop on Subsea
    Pipelines.* COPPE-UFRJ, Federal University of Rio de Janeiro.
    157–165.

7   Palmer, A. C. (1998) "A Radical Alternative Approach to Design and
    Construction of Pipelines in Deep Water." (OTC8670) *Proceedings of
    the 30th Annual Offshore Technology Conference.* Houston, TX. 4.
    325–331.

Appendix A

Glossary

Active Corrosion: Continuing corrosion, which if not controlled, could result in loss of integrity of the pipeline.

Adhesive joint: A joint made in plastic pipe using an adhesive material to form a bond between the mating surfaces without significant dissolution of the mating materials.

Alloy: A substance having metallic properties and being composed of two or more chemical elements of which at least one is an elemental metal.

Anaerobic: Free of air or uncombined oxygen.

Anchor Pattern: In corrosion contexts, the irregular peak and valley pattern created on the surface of a metal by the blasting medium used to remove the mill scale and oxides; in pipelaying contexts, the layout of the anchors of a moored barge.

Anion: A negatively charged ion. The ion may have more than one negative charge. Examples are Cl^-, $SO_4^=$, HS^-, $S^=$.

Anisotropy: The characteristic of exhibiting different values of a property in different directions with respect to a fixed reference system in the material. Pipeline steel exhibits anisotropy between longitudinal and transverse or circumferential mechanical properties and metallurgical morphology.

Anode: The electrode of an electrolytic cell at which oxidation is the principal reaction, usually seen as the site where the metal goes into solution (corrodes). Electrons are *generated* at the anode area and flow away from the anode in the external circuit.

Atmospheric zone: The section/zone of a platform or riser that extends upward from the splash zone and is exposed to sun, wind, spray, and rain.

Austenite: A solid solution of one or more elements in face-centered cubic iron. Unless otherwise designated (such as nickel austenite), the solute is generally assumed to be carbon. The symbol for austenite is gamma (γ). Carbon steels are austenitic above the transition temperature. The UNS S30000 series of chromium-nickel-iron stainless steels are austenitic and are non-magnetic.

Austenitic Steel: Alloy steel whose structure is normally austenitic at room temperature. Stainless steels are the most common form and the structure is determined by the addition of nickel. Manganese can also makes steel austenitic.

AUV: Autonomous underwater vehicle: small unmanned submarine, able to operate independently without an umbilical, in contrast to ROV (q.v.).

Ball Valve: A valve in which a pierced sphere rotates within the valve body to control the flow of the fluids.

Bauschinger Effect: Reduction of yield stress in compression as a consequence of previous plastic deformation in tension, and vice versa.

Beach Marks: Description of markings characteristic of fracture surfaces produced by fatigue crack propagation. These are also called *clamshell*, *conchoidal*, and *arrest marks*.

Bimetallic Corrosion: Also termed galvanic corrosion. Corrosion of two metals in electrical contact and immersed in the same electrolyte in which the metal of more negative corrosion potential is made to corrode at a higher rate by the metal of more positive corrosion potential.

Brass: An alloy consisting mainly of copper (>50%) and zinc to which smaller amounts of other elements may be added; most common alloying elements are arsenic (to reduce dezincification), iron, and lead.

Brinell Hardness Test: A test to determine the hardness of a material by forcing a hard steel or carbide ball of specified diameter into the material under a specified load. The result is expressed as the Brinell hardness number (BHN), which is the value obtained by dividing the applied load in kilograms by the surface area of the resulting impression in square millimeters. A Brinell impression may be used to produce a residual stress in a metal for stress corrosion cracking. A rough relationship between BHN and VHN over the range VHN 100 to 500 is BHN = 0.9345 (VHN) + 53.61

Carbon Steel: Steel is considered carbon steel if there is no minimum specification for aluminum, boron, chromium, cobalt, molybdenum, nickel, titanium, tungsten, vanadium, zirconium, or other alloying element to obtain particular properties and if manganese content is below 1.65% and copper content below 0.6%. Tramp elements arising from the raw materials (copper, nickel, molybdenum, chromium, etc) are present but not significant or reported.

Calcareous Deposits: A scale of largely magnesium hydroxide and calcium carbonate formed by precipitation from hard waters or on cathodically protected surfaces as a result of the change in pH at the surface. The rate and quantity of deposits formed depends on water temperature.

Cast Iron: Usually refers to grey cast iron, which is a ferrous material containing a high content of carbon. The major portion of the carbon occurs as flakes that are interspersed through the ferrous matrix.

Cathode: The electrode of an electrolytic cell at which reduction is the principal reaction. The cathode is the area where electrons are removed by cathodic reactions: oxygen reduction, hydrogen evolution, carbonic acid reduction, and hydrogen sulphide reduction. In some cases, cathodic reactions are the reduction of an element from a higher to lower valency state. Electrons flow towards the cathode in the external circuit. The cathode is also an item, e.g., the pipeline having cathodic protection.

Cathodic Disbondment: The loss of adhesion between a coating and the substrate metal caused by the products of a cathodic reaction, usually cathodic protection. For screening pipeline coatings, the cathodic disbondment is measured at operating temperature.

Cathodic Protection: A technique to reduce the corrosion rate of a metal surface by making it the cathode of an electrochemical cell.

Cation: A positively charged ion that migrates through the electrolyte towards the cathode. Usually metal atoms form cations at anodes, and the charge may range from single to six electron equivalents: Na^+, Fe^{++}, Fe^{+++}, Zn^{++}, Cr^{6+}.

Cavitation: The formation and rapid collapse within a liquid of vapor filled cavities or bubbles. The collapse of the bubbles on a metal surface results in strain hardening of the metal surface and enhanced corrosion damage. Cavitation occurs at areas of high turbulence and pressure change, e.g., pumps, orifice plates, and tees.

Chalk: Degradation of the binder component of a coating as a result of weathering and exposure to UV light releasing pigment particles.

Concentration Cell: Corrosion resulting from the potential differences generated on a metal surface by formation of areas of different concentration of chemical reactants. Typical examples are oxygen and salts.

Corrosion: The chemical reaction between a metal or alloy and its environment.

Corrosion Erosion: Also termed *erosion corrosion*. The conjoint action of corrosion and erosion resulting from the abrasive action of a fluid (either solution or gas) moving at high velocity, thus causing continuous removal of the protective films. Corrosion erosion should not be confused with conventional erosion that occurs when solids are present in the fluid. Impingement and cavitation are related phenomena.

Corrosion Fatigue: A cracking process caused by the combination of repetitive stress and corrosion. For undersea pipelines, the normal fatigue limit does not apply. The behavior of the pipeline steel is restored by adequate cathodic protection.

Corrosion Potential: The potential of a corroding surface in an electrolyte relative to a reference electrode. Also called the *rest potential, open circuit potential, free corroding potential.* See also *electrode potential.*

Couple: Term used to describe two metals associated with galvanic corrosion.

Crevice Corrosion: A type of concentration cell corrosion. Corrosion of a metal that is caused by the concentration of dissolved salts, metal ions, oxygen, or other gases in crevices or pockets remote from the principal fluid stream. The concentration cells result in differential potentials between the area of metal in the concentration cell and the bulk material usually leading to deep pitting. Typical examples are flange faces and valve seats.

Critical Cracking Potential: The electrochemical potential above which stress corrosion cracking occurs but below which it is not observed.

Critical Crevice Potential: The lowest value of potential at which crevice corrosion commences and continues; depends on the test conditions but is useful for ranking corrosion-resistant alloys.

Critical Pitting Potential: The lowest value of potential at which pits nucleate and grow; depends on the test conditions used but is useful as a ranking of corrosion resistant alloys.

Critical Pitting Temperature: The temperature at which pits nucleate and grow; depends on test conditions but is useful for ranking corrosion resistant alloys. Effects of critical pitting potential, temperature and chloride ion concentration, can be related in one diagram as the log of chloride concentration versus temperature for iso-potentials.

Crosslink: Chemical reaction linking two chains in the molecular structure of a coating that changes the final state of the coating; typical reaction in thermoset coatings which include epoxies, polyurethanes, phenolics.

CSE: Copper-copper sulphate electrode: a standard reference electrode fabricated from copper and saturated copper sulphate. It has a potential of +300 mV SHE and a temperature coefficient of 0.9 mV/°C The CSE is used for measurement of the cathodic protection potentials of items in soils; it is *not* suitable for use in seawater.

Current Density: The electrical current flowing to or from a unit area of an electrode surface.

De-alloying: The selective corrosion of one or more components of a solid solution alloy. Examples are dezincification, where zinc is removed from brass leaving behind a copper matrix, and graphitization, where iron is removed from cast iron leaving behind a graphite skeleton.

Depolarization: The removal of factors that hinder the flow of current in a corrosion cell.

Deposit Attack: A form of crevice corrosion resulting from a deposit of salt, silt, insoluble salts, corrosion products, etc. on a metal surface.

Defect: A blemish or imperfection in a material of sufficient size that the material is rejected as out of specification.

Destructive Testing: Testing in which part or the whole item under test is destroyed or damaged such that it is no longer fit for its purpose. Usually refers to tensile, fatigue, and impact testing.

Dezincification: A corrosion phenomenon resulting in the selective removal of zinc from copper-zinc alloys. Other selective de-alloying processes occur, e.g., graphitization.

Dielectric Shield: An electrically non-conductive material such as plastic or resin sheet or pipe that is placed between an anode and adjacent cathode to avoid current wastage and aid current distribution.

Differential Aeration Cell: Electrochemical corrosion resulting from a potential difference arising from variations in transfer of oxygen to a metal surface. The area deficient in oxygen becomes anodic and corrodes.

Doubler Plate: An additional plate or thickness of steel provided to confer extra strength at the point of an attachment to a pipeline or platform. Often used to avoid need for special welding for connection of sacrificial anodes.

DP: Dynamic positioning. Combination of navigation system and thrusters to enable a construction vessel to maintain position without using anchors.

Ductile Iron: A cast ferrous material of high carbon content in which the free carbon is present in spherical form, rather than flake form. Ductile iron has higher tensile and impact properties compared to grey cast iron.

Electrochemical Cell: Also termed *galvanic, electrolytic,* or *voltaic cell.* A combination of metals in which chemical energy is converted into electrical energy. Corrosion in aqueous solutions occurs by an electrochemical cell mechanism.

Electrochemical Equivalent: The weight of an element or alloy corroded at 100% efficiency by the passage of a unit quantity of electricity; typically reported as grams per coulomb.

Electrode: General term for anode and/or cathode.

Electrode Potential: The potential of an electrode in an electrolyte as measured against a reference electrode. The electrode potential does not include any resistance losses in potential in either the solution or external circuit. It represents the reversible work to move a unit charge from the electrode surface through the solution to the reference electrode. The international reference electrode (zero volts) is the standard hydrogen electrode (SHE). The relative electrode potentials of the metals and alloys form the Galvanic Series.

Electric Flash Welded Pipe: Pipe having a longitudinal butt weld formed by heating the junction with an electrical current followed by pressure after adequate heating is attained. Expulsion of metal occurs at the weld area. Recently applied to formation of girth welds.

Electric Fusion Welded Pipe: Pipe having a longitudinal butt weld formed by manual or automatic electric arc welding. The weld may be single or double and formed with or without addition of a filler metal.

Electric Resistance Welded Pipe: Pipe having a longitudinal butt weld formed by application of pressure after heating of the junction area by flow of electrical current through the pipe.

Electrolyte: A solution in which conduction of electrical current occurs by transport of dissolved anions and cations.

Electromotive Force Series (EMF Series): A listing of elements in the order of their standard electrode potentials; a series similar to, but different from, the Galvanic Series.

End Grain Attack: A form of exfoliation corrosion occurring in materials that show pronounced layering of different metallography that have dissimilar corrosion resistance. The exposed through-wall thickness of pipeline steels may be susceptible. This form of attack can lead to subsurface pitting in susceptible steels.

Endurance Limit: The maximum stress that a material can withstand for an infinitely large number of fatigue cycles. In seawater steels lose their endurance limit, but the limit is regained by application of adequate cathodic protection.

Environment: The conditions surrounding the item of interest: physical, chemical, and mechanical.

Environmental Cracking: Brittle fracture of a normally ductile material when the causative factor is a chemical environment. The term includes: *corrosion fatigue, high temperature hydrogen attack, hydrogen blistering, hydrogen embrittlement, hydrogen-induced cracking, stress corrosion cracking, sulphide stress cracking,* and *liquid metal cracking.*

Epoxy: General name for the resins containing the epoxy chemical group and includes epoxy esters, amine-cured epoxies, and epoxy plastics.

Equilibrium Potential: The potential of an unpolarized, reversible electrode.

Erichsen Test: A cupping test in which a piece of sheet metal restrained except at the centre is deformed by a cone-shaped, spherical-ended plunger until fracture occurs. The extent of deformation prior to fracture is a measure of the ductility.

Erosion: Destruction of metals or other materials by the abrasive action of moving fluids and is usually accelerated by the presence of solid particles or matter in suspension. When corrosion occurs simultaneously, the term *erosion corrosion* is often used.

Erosion Corrosion: Enhanced destruction of a metal or other material by the abrasive action of a moving fluid. The shear forces imposed by the fluid remove protective layers on the material surface.

ERW: Electric resistance welding, see chapter 5.

Exfoliation Corrosion: Localized subsurface corrosion in zones parallel to the surface that results in thin layers of uncorroded metal resembling pages of a book. In pipelines it is associated with end grain attack.

Exposed Line: A pipeline with the crown of the pipeline protrudes above the seabed in a water depth less than 5 m at mean low water.

Fatigue: The phenomenon leading to fracture under repeated or fluctuating stresses having a maximum value less than the tensile strength of the material. Fatigue fractures are progressive, beginning as minute cracks that grow under the action of the fluctuating stress. Spanning pipelines may suffer fatigue.

Fatigue Strength: The stress to which a material can be subjected for a specified number of fatigue cycles.

Ferrite: A solid solution of one or more elements in body-centered cubic iron. Unless otherwise designated, e.g., as chromium ferrite, the solute is generally assumed to be carbon. On some equilibrium diagrams there are two ferrite regions separated by an austenite area: the lower area is alpha ferrite and the upper delta ferrite. Unless otherwise designated alpha ferrite is assumed.

Filliform Corrosion: Corrosion that occurs under some thin film coatings in the form of randomly distributed thread-like filaments.

Firing Line: In lay barge pipelaying, line of rollers on which the pipe moves through a series of welding stations.

Flammable Range: The concentration range of a gas in air over which it will spontaneously ignite when exposed to an ignition source. Below the lower flammable limit (LFL), the gas is too lean to ignite while above the upper flammable limit (UFL), the mixture is too rich to burn.

Table A–1 Flammable Ranges for Oilfield Gases and Vapors

Gas/Vapor	Formula	SG air = 1	Ignition temp °C	LFL % gas	UFL % gas
Methane	CH$_4$	0.55	645	5.3	15.0
Natural Gas	Mixture	0.65	628	4.5	14.5
Ethane	C$_2$H$_6$	1.04	~564	3.0	12.5
Propane	C$_3$H^{*8}	1.56	~551	2.2	9.5
Butane	C$_4$H$_{10}$	2.01	~529	1.9	8.5
Hexane	C$_6$H$_{14}$	3.0	225	1.1	7.5
Gasoline	Blend	3–4	333	1.4	7.6
Carbon monoxide	CO	1.0	609	12.5	74.0
Hydrogen	H$_2$	0.1	500	4.0	75.0
Hydrogen sulphide	H$_2$S	1.2	260	4.0	44.0

Fouling: Accumulation of deposits.

Fracture Mechanics: The science of the cracking behavior of a cracked body under stress, taking into account applied stress, crack length, and specimen geometry.

Fretting Corrosion: Deterioration by corrosion at the contact area between metals subjected to small repeated relative displacements under load; can result in galling and wear.

Furnace Lap Welded Pipe: Pipe that has a longitudinal weld formed by the forge welding process in which the weld is formed by heating the pipe to welding temperature and passing the pipe over a mandrel and under rollers that compress the weld area to join the overlapping edges.

Galvanic Anode: Also termed sacrificial anode, a cast metal that provides electrical current to protect a metallic structure that is fabricated from a material more noble in the electrochemical series when in a suitable electrolyte.

Galvanic Current: The electric current flowing between two dissimilar metals or alloys in a galvanic couple. The size of the galvanic current is related to the potential difference between the metals and the ratio of the areas of the dissimilar metals.

Galvanic Series: A listing of metals and alloys arranged according to their electrochemical or corrosion potentials in a given environment.

Galvanizing: Coating of iron and steel with zinc using a bath of molten zinc. Sometimes the term is incorrectly used to describe zinc coatings on steel applied by electrodeposition.

Gate Valve: A full opening and closing valve that relies on deformation of the mating surfaces for sealing.

General Corrosion: A form of deterioration that is distributed more or less uniformly over a surface. Surveys indicate that general corrosion is about 30% of all corrosion, but it is not often observed in pipelines.

Globe Valve: A valve that seals the flow orifice in a spherical valve body with a plug moved by a stem driving the plug into a seat formed in the orifice.

Graphitization: A form of de-alloying in grey cast iron in which the iron corrodes leaving behind a graphitic matrix holding corrosion product. The metal appears uncorroded until grit blasted whereon the soft graphitized material is removed. Graphitization often occurs as localized plugs of materials and can occur in pipelines in water system valve castings.

Ground Bed: The buried collection of anodes that provide the path for the cathodic protection current to the protected item; a term usually reserved for on-land impressed current cathodic protection systems.

Half Cell: One of the electrodes with its immediate environment of an electrochemical cell.

Hardness: Measure of the deformation of a surface by an imposed force acting on a penetrating object, usually a sphere or pyramid. In pipeline engineering, Rockwell, Vickers, and Brinell hardness values are encountered. Hardness can be related to the tensile strength of a steel and is a suitable method for non-destructively testing materials.

HAZ: Heat affected zone, see chapter 5.

Heat: Quantity of metal produced by a single batch melting process.

Heat Analysis: Chemical analysis taken from material from one heat.

Heat Affected Zone (HAZ): The part of the base metal that was not melted during welding but where the microstructure and properties of the material were altered by the heat of the process. See also *sensitization.*

Holiday: A hole or discontinuity in a coating that exposes the metal surface to the environment.

Hoop Stress: The circumferential stress in a pipe wall resulting from the pressure of the contained fluid.

Hot Tap: The process of making a connection to an operating pipeline while the pipeline is under pressure.

Hydrogen Blistering: Voids produced within a metal by absorption of atomic hydrogen that forms molecular hydrogen at non-metallic inclusions within a low strength (usually) alloy and results in surface bulges.

Hydrogen Electrode: An electrode at which the equilibrium between hydrogen ions (H+) and atomic hydrogen is established at the surface of a platinum electrode. Under certain specified conditions, the potential of this electrode is given an arbitrary value of 0.00 V, and this Standard Hydrogen Electrode forms the reference potential for all other potentials.

Hydrogen Embrittlement: A condition of low ductility in metals, resulting from the absorption of hydrogen. Embrittlement can lead to fracture at stresses below the yield stress.

Hydrogen Induced Cracking (HIC): Hydrogen blistering in which the internal circumferential blisters link together by stepped cracking radial to the pipe wall possibly leading to a loss of integrity of the material. The blistering occurs at the mid-wall in SAW or UOE pipe and at the inner surface of seamless pipe. Also termed *Cotton cracking* and *hydrogen pressure induced cracking* and *stepwise cracking* (SWC); occurs in sour service pipelines.

Hydrogen Stress Cracking: The result of conjoint stress and hydrogen absorption that occurs in high strength alloys. The hydrogen may result from sour service or excessive cathodic protection though cases have occurred in seawater as a result of corrosion.

I D: Inside diameter.

Imperfection: Identifiable irregularity in a material that can be reliably detected by an inspection technique.

Impingement Attack: Corrosion associated with turbulent flow of liquids that may be accelerated by gas bubbles.

Immunity: Condition of non-corrosion of a metal or environment often used to describe the environmental conditions and potential zones where a metal or alloy does not corrode; a domain in an E-pH diagram.

Impressed Current: Direct current supplied by a power supply device external to the electrode system of a cathodic protection system.

Inclusion: A non-metallic phase in a metal. Examples are oxides, sulphides (manganese sulphide), and silicates.

Inhibitor: A chemical substance or combination of substances that prevent or reduce a chemical or physical reaction when in the proper concentration in an environment but that does not react with the components of the environment. Typical reactions requiring inhibition are corrosion, scaling, and wax crystal formation. Anodic inhibitors are substances that reduce the anodic reaction; cathodic inhibitors attach to cathodic areas and reduce the cathodic reactions. Most pipeline inhibitors are organic materials containing amines and/or other surface-active groups that attach to both anodic and cathodic areas.

Interaction: Also termed *interference* and *stray current corrosion;* corrosion of a metal structure caused by stray currents flowing from another structure. The current may arise from cathodic protection systems or other DC source, e.g., electric railway, tramway, earthing loops.

Interference Current: Also termed *interaction* or *stray current corrosion.*

Intergranular Corrosion: Preferential corrosion at or adjacent to the grain boundaries of a metal or alloy. Some stress corrosion processes are intergranular.

Ion: An electrically charged atom or group of atoms. See *anions* and *cations.*

J-lay: Version of lay barge pipelaying in which the pipe takes the form of a letter J. (See Figure 12–5.)

J-tube: A curved tubular conduit designed and installed on a platform to support and guide one or more pipeline risers, umbilicals, or cables.

Joint: The connection between to pieces of pipe or a length of pipe that will be joined to other sections of pipe to form the pipeline.

K_{ISCC} : The lowest stress intensity at which stress corrosion cracking has been observed for a given metal or alloy in a given environment. The units are stress times the square root of length: ksi $(in)^{0.5}$ or $MNm^{-3/2}$.

Knife-line Attack: Intergranular corrosion of an alloy along a line adjacent to a weld usually seen in sensitized stainless steels but has occurred in carbon steels in micro-aerobic waters.

Langelier Index: A numerical value of scale index predicting the likelihood of calcium carbonate scaling in natural waters. This index is of limited use in seawater.

Lay Barge: Anchored or dynamically positioned barge used to lay pipe.

Length: A piece of pipe as delivered from a pipe mill. The piece of pipe is termed a length independent of the actual dimension. Often also termed a joint.

Liftoff Point: In lay barge pipelaying, point at which the pipeline loses contact with the stinger.

Localized Attack: Common form of corrosion in which an area of metal surface is predominantly anodic and another area is predominantly cathodic, resulting in discrete loss of material at the anodic area.

Long-line Corrosion: Corrosion resulting from long-line currents that occur on welded pipelines as a result of the pipeline transiting different areas of soil, e.g., clay and sand.

Mandrel: A metal bar that serves as a former for casting, moulding, bending, or forging a metal to a predetermined geometry.

MAOP: Maximum allowable operating pressure: the maximum pressure at which a pipeline may be operated. The MAOP may be limited by either regulation or design code. MOP is the maximum actual operating pressure that occurs during normal operation during a year of service and is generally set by the operating company.

Martensite: (1) In an alloy, martensite is a metastable transitional structure intermediate between two allotropic modifications whose abilities to dissolve a given solute differ considerably, the high-temperature phase having the greater solubility. The amount of high-temperature phase transformed to martensite depends largely on the temperature attained in cooling, as there is a distinct beginning temperature.

(2) A metastable phase of steel formed by the transformation of austenite below the A_r temperature. It is interstitial super-saturated solid solution of carbon in iron having a body-centered tetragonal lattice. Its microstructure is characterized by an acicular (needle-like) pattern. Martensite is the structure responsible for the hardness and strength of traditionally quenched and tempered steels.

Mercaptan: An organic material containing sulphur groups that has a characteristic odor. Mercaptans may dissociate to release hydrogen sulphide and are often included in the total hydrogen sulphide concentration when evaluating the need to conform to NACE MR-0175.

Mesa Corrosion: Corrosion that leaves some sections of a surface relatively uncorroded, but exhibits extensive pits between sections where they are joined together (like the canyon and mesa topography of parts of the United States southwest).

Mill Scale: A heavy oxide layer formed on a metal during hot fabrication or heat treatment. On steels it is magnetic and may be laminar in nature. Mill scale must be removed before coating and is generally also removed from the inside of pipelines in corrosive service.

Mil: 1/1000 in.; used in corrosion rates as mils/year, mpy, as an alternative to mm/year.

Mixed Potential: The potential arising from two or more electrochemical reactions occurring simultaneously on one metal surface.

Mud Line: The ocean floor at the location of interest; not necessarily associated with actual mud.

Needle Valve: A small valve that is used to control flow by moving a pointed plug, or needle, within a flow orifice.

Nernst Equation: The equation describing the electromotive force of a cell in terms of the activities of products and reactants in the cell. The Nernst Layer is the diffusion layer defined as: $d = n\ F\ D\ DC/i$ where n is the valency, F is Faraday's Number, D is the diffusion coefficient, DC is the difference between surface and bulk concentration of the species, and i is the diffusion limited current density.

Noble: The positive direction of electrochemical potential. The term was once used to describe metals that are non-corrodible in normal environments, e.g., gold, platinum, and silver. A Noble Potential is a potential more cathodic (positive) than the standard hydrogen potential.

Nominal Wall Thickness: The pipeline wall thickness calculated from the maximum hoop stress formula. In certain circumstances, the pipe may be purchased to this wall thickness without adding an allowance to compensate for under-thickness tolerances permitted in specifications.

Non-Destructive Testing: Testing in which the item being tested is not damaged during the testing and remains serviceable.

Normalizing: Process of heating a steel to a temperature above the transformation range to produce austenite, holding the steel at that temperature for a period, and then cooling it in still air to a lower temperature to permit re-crystallization and tempering. The normalizing process is the minimum requirement for pipeline steels and evens out metallurgical variations in the steel resulting in more uniform properties.

OD: Outside diameter.

Open Circuit Potential: The potential of an electrode measured against a reference electrode when no current flows into or out of the electrode; usually used for defining the potential of an anode or corrosion resistant alloy prior to application of cathodic protection.

Overbend: Section of pipe in which the vertical profile is convex upward, as in an upside-down letter U; in lay barge pipelaying, the section of pipe over the stinger.

Oxidation: Loss of electrons by combination in a chemical reaction. *Oxidation* occurs at the cathode. Oxidation is also used to describe the combination of metals with oxygen at high temperatures.

Passive: (1) The state of the metal surface characterized by low corrosion rates in a potentially highly corrosive environment (i.e. a region that is strongly oxidizing for the metal). Passivity is generally due to the formation of thin, tight, oxide films invisible to the naked eye.

(2) A term used to describe sacrificial anodes that have ceased to be active because of the formation of a non-reactive layer on the surface.

Passivity: The state of being passive.

Passive-Active Cell: A corrosion cell in which the potential difference arises from the open circuit potentials of the same metal when some areas are active and others passive; usually occurs on corrosion resistant alloys that rely on passive films formed on the metal surface for their corrosion resistance.

Passivation: A reduction of the anodic reaction rate in corrosion.

pH: A measure of the hydrogen ion activity defined as $pH = \log_{10}[1/a_H]$ where a_H is the hydrogen ion activity and equals the molar concentration of hydrogen ions multiplied by the mean ion activity coefficient.

Pig: Piston-like device driven through a pipeline to clean the line, to separate different fluids, or to carry sensors to measure corrosion, wall defects, or the position of the pipe.

Pitting: Corrosion of a metal surface that is confined to a point or small area forming cavities. Pits on carbon steel tend to be hemispherical while pits on corrosion-resistant materials may have an adverse depth to area ratio.

Pitting Factor: The ratio of the depth of the deepest pit resulting from corrosion divided by the average penetration calculated from weight loss.

Polarization: The change of electrode potential as a result of reactions at the electrode surface; anodic polarization is a shift in potential to the noble direction (more positive), and cathodic polarization is a shift of the potential in the base direction (more negative).

Polarization Curve: A plot of the current density versus electrochemical potential for a specific electrode-electrolyte combination. Usually the current density is plotted as the logarithmic value. The shape of the curve gives information on the nature of the corrosion processes and their rates.

Polarization Resistance: The ratio of current to potential change over a limited portion of the polarization curve. It is also used to describe a method of measuring corrosion: LPRM for linear polarization resistance measurement.

Potential: Also termed *electrode potential*. Potential difference at an electrode/solution interface defined with reference to another specified reference electrode.

Potential-pH Diagram: A schematic representation of the domains of immunity, corrosion, and passivity as a function of potential and pH for a metal in a specific electrolyte; also termed *Pourbaix Diagrams* after their initiator.

Potentiodynamic: A corrosion measurement technique in which the metal or alloy potential is continuously altered, and the resulting current density measured; provides a potentiodynamic polarization curve.

Potentiostat: An electronic device that allows potentiodynamic measurements to be made. It includes a feed back loop to a reference electrode such that the current flow is altered to preserve the potential.

Poultice Effect: The concentration of corrosive chemical species under debris or other material forming a shield over the substrate metal. In pipelines the presence of poultices may also reduce the effectiveness of the corrosion inhibitor.

Precipitation Hardening: Hardening caused by the precipitation of a constituent from a super-saturated solid solution.

Primer: The first coat of paint applied to a surface. It is usually formulated to have good bonding and surface wetting characteristics, and it may contain inhibitive pigments. For pipelines, primer is often used to hold the quality of a blast prior to main coating.

Profile: The anchor pattern formed on a surface by abrasive blasting.

Redox Potential: The electrochemical potential measured on a platinum surface that expresses the equilibrium electrode potential of reversible oxidation-reduction reactions. It is used to evaluate the risk of differential aeration cells and of microbiological corrosion. Low Redox potentials represent reducing environments, e.g., anaerobic or sulphidic.

Reference Electrode: Sometimes termed *Standard Electrode*. A half-cell of reproducible and known potential by means of which an unknown electrode potential may be determined from the electromotive force (emf) of the cell composed of these two electrodes. The reference electrode must always be specified when quoting the potential. Commonly used reference electrodes are copper-copper sulphate, silver-silver chloride, calomel (mercury-mercurous chloride), and pure zinc.

Resin: A class or family of plastics or polymers used as a binder for coatings. Typical resins include alkyd resins, vinyl resins, ester resins, and epoxy resins.

Riser: The section of pipeline extending from the ocean floor up the platform.

Rockwell Hardness: A test for determining the hardness of a material by forcing a pyramid of a specified diameter into the metal surface under a specified load. The result is specified a Rockwell Hardness X where X specifies the specific test depending on load and pyramid dimensions. Rockwell Hardness is widely used in North America. The C test is most applicable to pipeline steels and is related to Vickers hardness by comparisons in ASTM E140 and BS 860. Rough relationships over the range from VHN 250 to 450 are:

$$HRC = 39.97 \ln(VHN) - 183.85 \text{ for ASTM E140 and HRC}$$
$$= 38.163 \ln(VHN) - 172.19 \text{ for BS 860.}$$

ROV: Remotely-operated vehicle: Small unmanned submarine connected to a mother ship by an umbilical cable.

Rust: Corrosion product of iron though may be used for corrosion products on other metals and alloys. Rust may be ferrous or ferric hydroxides or oxides and can range in color from white to green and brown and red, depending on the nature of the product and the mixture.

Sacrificial Protection: Reduction of corrosion of a metal by connecting it to a baser material in a galvanic cell. In cathodic protection, sacrificial protection is the attachment of galvanic alloys (magnesium, zinc, or aluminum alloys) to the buried or submerged structure. The term can also refer to the use of sacrificial coatings such as galvanizing.

Sagbend: Section of pipe in which the vertical profile is convex downward, as in a letter U. (See Figure 12–2.)

SAW: Submerged-arc welding. (See chapter 5.)

Scale: Thick, visible oxide (or sulphide) film formed during the high-temperature oxidation of a metal.

SCE: Standard calomel electrode. A standard reference electrode fabricated from mercury amalgam and mercuric chloride salts in a saturated, molar, or tenth molar KCl solution. The SCE is used for measurements of electrochemical potentials in laboratories and seawater. The SCE has a potential ranging from +241 to 334 mV SHE and a temperature coefficient of 0.22 to 0.59 mV/°C depending on solution concentration.

Sensitizing: Also termed sensitization. A heat treatment, which may be accidental or incidental (during welding), that causes carbide precipitation at grain boundaries and can cause the alloy to lose corrosion resistance at these areas; it may also result in stress corrosion cracking of the affected area.

SHE: Standard hydrogen electrode. The international reference electrode based on a platinum black electrode in 1N H_2SO_4 saturated with hydrogen gas at 1 atmosphere at 25 °C. The SHE has a defined potential of zero volts and a temperature coefficient of 0.67 mV/°C. Metals and alloys with positive potentials are sometimes termed noble; those with negative potentials are termed *base*.

Silver-Silver Chloride Electrode: A reference electrode that uses a silver chloride coated silver wire in conjunction with potassium chloride solution and seawater as the junction electrolyte; the most commonly used electrode for offshore cathodic protection evaluation.

Sigma Phase: An extremely brittle iron-chromium phase formed at elevated temperatures in Fe-Ni-Cr alloys.

Six o'clock Corrosion: Internal corrosion at the lowest point of the pipeline cross-section.

S-lay: Version of lay barge pipelaying in which the pipe takes the form of a letter S. (See Figure 12–2.)

Slip: A deformation process involving shear movement of a set of crystallographic planes.

Slow Strain Rate Testing: An experimental technique for evaluating the susceptibility of a material to stress corrosion cracking. A specimen of the material is strained to failure under uniaxial tension while exposed to the test environment. Subsequent to failure, the fracture surface is examined for evidence of stress corrosion cracking (SCC).

Solution Heat Treatment: Heating an alloy to a suitable temperature and holding it at temperature to allow one or more components to enter into solid solution then cooling rapidly to retain the constituents in solution.

Spalling: Breaking away or fragmentation of a surface coating; often also used to describe the failure of concrete cover.

Span: Section of pipeline not in contact with the seabed.

Splash Zone: The zone of a platform or riser that is alternately in and out of the water because of tides and waves. Excluded from this zone are surfaces that are wetted only during major storms.

SRB: Sulphate–reducing bacteria.

SSCE: Silver-silver chloride electrode. A standard reference electrode fabricated from pure silver plated with silver chloride and exposed to a normal or saturated KCl solution. SSCE is widely used for measurement of potentials in seawater and has potentials of +242 or 288 mV SHE and a temperature coefficient of 0.22 mV/°C, depending on solution concentration.

Stinger: Rigid or articulated structure attached to a lay barge to support the pipeline during laying. (See Figure 12–2.)

Strain-hardening: Increase of yield stress as a consequence of continued plastic deformation.

Stray Current Corrosion: Corrosion resulting from the direct current flow along paths other than those intended; results from DC traction systems, abnormal earths or cathodic protection systems. When resulting from within CP systems, it may be termed interaction.

Stress Corrosion Cracking (SCC): Cracking of a metal by the conjoint action of stress and a corrosive environment. The stress may be residual or applied or a combination of the two.

Sulphate-Reducing Bacteria (SRB): Species of gram-negative anaerobic bacteria that oxidize organic materials using the oxygen in the sulphate ion, consequently producing large quantities of sulphide and hydrogen sulphide. The SRB cause rapid localized corrosion of iron and steel in near-neutral soils and waters that are devoid of oxygen.

Sulphide Stress Cracking (SSC): Brittle failure by cracking by the combination of stress and the presence of water and hydrogen sulphide. The stress may be residual or applied or a combination. SSC was formerly called *sulphide stress corrosion cracking.*

Submerged Zone: The zone on a structure or riser that extends downwards from the splash zone and includes any sections below the mud line.

Tafel Slope: A linear portion of a polarization curve when the curve is plotted as log current to potential. Extrapolation of the slope to the rest potential gives the corrosion rate. The change in potential can be related to the current by the relationship:

$$h = \pm\, b \log (i/i_o),$$

where b and i_o are constants related to the metal environment and i is the current density.

Tensioner: (in lay barge pipelaying) Tracked or wheeled device that applies tension to a pipeline to control the curvature in the sagbend.

Thermoplastic: A plastic or coating capable of being repeatedly softened by heating and hardened by cooling. Coal tar and bitumen-based enamels are thermoplastic coatings.

Thermosetting: A material that undergoes a chemical reaction from the action of heat and pressure or catalysis or ultraviolet light leading to a relatively infusible state. Crosslinkage of the material is usually involved.

Throwing Power: The distance an electrode can affect the potential of an item. It is usually restricted to plating baths and cathodic protection.

Tie Coat: A special purpose intermediate coat used as a bridge between a primer and the finish coat to overcome incompatibility or application problems between the primer and the finish coat. Sometimes used to describe the adhesive used for polyethylene and polypropylene coatings.

Touchdown Point: In lay barge pipelaying, point at which the pipeline reaches the seabed.

Transpassive: The noble region of potential where an electrode exhibits a higher than passive current density; related to the behavior of corrosion resistant alloys.

Tuberculation: Formation of localized corrosion products as hollow mounds on the surface of metals. Usually describes corrosion products formed by biological action in water pipes.

Underfilm Corrosion: Corrosion occurring under organic films and termed *filiform corrosion*; also a term for corrosion under apparently intact coatings despite application of cathodic protection.

Urethane: A chemically cured coating consisting of vinyl, vinyl acrylic, or acrylic base reacted with an isocyanate converter to form a tough, durable, and glossy coating; usually used as a top coat.

Vickers Hardness (VHN): A test for determining the hardness of a material by forcing a hard pyramid into the material surface under a specified load. The result is expressed as Vickers Hardness Number X where X is the applied load and the numerical value is obtained by dividing the applied load by the diagonal of the impression. VHN is widely used in Europe. In sour service, the critical hardness for carbon steel is 248 VHN.

Voids: Holidays in coatings; holes within coatings, both organic and inorganic; shrinkage holes in castings and welds. Sometimes the term *void* is used to describe vacancies in metals subject to overstraining.

Vortex-induced Vibration (VIV): Oscillation of risers or spans induced by shedding of vortices.

Wear Plate: A sacrificial member attached to a structure or platform, usually in the splash zone, to protect it from anticipated corrosion and erosion from high velocity water, ice, or sand movement.

Weld Decay: Intergranular corrosion adjacent to welds usually observed on sensitized stainless steels in the HAZ but has been observed on carbon steels in micro-aerobic seawater.

Appendix B
Codes and
Standards

B–1 Background

Why are there codes? Where do they come from?

One can trace codes back in various directions. Roman engineers probably had rules of proportion for structures like aqueducts and roads. Medieval masons had rules for arches and vaults, not derived from structural mechanics but worked out slowly from experience of what stood up and what collapsed. Modern codes go back to the third quarter of the 19th century. There was a huge expansion of infrastructure, particularly railway bridges, and rapid industrial development, above all in boilers and ships. There were what is now regarded as an intolerable number of failures. Gies' book on bridges claims that of bridges built in the United States in the 1870s, there were 40 bridge failures a year. One bridge in four failed. There were some dramatic failures, among them the Ashtabula bridge collapse that killed 80 people and the Tay Bridge disaster in the UK.

A supposedly competent engineer carried out the design of the Tay Bridge, but he grossly underestimated the wind loads and made some padeyes of brittle castings. The Ashtabula Bridge was designed by a railroad manager who

had very limited design experience. The American Society of Civil Engineers (ASCE) said that the "construction of the [Ashtabula] truss violated *every canon of our standard practice*," but the standard practice had not been written down. That oversight led to pressure for standard practices to be recorded and codified in some formal way as a statement of minimum requirements for the protection of the community. It remained true that a design might meet the code requirements and still be a bad design: unnecessarily costly, or unsafe, or environmentally damaging, or ugly. That is still true today.

Who writes the codes? The answer to that question is complicated and variable. The responsibility is some national or international authority that has a hierarchy of committees of the great and the good. Often the committee subcontracts the original writing to a consultant or a group of consultants, then works on the draft, and sends it out for comment from the relevant industry. Approval may require more than one review cycle, and ultimately the committee issues the code document. Almost immediately some people complain and object, and after a few years of bringing pressure to bear for change, some revisions are carried out. The process is far from perfect.

Different countries have different codes, and they compete for attention in a way that has curious parallels with scientific communication. In Europe, for instance, there used to be many pipeline codes. In the UK there was the Institute of Petroleum IP6, which is exceedingly old-fashioned and reflects the notions of 50 years ago. About 20 years ago, there came a wave of new development, prompted by the offshore industry and led by the Norwegian organization Det norske Veritas, DnV, originally a ship classification society. It produced the 1981 code (a slightly uncritical assemblage of the collective wisdom of the time, but an appropriate response to the immediate need for something reasonably sensible for a new industry in Norway) then created the 1996 code (partly founded on intensive research on reliability theory), and latterly the 2000 code (which leaps forward in various directions).

Other countries have trailed along. The Dutch did some excellent work but did not promote it (and did not have it translated for a long time), and so their NEN3650 is unknown outside Holland. Similarly, the Germans produced the Germanischer Lloyd code but did not have it translated and the British rewrote BS 8010. There are three American codes, among them the relatively new API RP 1111, which reflects limit state design ideas. Then there is a quite different Canadian code. There also is International Standards Organization (ISO) code, ISO 13623, and a Eurocode. Naturally, it would be much better if there were fewer codes, ideally just one that everybody accepted and used and on which attention and research development would focus. There is no good reason not to have such a code, but one should not expect it to arrive any time soon. Nationalism and *not-invented-here* unfortunately play a part in the delay.

Codes are a product of imperfect and fallible human beings. They have mistakes and inconsistencies, and sometimes they are prey to special interests. BP had a wonderful chief metallurgist, Harry Cotton, who, unhappily, is no longer with us but who is famous for flying to Japan to buy 600 miles of 48-in. pipe for the Alaska pipeline before the design of the pipeline had been done. "When you think about it," Harry said once, "the minimum requirements for line pipe are going to be the best that can just be done by the least competent pipe mill represented on the code committee."

Inevitably there are mistakes. Here, for example, is the formula in the original version of BS8010 part III for the maximum hoop stress in an initially-unstressed thick-walled cylinder loaded by internal pressure p_i and external pressure p_o, inside diameter D_i and outside diameter D_o:

$$s_H = (p_i - p_o) \frac{D_o^2 + D_i^2}{D_o^2 + D_i^2} \qquad (B.1)$$

It was meant to be the Lamé formula for thick-walled cylinders and is supposedly more exact than the various versions of the Barlow formula. As can be seen immediately, the formula must be wrong. It says that if p_i and p_o are equal then the stress is zero. But in that case, the stress is uniformly compressive and equal to $-p_o$. The correct formula is as follows:

$$s_H = (p_i - p_o) \frac{D_o^2 + D_i^2}{D_o^2 + D_i^2} - p_o \qquad (B.2)$$

Similarly, the same code mistakenly added a factor of 1000 into a formula for shear stress, so that one component is a 1000 times larger than it should be.

It follows that codes need to be examined with a critical eye. But the opposite trap is to conclude too rapidly that some unexplained provision in a code is not a sensible one. It may in reality reflect some sound experience, possibly experience that cannot be quantified mathematically but is, nevertheless, important to take into account.

B–2 Alternative Code Philosophies

Some codes set rather general objectives, such as the objective that the probability of some kind of failure obeys the *as low as reasonably practicable* (ALARP) principle, which begs a number of questions. Other codes are far more prescriptive and impose quantified restrictions on design, such as the requirement that the hoop stress calculated according to a given formula be smaller than some specified fraction of the specified minimum yield stress, in turn defined in a closely-specified way. The modern trend in safety is goal-oriented, rather than prescriptive. Codes generally are prescriptive about some aspects of design but goal-setting about others. Different countries have different notions of what a code should be, and this difference is likely to remain a controversial issue as the industry slowly moves toward a universally accepted international code.

B–3 The Influence of Limit State Concepts

Early codes adopted an allowable-stress approach in the areas for which they were quantitatively prescriptive. For example, many codes limited the hoop stress under the maximum operating pressure to 0.72 of the specified minimum yield stress. The historical basis of that figure was that pipelines were pressure-tested at a pressure that induced a hoop stress 0.90 of specified minimum yield. It was judged reasonable to require a safety factor of 1.25 between the test pressure and the maximum operating pressure. The factor of 1.25 was not based on calculation, but on judgment and experience. There is no reason to suppose that it was not a safe number, though it is low by comparison with safety factors applied in other areas of structural engineering. It can be argued that a lower factor is justified by the fact that overload above the nominal maximum load is less likely to occur in pressurized systems protected against overpressure than it is to occur in structures loaded by more varied and less controllable loads such as wind, waves, and traffic.

Structural engineering has seen a strong trend away from allowable stress design and toward design based on the idea of a limit state. A limit state is a condition that directly threatens the continued safe operation of a system. In a steel-framed building, for example, exceeding a specified fraction of minimum yield stress is not a limit state because it does not necessarily threaten continued operation. On the other hand, partial collapse is a limit state because it is not consistent with continued operation. The concept of limit state design is to concentrate attention on the provision of margins of safety against limit states, rather than on indirect conditions such as certain stress levels. Continuing the example of a steel-framed building, one limit state is collapse, and the modern approach is to calculate the collapse load and to make sure that the largest applied load is no larger than the collapse load divided by a prescribed load factor greater than 1. Another limit state is excessive deflection to a limit that fractures cladding and finishes or makes the occupants feel unsafe. There the consistent approach is to restrict deflections to a limit deflection divided by some factor that reflects various sorts of uncertainty. The same limit state ideas have been incorporated in codes covering concrete and geotechnics, and, indeed, that it can be argued that geotechnics has always been based on limit states. When designing a slope or a retaining wall, the appropriate thing to consider is the factor of safety against collapse, not what the stresses are.

The following are among the limit states that marine pipeline design has to consider:

- Pressure containment

- Local buckling

- Large-scale buckling

- On-bottom stability

- High-cycle and low-cycle fatigue failure

- Fracture

- Excessive ovalization

The application of rigorous limit-state thinking to these different limit states is progressing at different rates and is not always consistent.

Limit-state design, sometimes more formally called load and resistance factor design (LRFD), is more rational than allowable-stress design. It can make possible valuable economies if it removes artificial constraints that have nothing to do with limit states or with real safety. The best example is the removal of the artificial limitation of longitudinal stress in constraint pipelines. All codes impose a limit on hoop stress, but traditional codes such as IP6 impose a second condition on an equivalent stress, which represents a combination of the hoop stress (always tensile) and the longitudinal stress (usually compressive) and is defined so as to be consistent with a yield condition (usually the von Mises condition). This becomes the governing condition for pipelines that operate at high temperatures, more than about 180 °F in a typical case, and it forces the designer to increase the wall thickness greatly. Indeed, at very high temperatures, more than about 300 °F in a typical case, it becomes impossible to design a constrained pipeline to meet the code conditions.

Allowable-strain design recognizes that reaching yield under a combination of longitudinal and hoop stress is not in itself a limit state and imperils the safety of a pipeline. It replaces the limit on equivalent stress with a much less restrictive limit on plastic strain, subject to certain conditions that are set out in the notes on structural analysis of pipelines.

A second thread in the argument is the influence of structural reliability analysis, which attempts to quantify the safety of structures by examining the statistics of the variability of loads and strength.[1] The most determined and ambitious effort to apply this to pipelines has been the submarine pipeline reliability-based design guideline (SUPERB) project, a joint industry project with 12 participants (Conoco, Elf, EMC, Exxon, HSE (UK), Norsk Agip, Norsk Hydro, NPD (Norway), Phillips, Shell, SNAM and Statoil) centered in Norway, which assembled databases of variability of pipe dimensions and yield stress. The yield stress database assembled more than 1000 data points from more than 20 projects, and from steels ranging from X60 to X80. The wall thickness database assembled more than 1000 data points from more than 17 projects.

An inherent difficulty with the application of reliability analysis is that pipelines and other civil engineering structures must be designed to have a very low probability of failure, and that failure is associated with the extreme tails of the statistical distributions, with some combinations of unusually low strengths and unusually high loads. Determining these tails with any degree of precision requires unimaginably large quantities of data, which are never going to be available.[2, 3] A way of evading the problem is to mix lots of data together. Rather than securing reliable statistics for apples and oranges separately, it is possible to mix apples from different orchards, oranges, melons and blueberries. This is unlikely to generate useful information. In fairness, it should be added that not all researchers share the writers' extreme skepticism.

Structural reliability analysis has different levels. Pure limit state design is level one. Level two is the application of statistics as an input into deterministic design, applying it to inform the choice of margins of safety. Most modern codes are influenced by level one. DnV 1996 and 2000 go further, and their development used some level two ideas, which are examined in section 5.

B–4 Risk

Risk is the product of probability and consequence. For example, the task of walking across a narrow and springy plank, with each end resting on a heap of sand three feet high carries a certain risk of falling off as one walks across the plank. However, if one does fall off, the consequence is probably minor, no more than a jolted knee (unless one is carrying a valuable piece of porcelain). In contrast is the example of walking across the same plank placed across a canyon a 100 ft deep. The probability of falling off is the same, but the consequences are more severe, and so the risk is higher. People apply this obvious idea a hundred times a day in their everyday lives.

Codes formalize the idea of risk by applying higher margins of safety to conditions where the consequences of failure are severe because they imperil life, lead to major secondary damage, or cause pollution. DnV 2000 addresses risk in three ways: by distinguishing between different fluids contained in the pipeline, by distinguishing between different locations, and by distinguishing between temporary and permanent exposure. Fluids fall into one of five categories (reference 4, table 2–1), identical to the classification in ISO 13283:

category A	Non-flammable water-based
category B	Flammable and/or toxic liquids at ambient temperature and atmospheric pressure (e.g., oil, methanol)
category C	Non-flammable and non-toxic gases at ambient temperature and atmospheric pressure (e.g., nitrogen)
category D	Non-toxic natural gas
category E	Flammable and/or toxic fluids, which are gases at ambient temperature and atmospheric pressure, and are conveyed either as gases or as liquids (e.g., hydrogen, propane, NGL, chlorine)

and location classes into two [reference 4, table 2–2]

location 1　　　　No frequent human activity

location 2　　　　"In the near platform are a or with frequent human activity" the extent to be determined by risk analysis or taken as 500 m from a platform

The 500 m figure is traditional, and there is obviously scope for argument about the risk analysis.

These classes are the combined into safety classes. Safety class *low* applies when failure implies low risk of human injury and minor environmental consequences. Safety class *normal* applies when failure implies risk of human injury, significant environmental pollution, or very high economic or political consequences and is stated to apply outside the immediate area of a platform. Safety class *high* applies when failure implies high risk of human injury, significant environmental pollution, or very high economic or political consequences and is stated to apply in location class 2 and, therefore, within the immediate area of a platform. There is again much room for argument and interpretation.

Location and fluid categories are combined into safety classes in the following way [reference 4, table 2–4] (Table B–1).

Table B–1 *Limit States, Safety Classes, and Failure Probabilities*

phase	fluid A, C location 1	fluid A, C location 2	fluid B, D, E location 1	fluid B, D, E location 2
temporary	low	low	low	low
permanent	low	normal	normal	high

Therefore, safety class determines safety class factors in calculations. The resistance factor γ_{SC} for pressure containment [reference 4, table 5–5] is 1.308 for safety class *high*, 1.138 for *normal*, and 1.046 for *low*. Factors stated to four significant figures give a misleading impression of scientific precision. They were arrived at by a calibration process intended to produce designs consistent with previous satisfactory experience. DnV 2000 makes clear that it is not the intention to alter the historic safety level, and the calibration took this into account.

These are not new ideas. Risers, for example, are customarily designed with a smaller design factor than pipelines well away from a platform.

Limit states are divided into ultimate limit states (ULS) that compromise the integrity of the pipeline and serviceability limit states (SLS) that make the pipeline unsuitable for normal operations. Two subsets of ULS are fatigue limit states (FLS) and accidental limit states (ALS) due to accidental loads. They can be linked to target acceptable failure probabilities and safety classes (SC) in the following way:

limit	failure probability per pipeline per year		
state	SC	SC	SC
	low	normal	high
SLS	10^{-2}	10^{-3}	10^{-3}
ULS, ALS, FLS	10^{-3}	10^{-4}	10^{-5}

and the conditions expressed in section 4 of ISO 13283 are supposed to achieve these failure probabilities, though the probabilities are rightly so low that it is not possible to prove this in a convincing way.

B–5 Influence of Level Two Structural Reliability Analysis

Level two structural reliability is beginning to influence code development. The idea is that if a parameter defining the material strength has a high level of variability, the probability of an unusually low value of strength is higher than it would be if the material had the same nominal strength but a lower level of variability. This can be recognized in some way by penalizing a high degree of variability.

OS-F101 applies this idea in the definition of the characteristic strength to be applied in strength calculations, formally defined [4, section 1, C208] as

> "the nominal value of the material strength to be used in the determination of the design strength. The characteristic strength is normally based upon a defined fractile in the lower end of the distribution function for strength."

The relationships between the characteristic strengths and the more commonly used specified minimum yield and ultimate strengths are

$$\text{characteristic yield strength } fy = (SMYS - f_{y,temp})\alpha_U \qquad \text{(B.3)}$$

$$\text{characteristic yield strength } fy = (SMTS - f_{u,temp})\alpha_U\alpha_A \qquad \text{(B.4)}$$

$f_{y,temp}$ and $f_{u,temp}$ allow for temperature derating (which kicks in at 20 °C (68°F) for chromium CRAs and at 50 °C (122 °F) for C-Mn steel). α_A is an anisotropy factor, which is 0.95 in the axial direction and 1 otherwise. α_U is a material strength factor: it is normally taken as 0.96, which is a way of recognizing the reference to the lower end of the distribution function mentioned above.

However, α_U may be taken as 1 if the material meets what is termed *supplementary requirement U*, which imposes stricter requirements on variability. U stands for high utilization, and the requirements are set out in [4, section 6, D500]. The test regime "…intend(s) to ensure that the average yield strength is at least two standard deviations above SMYS and the ultimate strength is at least three standard deviations above SMTS" and can be summarized as an algorithm in the diagram that follows (Fig. B–1):

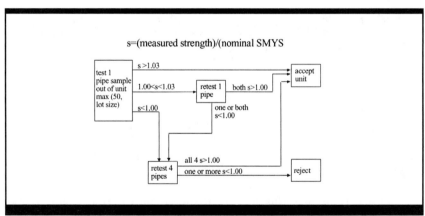

Fig. B–1 *Test Scheme for Supplementary Requirement U (reference 4, section 6, D500)*

This OS F101 scheme is more demanding than the corresponding requirement in API 5L, for example, which generates a high probability of a lot of pipe passing the yield stress test for a grade even though a substantial fraction of pipes in the lot have a strength less than the nominal strength of the grade.[4] Pipeline engineers often imagine that a length of X65 pipe necessarily has a strength higher than 65 ksi (448.2 MPa), but that is not the case. In the OS F101 scheme, the probability of passing when a substantial fraction of pipes in the lot have a strength less than the nominal strength is lower, but far from negligible.

An additional factor is that a length of pipe definitely does not have a single yield strength. Testing of many samples taken from the same pipe shows a high level of variability.[2] Indeed, it could be argued that variability between pipes is no higher than variability within one pipe. This is a subject that would benefit from further research.

References

1   Ellinas, C. P., A. C. Walker, A. C. Palmer, and C. R. Howard. *Subsea Pipeline Cost Reductions Achieved Through the Use of Limit State and Reliability Methods.* Proceedings of the Offshore Pipeline Technology Seminar, Amsterdam, IBC Technical Services, London (1989).

2   Palmer, A. C. *The limits of reliability theory and the reliability of limit-state theory applied to pipelines.* Proceedings of the 28th Annual Offshore Technology Conference, Houston, 4, 619–626, OTC8218 (1996).

3   Palmer, A. C., C. Middleton, and V. Hogg. *The Tail Sensitivity Problem, Proof Testing, and Structural Reliability Theory.* Structural Integrity in the 21st Century. Proceedings. 5th International Conference on Engineering Structural Integrity Assessment, Cambridge, 435–442 (2000).

4   Offshore Standard OS-F101. Submarine Pipeline Systems. Det norske Veritas, Hovik, Norway (2000).

Appendix C
Units

C.1 Introduction

Most of the scientific and technological world uses metric units for any serious purpose, generally and preferably in the Système Internationale (SI) version agreed at the Eleventh General Conference of Weights and Measures in Paris in 1961.[1] Unfortunately, the petroleum industry has only partially adopted SI units. This is because engineers from the United States played a predominant part, particularly in the early years, and their language and units were adopted generally. This situation is likely to persist for a long time in the United States, but the rest of the English-speaking world went over to the metric system 30 years ago. European practice in the petroleum industry is slowly changing. Engineering calculations are made in SI units, but measurements of volumes and weights are still in American units, though there is a trend towards SI units for volumes and weights. A further complication is that American units are not always the same as Imperial (British) units with the same name.

Mixed units and confusion about units are frequent sources of mistakes, sometimes very costly ones. The recommended approach is to use metric units in engineering calculations to the maximum extent possible or, alternatively, to use American units throughout, but not to mix the two. If input data are given in mixed units, for example, with volumes in barrels/day but lengths in km, it is straightforward to convert the input data, to carry out the calculation entirely in metric or American units, and, if necessary, to convert the output back to other units.

One of the advantages of metric units is that they are strictly decimal, so that different units for the same quantity are always related by multiples of 10. There are no multiples such as 12 (the ratio between 1 foot and 1 inch) or 2240 (the ratio between 1 Imperial ton and 1 pound) or 2000 (the ratio between 1 American ton and 1 pound). The preferred submultiples go up and down in powers of 10^3: 1 mm is 10^{-3}m and 1 km is 10^3 m. It is good practice to use only submultiples related by factors of 10^3 so that length, for example, should be measured in m or km or mm, but not cm. Not everyone follows this methodology.

Non-metric units are likely to persist longest in the description of pipe diameters (section C.2), oil and gas volumes (C.3), forces (C.4), pressure, (C.5) and density (C.6).

C.2 Lengths

SI measures length in m, km or mm.

Pipe diameters are often measured in inches. 1 in. = 25.4 mm exactly. Care needs to be taken with nominal diameters, which regrettably are not always exact conversions. A nominal 10-in. pipe has an outside diameter of 273.05 mm (10.75 in.), not 254 mm.

Distances are sometimes measured in miles. 1 mile = 1.60934 km. However, 1 nautical mile is different and is 1.85318 km in the UK but 1852 m everywhere else.

C.3 Volumes

Oil volume is usually measured in *barrels*, abbreviated bbl or sometimes b. One barrel is defined as 42 United States gallons (smaller than Imperial gallons) and is 0.158987 m³. Volume changes slightly with temperature and pressure, and the standard conditions are *stock tank conditions*, 60 °F (15.55 °C) and 1 atmosphere (101.325 kPa). There is a tendency, particularly in Europe and the FSU, to shift to metric volume measurements or instead to mass that is measured in metric tons (tonnes). 1 metric tonne = 1000 kg, significantly larger than an American ton, which is 2000 lb (907.2 kg), which is different again from an Imperial ton, which is 2240 lb.

Gas is often measured in *standard cubic feet*, abbreviated scf (sometimes pronounced as *scuff*). Multiples are Mscf (thousands), MMscf (millions), Bcf (10^9), Tcf (10^{12}). Standard conditions are generally 60 °F (15.56 °C) and 1 atmosphere (101.325 kPa), but sometimes 15 °C and 1 bar (100 kPa). The difference is clearly small. Metric measurements are in *normal cubic meters* (Nm³), referred to as *normal* conditions 0 °C and 1 bar, slightly different from standard conditions.

C.4 Force

SI measures force in N (newtons) and multiples kN and MN. 1 N is about the weight of one apple, appropriately, the weight of approximately 0.1 kg of mass. 1 kN is about the weight of a heavy person with a mass of 100 kg. 1 MN is about the weight of a railway locomotive with a mass of 100,000 kg (100 tonnes). The take-off weight of a Boeing 747 is about 3.5 MN.

Another advantage of the SI system is that it is consistent. One unit of force applied to one unit of mass generates one unit of acceleration. 1 N applied to 1 kg generates an acceleration of 1 m/s². The gravitational acceleration, 9.80665 m/s² in SI, only enters calculations when gravity itself is significant, not as a conversion factor from force units based on weight.

A 1 kgf (kilogram force) is 9.80665 N under the standard gravitational acceleration of 9.80665 m/s². The weight of 1-pound mass (0.45359237 kg) is 4.4482 N under the standard gravitational acceleration of 9.80665 m/s².

C.5 Pressure

SI measures pressure in Pa (Pascals). 1 Pa is 1 N/m², which is 0.00014504 lb/in². Because 1 Pa is such a small pressure, the multiples kPa and MPa are more widely used: 1 kPa = 0.14504 lb/in2. 1 MPa = 145.04 lb/in2.

Two non-SI metric units of pressure, the bar and the kg/cm², are still widely used in countries that went metric before SI was adopted. 1 bar = 100 kPa=14.504 psi. 1 kg/cm² = 98.0665 kPa = 14.22 psi.

C.6 Density

SI measures density in kg/m³. 1 kg/m³ is 0.06243 lb/ft³. Density often is described by a scale called American Petroleum Institute (API) *gravity*, defined by the following:

$$API = \frac{141.5}{SG_{60}} - 131.5$$

where SG_{60} is the specific gravity (relative density) relative to water at 60 °F (15.56 °C) and 1 atmosphere (101.325 kPa). This odd scale was chosen because it gives a linear scale on a hydrometer. Note that this scale is potentially confusing because heavier oil has a lower API gravity. Most oil has a gravity in the range from 20 to 45.

The density at 15.56 °C can be calculated from the API gravity using the following formula:

$$\text{density} = \left(\frac{141.5}{131.5 + A}\right) 999.01 \text{ kg/m}^3 = \left(\frac{141.5}{131.5 + A}\right) 62.37 \text{ lb/ft}^3$$

Reference

1   Changing to the Metric System: Conversion Factors, Symbols and Definitions. Her Majesty's Stationery Office (1967).

# Index

A

C

G

L

M

P